Series on
Reproduction and Development in Aquatic Invertebrates

Volume 3

Reproduction and Development in Echinodermata and Prochordata

W0113092

Series on

Reproduction and Development in Aquatic Invertebrates

Volume 3

Reproduction and Development in Echinodermata and Prochordata

T. J. Pandian

Valli Nivas, 9 Old Natham Road
Madurai-625014, TN, India

CRC Press is an imprint of the
Taylor & Francis Group, an **informa** business

A SCIENCE PUBLISHERS BOOK

Cover page: Representative echinoderms and prochordates. Row 1 (from left to right) crinoid *Antedon bifida* (Public domain, British Museum), *Holothuria scabra* (from Asha et al., 2015), sea urchin (from dreamstime. com), sand dollar (free hand drawing), Row 2 *Ctenodiscus crispatus, Peltaster placenta* (from Downey, 1973), star fish (from dreamstime.com), *Asternyx excavata* (free hand drawing from Hyman, 1955), Row 3 *Ptychodera* (Public domain, Cambridge Univ), *Cephalodiscus* (free hand sketch from Hyman, 1955), Amphioxus (from dreamstime.com), Row 4 Oikepleura (free hand sketch from Alldredge, 1976, meerhome.org), Oozoid and blastozoid of *Thalia democratica* (Hereu and Suarez-Morales, 2012, Courtesy: Zootaxa), *Ciona intestinalis* (free hand sketch from Millar, 1970), *Botryllus* (free hand sketch from Marine Collection, Stanford University)

CRC Press

Taylor & Francis Group

6000 Broken Sound Parkway NW, Suite 300

Boca Raton, FL 33487-2742

First issued in paperback 2020

© 2018 by Taylor & Francis Group, LLC

CRC Press is an imprint of Taylor & Francis Group, an Informa business

No claim to original U.S. Government works

ISBN-13: 978-0-8153-6472-6 (hbk)

ISBN-13: 978-0-367-78134-7 (pbk)

Library of Congress Cataloging-in-Publication Data

Names: Pandian, T. J., author.
Title: Reproduction and development in echinodermata and prochordata / T.J. Pandian.
Description: Boca Raton, FL : CRC Press, Taylor & Francis Group, [2018] | Series: Reproduction and development in aquatic invertebrates ; volume 3 | "A science publishers book." | Includes bibliographical references and indexes.
Identifiers: LCCN 2017057110 | ISBN 9780815364726 (hardback : alk. paper)
Subjects: LCSH: Echinodermata--Reproduction. | Echinodermata--Development. | Protochordates--Reproduction. | Protochordates--Development.
Classification: LCC QL381 .P36 2018 | DDC 593.9--dc23
LC record available at https://lccn.loc.gov/2017057110

Visit the Taylor & Francis Web site at
http://www.taylorandfrancis.com

and the CRC Press Web site at
http://www.crcpress.com

Preface to the Series

Invertebrates surpass vertebrates not only in species number but also in diversity of sexuality, modes of reproduction and development. Yet, we know much less of them than we know of vertebrates. During 1950s, the multi-volume series by L.E. Hyman accumulated bits and pieces of information on reproduction and development of aquatic invertebrates. Through a few volumes published during 1960s, A.C. Giese and A.S. Pearse provided a shape to the subject of Aquatic Invertebrate Reproduction. Approaching from the angle of structure and function in their multi-volume series on Reproductive Biology of Invertebrates during 1990s, K.G. Adiyodi and R.G. Adiyodi elevated the subject to a visible and recognizable status.

Reproduction is central to all biological events. The life cycle of most aquatic invertebrates involves one or more larval stage(s). Hence, an account on reproduction without considering development shall remain incomplete. With passage of time, publications are pouring through a large number of newly established journals on invertebrate reproduction and development. The time is ripe to update the subject. This treatise series proposes to (i) update and comprehensively elucidate the subject in the context of cytogenetics and molecular biology, (ii) view modes of reproduction in relation to Embryonic Stem Cells (ESCs) and Primordial Germ Cells (PGCs) and (iii) consider cysts and vectors as biological resources.

Hence, the first chapter on Reproduction and Development of Crustacea opens with a survey of sexuality and modes of reproduction in aquatic invertebrates and bridges the gaps between zoological and stem cell research. With capacity for no or slow motility, the aquatic invertebrates have opted for hermaphroditism or parthenogenesis/polyembryony. In many of them, asexual reproduction is interspersed within sexual reproduction. Acoelomates and eucoelomates have retained ESCs and reproduce asexually also. However, pseudocoelomates and hemocoelomates seem not to have retained ESCs and are unable to reproduce asexually. This series provides possible explanations for the exceptional pseudocoelomates and hemocoelomates that reproduce asexually. For posterity, this series intends to bring out six volumes.

August, 2015   T. J. Pandian
Madurai-625 014

Preface

KNOWN for their amazing potential for regeneration and/or clonal reproduction, echinodermates and prochordates have attracted much attention. The accumulated publications including those, which fetched the Nobel Prize, have formed the base for many books. In general, these books are concerned with one or other of these taxons or on a specific theme alone. This book is a comprehensive synthesis covering almost all aspects of Reproduction and Development in Echinodermata and Prochordata from the crinoid *Antedon bifida* to the benthic colonial ascidian *Botryllus schlosseri*.

It is organized in 10 chapters, of which the first seven are devoted to Echinodermata, and of the remaining three, one each to Hemichordata, Cephalochordata and Urochordata. The fairly long first chapter commences with a brief introduction to taxonomy, structural diversity and geographical distribution of echinoderms. Echinoderms are unique, as their symmetry is bilateral in larva but pentamerous radial in adults. This has eliminated the development of an anterior head and lateral appendages. Consequently, they are sessile/sedentary or low motile. The obligate need for acquisition of food by substratum-faced oral surface in 74% of echinoderms has also eliminated them from colonizing pelagic realm as plankton and entire water column as nekton. In sea urchins, *Laminaria japonica* facilitates a greater allocation for commercially valuable gonad, the roe but *Rhodemenia palmata* somatic growth. Depth has profound negative effects on fecundity. In lecithotrophic ophiuroids and asteroids, egg size increases with increasing depth up to 2,000–5,000 m but remains constant in their planktotrophs, albeit at different levels. At the interspecific level, fecundity of lecithotrophs decreases with increasing depths. A striking contrast is that ophiuroids larger than 18 mm size do not produce planktotrophic eggs but asteroids larger than 110 mm size generate planktotrophic eggs alone.

The second chapter presents a brief account on the smaller sized capture and aquaculture fisheries; it points out the scope for (i) application of prevsiceral coelomic fluid to induce spawning in holothuroids, (ii) profitable restoration of abandoned shrimp farms by holothuroids and (iii) utilization of abundant benthic sea grasses as ingredients for synthetic feed for echinoids.

The third chapter is concerned with sexual reproduction. Incidentally, injection of potassium chloride and L-methyladenine induces successful spawning in sea urchins and asteroids, respectively. In them, thyroxine regulates larval duration, offering considerable scope to shorten the critical larval duration. Its role in holothuroids remains to be tested. Echinoderms display one of the simplest reproductive systems. In the absence of an intromittent organ, many brooding holothuroids remain either unfertilized or with low (48%) fertilization success.

In echinoderms, hybridization, karyotyping and ploidy induction have not provided evidence for the existence of heterogametism. But the limited variations in sex ratio indicate that sex is irrevocably determined at fertilization. Two models of steroid functions have become apparent: (i) the carnivorous asteroids with pyloric caeca and (ii) the herbivorous echinoids with nutritive phagocytes but without storage organ. Echinoids and larvacea have economized the costs of hormones and translocation of nutrients from the storage organ to the gonad by harboring germ cells and nutritive cells within the gonad itself. For example, the E_2 level is high as 4–460 pg/g gonad in asteroids, in comparison to 5–160 pg/g in echinoids. Endocrine disruptors alter the 'hormonal climate' and interfere with the regenerative process. Parasite castration is not common among echinoderms and when it occurs, it induces partial sterility alone.

From an incisive analysis of available literature on clonal reproduction in echinoderms, the fourth chapter brings to light for the first time the following: 1. Barring crinoids, ~ 2.2% of other echinoderms clonally reproduce by fission (87%), larval cloning (9%) and clonal autotomy (4%). These clonal reproductive modes mutually exclude each other. 2. Clonal and sexual reproduction is distinctly separated by season and/or body size. 3. With increasing interval between successive cloning, cloning potency is decreased in the following order: ophiuroids < holothuroids < asteroids. The presence of > 5 regenerating arms inhibits the ensuing clonal reproduction in asteroids and possibly ophiuroids. 4. Cloning is triggered by accumulation of nutrient resource in the body, i.e. food availability rather than population density triggers fission. 5. Evidences are adduced for the role played by multipotent stem cells and primordial germ cells.

The fifth and the last three chapters highlight the autoregulative potential of echinoids and solitary ascidians. In echinoids, planktotrophic (80 μm) and lecithotrophic (144 μm) eggs can autoregulate to yield viable twins and quadruplets, respectively. In solitary ascidians, however, one of the two blastomeres develops into a half larva, which autoregulates at metamorphosis to produce a complete viable juvenile. This delayed autoregulation is relevant to biomedical applications. In the epimorphic crinoids, holothuroids and ophiuroids, and morphallaxic asteroids and echinoids, regeneration of missing tissues/organs/systems is accomplished by multipotent stem cells like coelomocytes and the like, and involves de-, re- and trans-differentiation processes. Despite the amazing regenerative potential, the occurrence of regeneration is surprisingly limited to 204+ species, i.e. about 3% of all echinoderms. Nevertheless, its prevalence within each species is so high that ~ 300 mt ophiuroid arm biomass is suggested to be injected to the trophodynamics. Incidentally, the cephalochordates, known for prolific breeding, inject a few million tons of larvae into the trophodynamics.

Natural and/or induced fission produces bidirectional cloning in echinoderms but unidirectional cloning in manipulated enteropneusts and solitary ascidians. In pterobranchs, thaliaceans and ascidians, the repeated and rapid budding leads to colonial formation. Coloniality imposes reductions in species number and body size, generation time and life span, gonad number and fecundity as well as switching from gonochorism to simultaneous hermaphorditism and oviparity to ovoviviparity/viviparity.

This book is a comprehensive synthesis of > 662 publications carefully selected from widely scattered information from 194 journals and 66 other literature sources. The holistic approach and incisive analysis have led to of several new findings and ideas related to reproduction and development of echinoderms and prochordates. Hopefully, the book serves as a launching pad to further advance our knowledge on these deuterostomes.

July, 2017   T. J. Pandian
Madurai-625 014

Acknowledgements

It is with pleasure that I thank Drs. P.S. Asha, T. Balasubrmanian, E. Vivekanandan for critically reviewing parts of the manuscript of this book and offering valuable suggestions. In fact, I must confess that I am only a visitor to the theme of this book but my earlier books in this series and editorial service on energetic of echinoderms and prochordates (Pandian, 1987, *Animal Energetics*, Academic Press) emboldened me to author this book. The Central Marine Fisheries Research Institute (CMFRI), Kochi provides the best library and excellent service to visitors. I wish to place on record my sincere thanks to Dr A. Gopalakrishnan, Director, CMFRI and his library staff especially Mr. V. Mohan. Thanks are due to the American College, Madurai for providing valuable books. I also thank Drs. R. Jeyabaskaran, G. Kumaresan, N. Munuswamy, P. Murugesan, B. Senthilkumaran and K.K. Vijayan for helping with many publications. The manuscript of this book was prepared by Mr. T.S. Balaji, M.Sc. and I wish to thank him profusely for his competence, patience and co-operation.

Firstly, I wish to thank many authors/publishers, whose published figures are simplified/modified/compiled for an easier understanding. To reproduce original figures from published domain, I gratefully appreciate the permissions issued by *Biological Bulletin*, CMFRI, International Ecology Research Institute (*Marine Ecology Progress Series, Diseases of Aquatic Organisms*), Smithsonian Institute and Magnolia Press (*Zootaxa*). For permissions issued to reproduce their original figures, I remain thankful to Drs. P.S. Asha, J.M. Lawrence and S. Stohr. Special thanks are due to Dr. I.Y. Dolmatov for sending me relevant Russian literature. For advancing our knowledge in this area by their rich contributions, I thank all my fellow scientists, whose publications are cited in this book.

July, 2017   T. J. Pandian
Madurai, 625014

Contents

Section I
Non-Chordate-Deuterostomia

Section Ia
Echinodermata

1

Introduction

The echinoderms comprise feather stars and sea lilies (Crinoidea), sea cucumbers (Holothuroidea), sea urchins, pencil urchins (cidaroids), heart urchins (spatangoids), sand dollars and sea biscuits (clypeasteroids) (Echinoidea), sea stars (Asteroidea) as well as basket, brittle (euryalids with branching arms) and snake (euryalids with non-branching arms) stars (Ophiuroidea) (Fig. 1.1). Conspicuously different from all other invertebrates, they exhibit pentamerous radial symmetry, mesodermal endoskeletal and water-vascular systems, which facilitate slow motility, food acquisition, respiration, excretion and sensory perception (Rinkevich et al., 2009, Reinardy et al., 2015). In echinoderms, the water-vascular system with numerous projections (podia) facilitate communication with external *mileu*, except in crinoids, in which direct opening to the exterior is missing (Table 1.1). Being the only taxon saluted by L.H. Hyman, they are "a noble group, especially designed to puzzle zoologists" (Hyman, 1955). Although invertebrates yield lower landings in comparison with finfishes, they account for 40% of the world fishery trade, due to the high price fetched by shellfishes and cephalopods (see Pandian, 2017). Among echinoderms, sea urchins are harvested for roe and sea cucumbers for the body, longitudinal muscles and viscera. Holothuroids are commercially fished mostly in the Indo-Pacific areas (Jangoux, 1990). In the Gulf of Mannar and Palk Bay of India, for example, only seven out of 39 holothuroid species are commercially valuable (Asha et al., 2015). They are consumed raw, boiled or pickled and their consumption is restricted to countries like China and Japan. In other countries like India, the fishing and trading of sea cucumbers either alive or dried is banned. However, there is no restriction for fishing and trading sea urchins. The star fish *Echinaster guyanensis* is economically important, as they are harvested in great quantities to make handicrafts (Hadel et al., 1999). Briefly, the echinoderm fisheries are small in scale (Conand and Byrne, 1993, Keesing and Hall, 1998, however see Chapter 2).

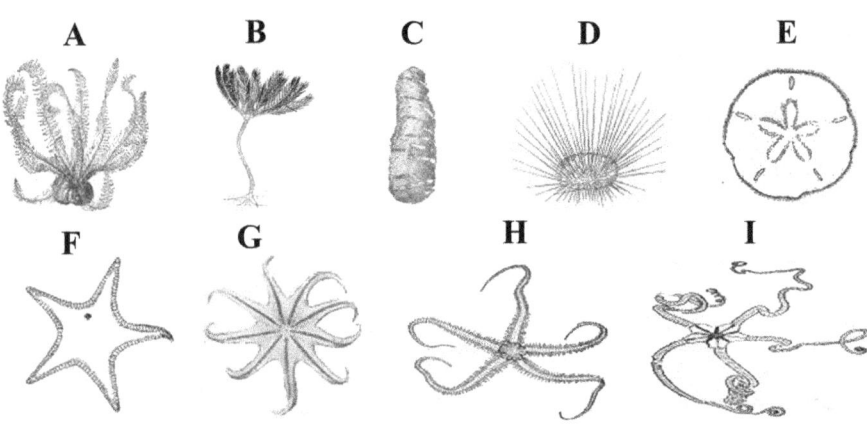

FIGURE 1.1

A. Comatulid crinoid *Antedon bifida*, B. Stalked crinoid (from an unknown Google source), C. *Holothuria scabra* (from Asha et al., 2015), D. Sea urchin *Diadema antillarum* (side view), E. Keyhole sand dollar *Mellita sexiesperforata* (aboral view), F. Mud star fish *Ctenodiscus crispatus*, G. Oral view of *Solaster endeca*, H. *Ophiocoma* with spiny arms, I. *Asternyx excavata* with simple arms (all others are modified free hand drawings from Hyman, 1955).

Nevertheless, echinoderms together with primitive deuterostomes constitute a basal group of non-chordate invertebrate deuterostomes and occupy a key position in evolution (Rinkevich et al., 2009). The genomes of echinoderms and vertebrates have revealed 70% homologies of sea urchin with human (Cameron et al., 2000, Sea Urchin Genome Sequencing Consortium [SUGSC], 2006). Further, many echinoderms serve as excellent tractable, inexpensive models and are amenable to all sorts of manipulations in stem cell and regeneration research but warrant no ethical, political and moral concerns. Not surprisingly, sea urchins have served as a model organism in embryology and developmental biology for more than a century (Lessios, 2007). Notably, it is from the sea urchin that fundamental contribution has been made on the cell cycle including the discovery of cyclin protein that regulates the progression of cell cycle (Evans et al., 1983). It is from the sea star, Metchinikoff (1845–1916) has first demonstrated phagocytosis of microbes (see Tan and Dee, 2009). Both these findings have been awarded Nobel Prize (Kondo and Akasaka, 2012). Other important discoveries like (i) fertilizin, a chemoattractant released from the eggs that attract conspecific sperm (e.g. Hertwig, 1876), (ii) chromosomes as carriers of hereditary information (see Maderspacher, 2008) and (iii) gene regulating network (Davidson et al., 2002) have all arisen from echinoderm research. The echinoderms are now considered as model animals for ageing research (see Francis et al., 2006), as some of them like *Echinometra lucunter* and *Strongylocentrotus franciscanus* are estimated to live for 75 and 200 years, respectively (Ebert, 2008, Ebert et al., 2008) and yet do not undergo senescence (Bodnar, 2015). It is in this

TABLE 1.1

Structural diversity in echinoderms (collated from different sources)

Structure	Crinoidea	Holothuroidea	Echinoidea	Asteroidea	Ophiuroidea
Morphology	Stalked or stalkless cup-like body with branched or unbranched pinnated arms and mouth facing upwards	Elongated body on oral-anal axis; modified podia encircle the mouth	Globose, oval or discoid body moving on oral surface	Pentagonal oral-aborally flattened disk with ray-like arms; moving on oral surface	Central stellate, oral-aborally flattened disk with ray-like arms; moving on oral surface
Motility	Mostly sessile	Mostly sedentary	Slow motile by podia and/or spines	Motile by tube feet	Motile by tube feet
Anatomy	Water-vascular system lacks direct exterior opening. Anus opens aborally	Respiratory tree. Anus opens aborally	Mastigating Aristotle's lantern in sea urchins. Anus opens aborally	External digestion by everted cardiac stomach. Anus may be absent	Anus absent. Unlike in asteroids, ambulacral groove is closed
Reproduction	Discrete gonad is absent. Sexual only but autotomic	Median ovary/ovotestis open at the oral end. Asexual by transverse fission	>70% of spacious coelom occupied by 5 or more lobulated gonads terminating in aboral sinus. Sexual only	Spacious coelom occupied by tubular lobulated/saccular gonad projecting into the arms. Asexual by fission and clonal autotomy, autotomic	Saccular gonad attached to coelomic bursae; through bursal slit, gametes released. Asexual by fission. Autotomic
Regeneration	Autotomized arms and pinnules are readily regenerated. Induced evisceration in *Antedon*	Regeneration of auto-eviscerated gut is common	Spines, pedicellariae and podia are readily regenerated	Autotomized arms and disks are readily regenerated	Autotomized arms and disks are readily regenerated

context, studies on reproduction and development of echinoderms have gained much importance during recent years, especially in exploration of their genes responsible for extra-ordinary potency for regeneration and agametic cloning. With the advent of molecular biology and arrival of more sophisticated tools and techniques, publications in this area are covered by an increasing number of high profile journals, to which conventional zoologists have no immediate access. An objective of this book is to bridge the widening gap between conventional zoologists and molecular biologists.

1.1 Taxonomy and Structural Diversity

The echinoderms consist of ~ 7,000 living species grouped into five classes namely Crinoidea (700 species), Holothuroidea (1,100 species), Echinoidea (~ 1,000 species), Astroidea (1,800 species) and Ophiuroidea (2,064 species). Table 1.2 shows that there are more extinct species than living ones. For example, > 5,000 species of crinoids have become extinct, leaving only about 700 living species (see Hyman, 1955). On the whole, of ~ 20,000 echinoderm species, over 13,000 species have become extinct (Pawson and Miller, 2012). Over the years, due to the discovery and erection of more and more new species, the taxonomy of echinoderms has remained dynamic. For example, from the mid-18th century, the discovery rate of ophiuroid species has accelerated and remained relatively high for about a century and has now levelled-off (Fig. 1.2). At present, Ophiuroidea is the largest class of Echinodermata and includes 16 families and 270 genera and is rich with 2,064 known species. At the same time, the taxonomy has also remained fluid. For example, Mortensen's Monograph of the Echinoidea (1928–1951), recognized as the major treatise on the group (e.g. Emlet, 1995), has and continues to be valid for the alpha taxonomy of echinoids. *Echinometra mathaei* is recognized as one

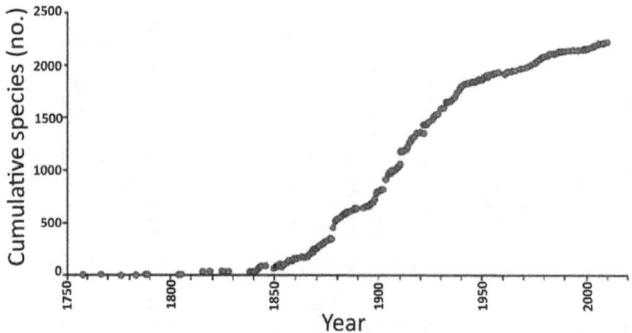

FIGURE 1.2

Discovery rate of ophiuroid species (from Stohr et al., 2012).

TABLE 1.2

Systematic resume of echinodermates (~ 7,000 living species) (from Hyman, 1955 and others), † = extinct, * = living, >†,<* = mostly extinct but a few are living

Subphylum	Pelmatozoa
Class	Heterostelea†
	Cystidea†
	Blastoidea†
	Crinoidea†* (700 species)
Order	Inadunata†, 53 families, > 300 genera
	Flexibilia†, 10 familes, > 47 genera
	Camerata†, 29 familes, > 129 genera
	Articulata†*, 630 species
Class	Edrioasteroidea†
Subphylum	Eleutherazoa
Class	Holothuroidea* (1,100 species)
Order	Aspidochirota, *Stichopus badionotus*
	Elasipoda, *Kolga hyalina*
	Dendrochirota
	Molpadonia, *Paracaudina chilensis*
	Apoda, *Synaptula hydriformis*
Class	Echinoidea
Sub class	Bothriocidaroidia†
Sub class	Regularia
Order	Lepidocentroida†
	Melonechinoida†
	Cidaroidea†*, *Ctenocidaris*
	Aulodonta†*, *Diadema*
	Stirodonta†, *Arbacia stellata*
	Camarodonta*, *Strongylocentrotus droebachiensis*
Sub class	Irregularia (900 species)
Order	Holectypoida†
	Cassiduloida >†,<*, *Cassidulus caribaearum*
	Clypeastroida*, *Clypeaster rotundus*
	Spatangoida*, *Lovenia cordiformis*
Class	Asteroidea (1800 species)
Order	Platyasterida†
	Hemizonida†
	Phanerozonia*, *Astropecten aranciacus*
Sub order	Postulosa†
	Cribellosa*, *Porcellanaster*
	Paxillosa*, *Ctenodiscus crispatus*
	Notomyota*, *Pectinaster*
	Valvata*, *Pseudoarchaster pisillus*
	Spinulosa*, *Crossaster papposus*
	Forcipulata*, *Neomarphaster*
Class	Ophiuroidea (2,064 species, Stohr et al., 2012)
Order	Ophiurae*, *Amphipholis squamata*
	Euryalae*, *Asteronyx loveni*
Class	Ophiocistioidea

nominal species by Mortensen. However, recent studies on egg-sperm compatibility (Metze et al., 1991) and karyotype (Uehara et al., 1991) have suggested that *E. mathaei* is a complex of four species. Yet, no modern method like barcoding seems to have been used to conclusively show that four species can be erected from *E. mathaei*. In the asteroid genus *Echinaster*, two morphs, distinguishable by coloration, skeletal structure and reproductive biology, occur in the Gulf of Mexico. Their taxonomy remains unresolved, although they have earlier been referred as *E. spinulosus, E. echinophorus, E. modestus* by previous authors (see Scheibling and Lawrence, 1982). Larval taxonomy remains woefully incomplete; for example, at least 48 ophiopluteus species are known but they have not yet been assigned to respective adults (see Stohr et al., 2005). Based on morphology, ecology and reproductive biology, *Holothuria scabra* is being divided into two subspecies namely *H. scabra* and *H. scabra versicolor* (Hamel et al., 2001). On the other hand, many new species have been brought under one species and a few are renamed (e.g. *Antedon* [*rosacea*] *adriatica,* Hyman, 1955).

Echinoderms are unique and distinctly differ from all other animal taxons in that their embryos/larvae display bilateral symmetry but the adult pentamerous radial symmetry. However, the manifestation of pentamerous radial symmetry in echinoderms has eliminated the scope for the formation and development of an anterior head and lateral appendages. Consequently, the echinoderms are either sessile/sedentary or slow motile (Table 1.1). The obligate need for locomotion and acquisition of food by facing the oral surface over the substratum in echinoids (19%), asteroids (26%) and ophiuroids (29%) has unbreakably tied the distribution of > 74% of echinoderms to benthic habitats; others are either sessile (crinoids 10%) or sedentary (holothuroids 16%). As an immense consequence, the echinoderms have neither colonized the entire column of sea water as nekton nor the pelagic realm as plankton. A couple of planktonic holothurians *Planktothuria natarix* and *Enypiastes ecalcarea* are exceptions (Hyman, 1955). The presence of capacious water-vascular system has also reduced the need for the formation of discrete excretory and respiratory systems; a single exception is the presence of the respiratory trees in holothuroids. A striking feature of echinoderms is their extraordinary ability for regeneration. For example, arms and pinnules in crinoids, spines and pedicellaria in echinoids, arms and disk in asteroids and ophiurods as well as the entire alimentary canal in holothuroids are readily regenerated. However, asexual reproduction by fission and/or clonal autotomy is limited to holothuroids, asteroids and ophiuroids.

Except for the consistent manifestation of radial symmetry, the morphology (Fig. 1.1) and anatomy (Table 1.1) of the five classes of echinoderms are so diverse that Linnaeus has placed *Asterias, Echinus* and *Holothuria* with the naked, warty/spiny molluscs (see Hyman, 1955). Most remarkable is the presence of the mouth facing the substratum in echinoids, asteroids and ophiuroids, a unique feature not encountered in any other invertebrates. The shape of the main body is cup-like in crinoids, cylindrical in holothuroids and stellate in asteroids and ophiuroids. The body is globose in regular echinoids but a flattened disk in irregular echinoids. In asteroids, the flattened central stellate disk is continued with ray-like arms; the number of arms is usually five but it can be nine in *Solaster endeca* (Fig. 1.1G), 13 in *Crossaster papposus*, 16 in *Rathbunaster californicus*, 22 in *Heliaster kubiniji*, 22 in *Pycnopodia helianthoides* and as many as 23–42 in *Labidaster radiosus* (see Hotchkiss, 2000). The arm length in relation to disk diameter of ophiuroids varies from about two-three times to 20 times or more (e.g. *Macrophiothrix*). Ophiuroids lack an anus and madriporite that connect the water-vascular system with the surrounding sea water (Stohr et al., 2012). In echinoderms, the gonadal morphology ranges from the absence of discrete gonad in crinoids to a single bunch of median gonad in holothuroids, five or more lobulated gonads in echinoids, tubular/saccular gonad, which may project into the arms in asteroids and to saccular gonad attached coelomic bursae in ophiuroids.

1.2 Distribution, Locomotion and Dispersal

Geographic range is the horizontal and vertical areas, in which populations of a species are distributed. It is a species specific trait with important evolutionary consequences (Emlet, 1995) inclusive of determination of life span (short in brooders, Strathmann and Strathmann, 1982, Strathmann, 1985), and exposure to different environmental and biological factors. The comatulid crinoids reach their greatest abundance in shallow waters of the Indo-Pacific region. However, their paucity in the Atlantic and eastern Pacific is a striking contrast (Hyman, 1955). The holothuroids inhabit all seas upto a depth of 5,000 m. As in crinoids, the Indo-Pacific harbors the richest holothuroid fauna. For example, the commercially important *Holothuria scabra* is distributed between 30° N and 30° S in the Indo-Pacific region at depths of 5–25 m and occurs at densities of 1–7 cucumbers/m² (Hamel et al., 2001). Asteroids are more prolific in species and numbers in the west coast of North Pacific than in any part of the world (Hyman, 1955). This region is also known for endemicity of ~ 70 species belonging to 7 genera: *Thrissacanthias, Nearchaster, Myonotus, Gephyreaster, Dermasterias, Cryptopeltaster* and *Heterozonias*. In contrast, the Indo-Pacific is not as rich as the North Pacific but is conspicuous by their brilliantly colored species. *Henricia sanguinolenta* is circumboreal in distribution. Similarly, *Ceramaster patagonicus* is known for its 'bipolarity', i.e. it occurs in both north and south Polar Regions (Hyman, 1955). Among the littoral ophiuroids, *Amphipholis squamata* is cosmopolitan and descends upto 250 m depth. *Ophiactis savignyi* is circumtropical, *Ophiacantha bidentata* circumpolar and *Ophiopholis aculeata* circumboreal. Again, the Indo-Pacific is richest for ophiuroids. From her survey, Hyman (1955) "gained an impression that ophiuroids have been more successful than the other existing echinoderm classes in spreading over the sea."

TABLE 1.3

Species richness and endemism in described ophiuroids across 12 regions and four depth strata (modified from Stohr et al., 2012)

Region	Species (no.)	Endemic Species (%)	Species in Each Depth Stratum (no.)				Area (million km²)				
			Shelf (0–200 m)	Bathyal (200–3500 m)	Abyssal (3500–6500 m)	Hadal (> 6500 m)	Shelf (0–200 m)	Bathyal (200–3500 m)	Abyssal (3500–6500 m)	Hadal (> 6500 m)	Total
Arctic	73	8.2	36	60	7	0	6.9	8.9	1.6	0.0	17.4
Antartic	126	36.5	72	101	27	1	1.6	7.4	12.8	0.0	21.8
N. Pacific	398	50.8	262	229	20	2	2.8	4.5	18.3	0.7	26.3
N. Atlantic	241	23.7	138	180	30	0	4.8	8.1	18.9	0.0	21.8
E. Atlantic	118	39.8	73	63	17	2	0.6	3.7	21.0	0.1	25.4
W. Atlantic	335	60.6	217	229	16	0	2.2	4.5	11.0	0.2	18.0
E. Pacific	186	62.9	92	111	28	1	0.4	6.0	13.5	0.0	19.9
S. America	124	24.2	79	102	17	1	1.5	2.8	11.1	0.2	15.6
S. Pacific	355	22.8	235	259	21	0	0.9	9.9	33.8	0.0	44.6
Indian	316	25.6	222	160	19	1	1.7	8.9	18.0	0.0	28.5
Indo-Pacific	825	47.5	551	507	31	6	6.8	21.2	70.3	1.0	99.3
S. Africa	201	21.9	152	135	20	1	0.2	7.9	22.0	0.0	28.1

For the first time, Stohr et al. (2012) have brought out a model picture of the geographic distribution of ophiuroids in 12 defined areas as a function of available areas at the shelf, bathyal, abyssal and hadal depths (Table 1.3). Expectedly, the number of ophiuroid species decreases in the following order: shelf < bathyal < abyssal < hadal. Not surprisingly, within the areas of 30.4 million km², the shelf provides more habitats for the ophiuroids than the abyssal with > 252.3 million km² areas. Surprisingly, > 60% of the ophiuroids are endemic in West Atlantic and East Pacific. Only seven ophiuroid species have been reported from the hadal.

Brittle stars have also been reported from hydrothermal vents (e.g. *Ophioctnella acies*, Tyler et al., 1995, *Spinophiura jolliveti*, *Ophiolamina eprae*, Stohr and Segonzac, 2006) and methane cold seeps (e.g. *O. acies*, *Ophienigma spinilimbatum*, Stohr and Segonzac, 2005). In thermal vents and cold seeps, the vesicomyid and mytilid bivalves symbiotically engage thiotrophic and methanotrophic autotrophic microbes to draw required energy (Pandian, 2017). However, studies on the ability of ophiuroids to symbiotically engage and draw energy remain to be made.

As is obvious from the present description, vertical expansion especially into the abysmal depths encounters more environmental challenges than that of horizontal distribution. Like many abyssal fauna, echinoderms have also descended upto 6,035 m depth. The maximum depths, to which they could descend is 4,790 m for echinoids, 5,000 m for the elasipod holothuroids and upto 6,000 m for the antedonid crinoids (Table 1.4). The greatest depth of 6,035 m, at which an asteroid and a couple of ophiuroids were collected, was from the eastern Atlantic by the Monaco yacht Princesse Alice. Incidentally, buccinid and prosobranch molluscs could be collected from even greater depths of 7,200 and 9,050 m from the Kuril Trench (see Pandian, 2017).

Table 1.5 lists information on motility of representative echinoderms. Hyman (1955) indicated the then available information on the speed of many asteroids, a few echinoids and a couple of ophiuroids. For asteroids, more information is reported by Mueller et al. (2011) and Montgomery and Palmer (2012). Asteroids and echinoids exhibit mostly direction-limited motility. To arrive at a 9-point destination, a star fish takes a deviated route of 13 points (Mueller et al., 2011). To reach a food source kept at ~ 50 cm distance, a regular urchin goes through a circuitous route but a direct one, when it is kept at ~ 40 cm distance (Hyman, 1955). Comatulid crinoids can swim, crawl or creep (Hyman, 1955). By alternate muscular contractions, holothuroids move more like a 'gigantic caterpillar' (Hyman, 1955). Echinoids, asteroids and ophiuroids move using tube feet aided by the water-vascular system. The nature of substratum plays an important role on surface motility of holothuroids. Understandably, the presence of edible organic matter slows down the speed of *Holothuria scabra* (Table 1.5). The distance travelled by *Paracentrotus lividus* may reach 40 cm per hour (h) but effectively just 4.8 m/day (d), as the travel is interrupted by

TABLE 1.4

Vertical distribution of echinoderms (compiled from Hyman, 1955, Emlet, 1995)

Species/Taxon	Depth (m)	Location
Crinoidea		
Annacrinus wyville-thomsoni	2,000	Portugal
Antedonids	up to 6,000	
Holothuroidea		
Kolga hyalina	2,000	Norway
Elasipods	2,000–5,000	Challenger
Echinoidea		
Aporocidaris milleri	48–3,875	Alaska
Hygrosoma petersii	200–3,800	Bahamas-Iceland
Urechinus naresianus	770–4,400	Atlantic-Greenland
Notocidaris hasata	2,450–2,725	Antarctic
Helicocystis carinata	3,656–4,790	Antarctic
Hemiphormosoma paucispinum	4,200	Philippines
Asteroidea		
Crossaster papposus	up to 1,200	North America
Ctenodiscus crispatus	up to 1,860	North Atlantic
Hymenaster pellucidus	15–2,800	Arctic-Boreal
Solaster borealis	400–1,800	North Pacific
S. abyssicola	1,500–3,800	North Atlantic
Fryella sexadiata	4,000–5,000	North Atlantic
Plutonaster gracilis	3,000	Indo-Pacific
Dyaster gilberti	4,000	Panama
Albatrossaster richardi	6,035	Cape Verde Islands
Ophiuroidea		
Ophiura flagellata	96–2,330	Indian, Pacific, Atlantic
Ophiernus adspersus	290–3,600	Indian, Pacific, Atlantic
Ophiomuscium lymani	700–4,000	Almost cosmopolitan
O. planum	5,000	Indo-Pacific
Amphiophiura obdita	6,035	Cape Verde Islands
Ophiacantha opercularis	6,035	Cape Verde Islands

intermittent rest (Boudouresque and Verlaque, 2001). Vertical burrowing rate is also regulated by soil texture, as in *Leptosynapta* and *Molpadia*. The urchin *Centrostephanus rodgersii* emerges from its burrow to feed between dusk and dawn (Fig. 1.3A). In regular echinoids, 'walking' on spine tips is more rapid. Figure 1.3B shows the emerging trends for minimum and maximum speed displayed by representative species belonging to the major classes of motile echinoderms. Accordingly, the speed decreases in the following order: Ophiuroidea > Asteroidea > Echinoidea > Holothuroidea > Crinoidea. Briefly, the short and stout tube feet of starfishes and brittle stars facilitate faster

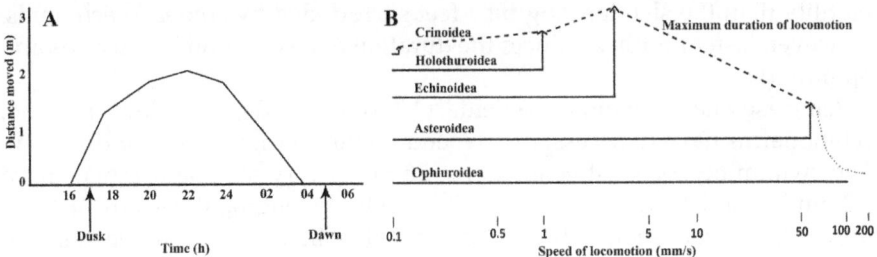

FIGURE 1.3

A. Nocturnal movement by *Centrostephanus rodgersii* from its crevice as a function of time (redrawn from Jones and Andrew, 1990). B. Speed and maximum duration of locomotion of selected representative echinoderm species belonging to the major classes. The trends are drawn using data reported in Table 1.4.

TABLE 1.5

Speed and duration of locomotion in echinoderms

Species	Speed (mm/s)	Duration (m/d)	Reference
Crinoidea			
Crawling isocrinids	0.1		Baumiller and Messing (2007)
Holothuroidea			
Holothuria scabra	0.2		Lokani (1996)
on shells and corals		0.8	Mercier et al. (2000)
on sea grass		0.5	
on organic matter		0.4	
Leptosynapta	0.76		Hyman (1955, p 209)
Molpadia	1.0		
Echinoidea			
Strongylocentrotus droebachiensis	0.03		Montgomery and Palmer (2012)
Paracentrotus lividus	1.1	0.5	Boudouresque and Verlaque (2001)
Arbacia punctulata	0.4		Hyman (1955, p 550)
Centrostephanus longispinus	3.0		
C. rodgersii		2.0	Jones and Andrew (1990)
Abatus ingens	0.54		Thompson and Riddle (2005)
Asteroidea			
Asterina gibbosa	0.06		Hyman (1955, p 351)
Astropecten spinulosus	60.0		
Asterias vulgaris		0.7	
Marthasterias gracilis	0.5		Montgomery and Palmer (2012)
Archaster typicus	12.7		
Linckia laevigata	13.5		Mueller et al. (2011)
Acanthaster planci	58.8		
Ophiuroidea			
Ophionereis reticulata	83.3		Hyman (1955, p 660)
Ophiocoma	100		Shaeffer (2016)

motility than the slender long tube feet surrounded by spines in echinoids. However, faster motility reduces the duration of locomotion in asteroids and ophiuroids.

Being sessile (crinoids), sedentary (holothuroids) or slow moving, echinoderms have to necessarily depend on the larval stage(s) for dispersal. The swimming speed of echinoderm larvae is very slow and ranges from 0.2 µm/seconds(s) to 0.7 µm/s (= 0.2 to 2.4 body length[s]) (McEdward and Miner, 2001a). Consequently, the dispersal of pelagic larvae is at the mercy of waves and currents, which, in turn, depends on the larval duration and it has a potential to influence geographic distribution. The pelagic planktotrophic larval duration ranges from 5 days to 100 days in echinoids (McEdward and Miner, 2001a) and for asteroids from > 28 days (*Linckia laevigata*, see Poulin and Feral, 1996) to > 65 d (*Pisaster ochraceus*, Vickery and McClintock, 2000). The larval duration lasts for 5–35 days and 4–30 days in facultative feeding and non-feeding lecithotrophic echinoid larvae, respectively. The one year stay of non-feeding larva of the star fish *Mediaster aequalis* in pelagic realm is more an exception than rule (Birkeland et al., 1971). Larvae of brooders spend little or no time in the water coloum (McEdward and Miner, 2001a).

A majority of echinoderms display an indirect life cycle involving one or more larval stages. For example, 114 out of 175 echinoids belonging to 26 of the 31 families and 9 of the 11 orders pass through pelagic planktotrophic pluteus stage. Facultative planktotrophy is limited to two species: the clypeasteroid *Clypeaster rosaceus* and the spatangoid *Briaster latifrons*. Non-feeding lecithotrophs occur in 9 species (5 families and 4 orders) and brooders in 49 species (8 families and 7 orders) out of 175 species (McEdward and Miner, 2001a). Briefly, 65, 6 and 28% of echinoids are pelagic planktotrophs, lecithotrophs and brooders, respectively. In other classes of echinoderms, the proportions of non-feeding lecithotrophs and brooders are less. In ophiuroids, the brooders account for < 3% (see Stohr et al., 2005).

Quantitative evaluations of the relationship between development mode and geographic range have been investigated only for a few gastropods (e.g. Scheltema, 1971). Perron and Kohn (1985) have found that *Conus* sp, which are widely distributed among oceanic islands, have longer planktonic periods. As a consequence, the longer planktonic periods do ensure genetic continuity between geographically isolated populations. Similar studies are available for asteroids and echinoids. The asteroid *L. laevigata* with 28-days planktonic larval stage has a geographic range of 1,000 km. Conversely, the range of brooding asteroids *Leptasterias epichlora* and *L. hexactis* with 0-day planktonic stage is limited to 25 km only (see Poulin and Feral, 1996). Emlet (1995) has made a global survey of 215 species belonging to regular echinoids that are planktotrophs (149 species), lecithotrophs

(51 species) and brooders (15 species) (Table 1.6). In these echinoids, the larval duration lasts for 5–100 days, 4–30-days and < 10-days in the planktotrophs, lecithotrophs and brooders, respectively (McEdward and Miner, 2001a). Expectedly, the geographic range runs to 6,969 km for planktotrophs, 3,850 km for lecithotrophs and 4,172 km for brooders. Figure 1.4 shows the relation between the larval duration and geographic range for 33 species of regular echinoids with feeding and non-feeding larvae. Examining the development patterns of echinoids as functions of latitude and depth, Emlet (1995) have shown that 80% of echinoids display planktotrophic development in shallow water (< 100 m depth); their range of ~ 7,000 km is maintained in echinoids in general, and in cidaroids and temnopleuroids, the largest taxa of echinoids. However, this proportion decreases to 9% in species inhabiting waters deeper than 100 m. On this basis, Emlet (1995) has considered 100 m depth as the transition level from planktotrophy to lecithotrophy. It could have been more enlightening, had Emlet taken 200 m depth as transition level, as has been considered by Stohr et al. (2012) for ophiuroids. Incidentally, no information is yet available on the buoyancy rate of eggs and embryos from benthic to pelagic as well as sinking rate of settling larvae from pelagic to benthic.

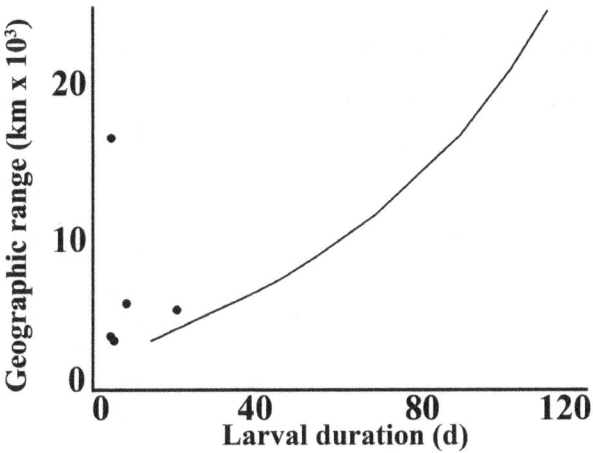

FIGURE 1.4

Relation between larval duration and geographic range in 28 species of regular echinoids with feeding larval stage. Values reported for the echinoids with non-feeding larval stage are indicated by filled circles (redrawn from Emlet, 1995).

TABLE 1.6

Geographic range of echinoid species inhabiting different depths (modified from Emlet, 1995). P = Pelagic planktotrophs, L = Pelagic non-feeding lecithotrophs, B = Brooders

Depth	Species (no.)			Geographic Range (km)		
	P	L	B	P	L	B
All depths						
Echinoids	149	51	15	6969	3850	4172
Cidaroids	24	23	13	7418	3559	4774
Temnopleuroids	22	17	1	7163	1862	–
Shallow waters (< 100 m depth)						
Echinoids	102	22	6	7926	3482	2214
Cidaroids	12	5	5	10467	5334	2629
Temnopleuroids	18	14	0	6535	2107	–
Deep waters (> 100 m depth)						
Echinoids	47	29	9	4893	4128	5478
Cidaroids	12	18	8	4369	3066	6115

1.3 Population Density and Microhabitats

Echinoderms play a key role in structuring many marine ecosystems and are notorious for large scale variations in population density (Uthicke et al., 2009). Not surprisingly, voluminous literature on their density is available (e.g. crinoids: Meyer et al., 2008; holothuroids: Al-Rashdi et al., 2007; echinoids: Lawrence, 2001a; asteroids: Britt et al., 2002; ophiuroids: Fujita and Ohta, 1990). However, this account is limited to microhabitats that offer protection for smaller echinoderms to escape from predators and facilitate reproduction, as well. The fact that in *Strongylocentrotus droebachiensis*, Fertilization Success (FS), for example, increases from 18% at 3 no./m^2 to 62% at 160 no./m^2 (Wahle and Peckham, 1999) reveals the importance of density for reproduction. In other words, the FS frequency increases from 15% at distances > 20 cm between a male and a female to 95% with progressively decreasing distances of < 20 cm between them (Scheibling and Hatcher, 2001). More specifically, FS remains at the level of 0.1, 7.3, 18.0 and 45.0% at the male density of 0, 1, 4 and 16/m^2 (Levitan, 1991).

Microhabitats called refuge-substrata such as the alga *Halimeda* may repel predators, and the sub-tidal zone provides the minimal exposure and stress-free habitat. Sponges too offer structural and chemical protection from predators of brittle stars. Boffi (1972) has made a year long study on the dispersion of nine ophiuroid species amidst 28 algal species, four sponge species and one colonial bryozoan *Schizoporella unicornis* in the northern coast of Sao Paula, Brazil. The preferred microhabitat for these ophiuroids is calcareous over non-calcareous algae, phaeophytes over rhodophytes and chlorophytes, and habitats in the subtidal zone, which are exposed only during the lowest tide. Similarly, the sponges *Mycale* sp and *Haliclona* sp provide a more protective microhabitat than *Tedania* sp. Table 1.7 lists the density of the ophiuroids on the algae and sponges from the coasts of Brazil, Jamaica, Belize and Bermuda. Besides listing the density of two ophiuroids in the favored algal and sponge microhabitats, Mladenov and Emson (1988) have noted that the microhabitats are more preferred by small fissiparous ophiuroids than those that reproduce sexually. Within sponges, mangrove sponges provide a more stable and protected microhabitat than shallow-water algal turf. However, the fragments of *Halimeda* produced by waves can be transported by currents and may provide scope for dispersal of the fissiparous clonal aggregations of these ophiuroids.

Hendler and Littman (1986) made a more meaningful analysis of the microhabitat-dwelling ophiuroids and their reproductive modes. Their analysis included 33 species, of which the smallest were found amidst *Halimeda opuntia* and the largest in corals *Porites porites, Madracis mirabilis* and *Agaricia tenuifolia*. More specifically, the density of smaller (disk diameter < 4 mm) brooding and fissiparous ophiuroids amidst *Halimeda* was > 19–21

TABLE 1.7

Maximum density of algae- and sponges-inhabiting ophiuroids in the northern coast of Sao Paulo, Brazil (compiled from Boffi, 1972) and Jamaica, Belize and Bermuda coasts (compiled from Mladenov and Emson, 1988)

Ophiuroid Species	Algal/sponge Species	Density (no./100 ml algae or no./100 g dry sponge
Ophiactis lymani	*Dictyopteris delicatula*	1510
Axiognathus squamatus	*Jania rubens*	500
Hemipolis elongata	*J. rubens*	20
O. savignyi	*Galaxaura fructens*	8
Amphipholis januarii	*Hypnea musciformis*	8
Ophiothrix angulata	*Spatoglossum schroederi*	3
Ophiactis savignyi	*Halimeda* sp	3.9*
O. savignyi	*Amphiroe* sp	7.1*
Ophiocomella ophiactoides	*Halimeda opuntia*	11–26*
O. ophiactoides	*Gracilaria* sp	11*
O. savignyi	*Haliclona* sp	49–1892**
Ophiothrix angulata	*Haliclona* sp	0.5–16.3**
Ophiactis lymani	*Haliclona* sp	1.9–15.8**
Axiognathus squamatus	*Haliclona* sp	0.5**
O. savignyi	*Amphimedon viridis*	45–150*
O. savignyi	*Ulosa ruetzleri*	16–27*
O. savignyi	*Tedania ignis*	3**

*(no./l), ** (no./sponge)

no./l, in comparison to 5 and 9 no./l for lecithotrophic and planktotrophic ophiuroids, respectively (Table 1.8). On offer of three microhabitats, the smaller brittle stars select within 12 h, *Halimeda* in preference over *Porites* and *Agaricia*. The analysis and experiment confirmed that the *Halimeda* offers the safest refuge microhabitat for the smallest brooding and fissiparous ophiuroids.

TABLE 1.8

Density of ophiuroids of different reproductive modes from different microhabitats of the Belize Barrier Reef in the Caribbean Sea (data calculated from Hendler and Littman, 1986)

Group	Species (no.)	Disk Size (mm)		Density (no./l)		
		Mean	Range	*Halimeda*	*Porites + Madracis*	*Agaricia*
Planktotrophs	9	12.42	3.3–21.1	9.0	2.9	1.5
Lecithotrophs	8	13.51	5.7–21.8	4.9	1.1	0.6
Brooders	5	3.92	2.4–5.5	21.1	1.4	0.7
Fissiparous	4	3.55	1.9–5.2	18.7	1.3	0.9

In echinoids, the fleshy algal cover, besides providing refuge microhabitat, may greatly reduce per capita food availability and gonad mass but increase FS with increasing density. To know whether increasing density ultimately increases zygote production also, Wahle and Peckham (1999) investigated the density of *S. droebachiensis* in seven sites at 5 and 15 m depth in the Gulf of Maine. Due to heavy predation, the coverage of fleshy non-coral macroalgae declined to 23 and < 3% at densities of 150 and 250 urchins per m² at 5 and 15 m depth, respectively. The urchin density ranged from 0.1 no./m² to 250 no./m² (= 42–6,145 g/m²) at 5 m depth. With increasing density, both per capita availability of food and gonad mass were reduced (Fig. 1.5); the decline of gonad mass was more pronounced at 15 m depth. With increasing density, FS, however, increased almost linearly, despite that the eggs remained viable for at least 8 hours after release and the fertilizing ability of sperm decreased to < 5% within 2 hours after milting. Consequently, the per capita zygote production, i.e. zygote index also decreased. As a result, the reproductive benefit, i.e. increased FS due to increased density and greater aggregation, was neutralized by reduction in gonad mass (Fig. 1.5).

1.4 Energy Budget and Reproduction

Energy budget: In animals, energy budget is assessed by estimations on C = F + U + R + P, where C is the food energy consumed, F, U and R, the

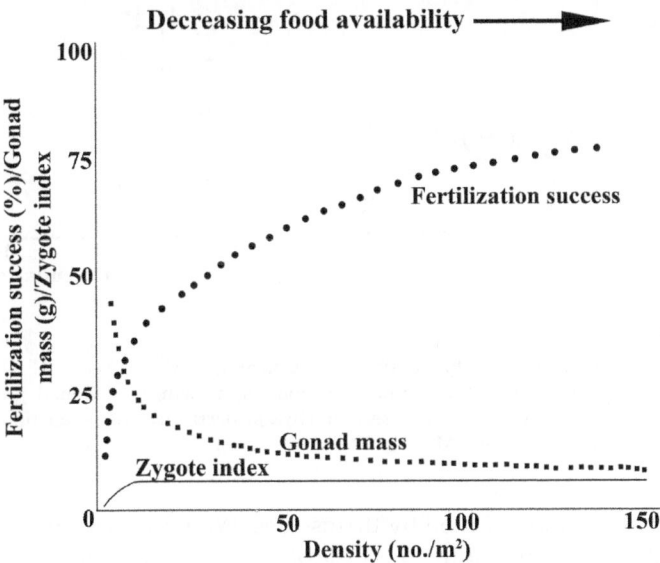

FIGURE 1.5

Rough sketch to show the relations between fertilization success/ gonad mass/zygote index and population density in *Strongylocentrotus droebachiensis* (compiled and redrawn from Wahle and Peckham, 1999 and Levitan, 1991). An arrow indicates decreasing food availability with increasing density.

energy lost on feces, urine and metabolism, respectively and P, the energy gained due to growth and/or reproduction (e.g. Pandian, 1987, Fig. 1.6A). In ecological energetics, one or more of these components have been estimated at population level in a few holothuroids. The physiological energetics is associated with relatively more precise estimates of these components (e.g. McBride et al., 1999). Available information on energetics of echinoderm is more or less limited to echinoids and holothuroids (Hyman, 1955), as (i) other echinoderms are not amenable to rearing in laboratory (however, see also Lawrence, 1987), (ii) almost all crinoids, ophiuroids and asteroids are of no direct economic importance, and hence, the need for rearing them has not arisen, (iii) C estimates are not possible in starving females during the breeding period; for example, the starvation lasts for 2 months in *Leptasterias hexactis* (Menge, 1975), (iv) ~ 3% of all echinoderms are capable of regenerating one or more tissues, organs and systems (Table 5.5) and (v) ~ 2% of holothuroids, asteroids and ophiuroids frequently undergo fission or clonal autotomy (Table 4.4); they make the precise estimates of energetics components a more difficult task.

No echinoderm is thus far reported to acquire nutrients engaging photoautotrophic or chemoautotrophic microbes. Hence, they necessarily

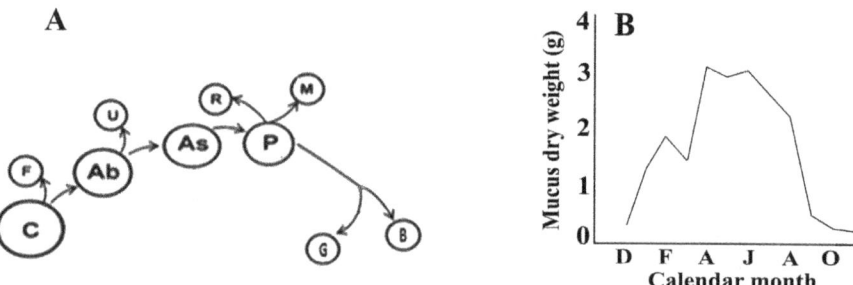

FIGURE 1.6

A. Energetics components in echinoderms. C = Consumption, F = Feces, U = Urine, R = Respiration, M = Mucus, P = Production, B = Somatic growth, G = Gonad growth, Ab = Absorption, As = Assimilation. B. Mucus secretion by *Cucumaria frondosa* as function of calendar month (redrawn from Hamel and Mercier, 1999).

depend on acquisition of food by themselves. With no need for estimates on F in the absence of the anus, the task of estimating energetic components becomes less in ophiuroids. Holothuroids are readily amenable for estimation of F. Intertidal sediments, in which particle size is ~ 0.1 mm, contain 0.02% organic nitrogen and 0.2% carbon. Hence, they support a rich fauna of microbes. Consuming the sediments with organic substances, holothurians serve as 'bioturbulants'. For example, an individual sea cucumber of *Paracaudina chilensis* consumes ~ 168 g/d or 62 kg/y. Population of *Stichopus* sp, inhabiting 1.7 km² areas in the Bermuda region, turbulates 500–1,000 t/y (see Pandian, 1975). Incidentally, *H. scabra* consumes 41 kg/cucumber/y (Mercier et al., 1999) and *H. atra* ~ 70 kg/cucumber/y (Bonham and Held, 1963). In the presence of capacious water-vascular system, no discrete excretory organ is present in echinoderms. Nevertheless, there are estimates on quantities of ammonia and urea excreted by a few of them. For example, 9 μg of ammonia/g dry weight/d is excreted by *Strongylocentrotusdroebachiensis* (see Pandian, 1975). The proportion of ammonia, urea and purine excretion amounts to 36, 14 and 7% in *Asterias rubens*, and 39, 6 and 12% for *Holothuria tubulosa*, respectively (see Pandian, 1975). Like most aquatic animals, echinoderms secrete mucus to (i) chemically clean themselves and (ii) collect small particles (Hyman, 1955). However, they do not secrete mucus, as profusely as molluscs. A rare publication by Hamel and Mercier (1999) indicates that mucus secreted by *Cucumaria frondosa* increases from ~ 0.2 g dry weight in December to 2.8 g in July and subsequently decreases (Fig. 1.6B). Besides serving as a chemical mediator for aggregation, more active locomotion and feeding during summer may also involve greater secretion of mucus. In fact, the seasonal changes in mucus synthesis are correlated with gametogenesis and Gonad Index (GI).

Gonad index and growth: Relatively more information is available on C of many sea urchins feeding on a range of algae. Bringing to light many Japanese publications, Agatsuma (2001a, b) has named a few chemicals secreted by algae that either stimulate or deter grazing by the urchins. Using the criteria of feeding responses of animals (Pandian, 1975), these chemicals are listed as repellants, suppressants, deterrents and stimulants (Table 1.9). Vadas (1977) has made a detailed investigation on food preference and absorption efficiency (Ae = A/C × 100) in the context of GI of *Strongylocentrotus* spp. From Table 1.10, the following may be inferred: 1. Of the chosen 8 algae *Nereocystis luetkeana*, *Laminaria saccharina* and *Costaria costata* are digested and absorbed ~ 2 times more efficiently by the urchins *S. franciscanus*, *S. purpuratus* and *S. droebachiensis* than those of *Agarum fimbriatum*, *A. cribrosum* and *Monostroma fuscum*. 2. The efficiency of the urchins decreases in the following order: *S. franciscanus* < *S. droebachiensis* < *S. purpuratus*. Interestingly, using chromic oxide (Cr_2O_3) as a marker, Sun et al. (2004) have estimated Ae of the sea cucumber *Apostichopus japonicus* as a function of diets consisting of different levels of proteins. With increasing protein content from 14.7 to 21.5%, Ae also linearly increases from 41 to 64%. Explaining the advantages and limitation of using Cr_2O_3 as a marker in estimation of Ae in fishes (Pandian and Marian, 1985a) and polychaetes (Pandian and Marian, 1993), Pandian and Marian (1985b) have shown that nitrogen content of food serves as a better marker to estimate Ae in these animals. Nitrogen content of food proves to be a good marker to assess/estimate Ae in urchins and cucumbers fed on different feeds and diets. Of course, values for some algae are scattered, as they may contain suppressants.

TABLE 1.9

Chemicals secreted by algae that act as repellants, suppressants, deterrents and stimulants to *Strongylocentrotus* spp (compiled from Agatsuma, 2001a, b)

Repellants

Phlorotannins secreted by *Ecklonia stolonifera*, *E. kurone*, *E. cava* and *Eisenia bicyclis* to repel *S. nudus*

Suppressants

Aplyisaterpenoid A and Macrocyclic-r-pyrene secreted by red algae *Phacelocarpus labillardieri* and *Placomium leptophylum* suppress feeding by *S. intermedius*

Deterrents

Bromophenol secreted by *Odonthalia corymifera* deters continuation of feeding by *S. intermedius*

Stimulants

Digalactosyldiacyl glycerol, 6-sulfoquinovosyldiacyl glycerol and Monogalactosyldia glycerol secreted by brown alga *Eisenia bicyclis* stimulates continuation of feeding by *S. intermedius*

TABLE 1.10

Food preference rank, absorption efficiency and gonad index of *Strongylocentrotus* spp fed on different algae (compiled from Vadas, 1977). * Values are for *S. droebachiensis*

Alga	Preference Rank	Absorption Efficiency (%)			Gonad Index*
		S. franciscanus	*S. purpuratus*	*S. droebachiensis**	
Nereocystis luetkeana	1	91	85	84	25.0
Laminaria saccharina	2	78	64	77	17.5
Costaria costata	3	83	77	78	–
Opuntia californica	4	–	–	–	–
L. groenlandica	5	–	–	–	–
Monostroma fuscum	6	48	28	56	–
Agarum fimbriatum	7	52	36	46	10.2
A. cribrosum	8	56	44	40	10.5

More importantly, the algae that are preferred by *S. droebachiensis* are not only digested and absorbed more efficiently but also allocated two times more for GI. The proportion of P allocated for gonad growth of *S. droebachiensis* of the Barent's Sea increases from 20% in young (3 years old) to 45 and 33% old (6 years old) and oldest (> 8 years old) females, respectively (see Scheibling and Hatcher, 2001). The Japanese experiments have confirmed that the allocation for gonad growth increases from ~ 5% in young (19 mm) to > 20% in mature (55 mm) *S. intermedius* females fed on *L. japonica*. Simultaneously, the allocation for somatic growth is decreased from 18.4 to 2.4% (Fig. 1.7 Left panel D). Evidently, gonad growth occurs at the cost of somatic growth. This negative relation also occurs in this Japanese urchin fed on different algae and in each alga with different rations. With regard to quality of feed, the algae that facilitate a greater allocation for gonad growth decrease in the following order: *Ulva pertusa* > *L. japonica* > *Alaria crassifolia* (Fig. 1.7 Left panel A, B, C). More interestingly, with increasing ration, the allocation is switched from somatic growth to gonad growth at a specific ration for these algae. Accordingly, the switch occurs at ~ 40, 70 and 190 mg dry feed/d for *U. pertusa, L. japonica* and *A. crassifolia*, respectively (Fig. 1.7 Left panel A, B, C, see the arrows). However, it is not clear, whether the provision of a wider range of ration/consumption shifts the switching point to higher ration. Secondly, higher ration of *A. crassifolia* facilitates greater allocation for gonad growth, in comparison to the limited ration/consumption range of *U. pertusa*. Figure 1.7 Right panel shows that the algae like *Scytosiphon lomentaria, L. japonica* and *Pachymeniopsis yendoi* facilitate a greater allocation for gonad growth than for somatic growth. But *U. pertusa, Rhodymenia palmata* and *Agarum cribrosum* allocate more for somatic growth than for gonad growth. This account indicates that the algae *S. lomentaria* and *L. japonica* increase the commercially important roe size.

As availability of algal feed decreases with increasing depth, depth has profound effects on population density and biomass released as gametes.

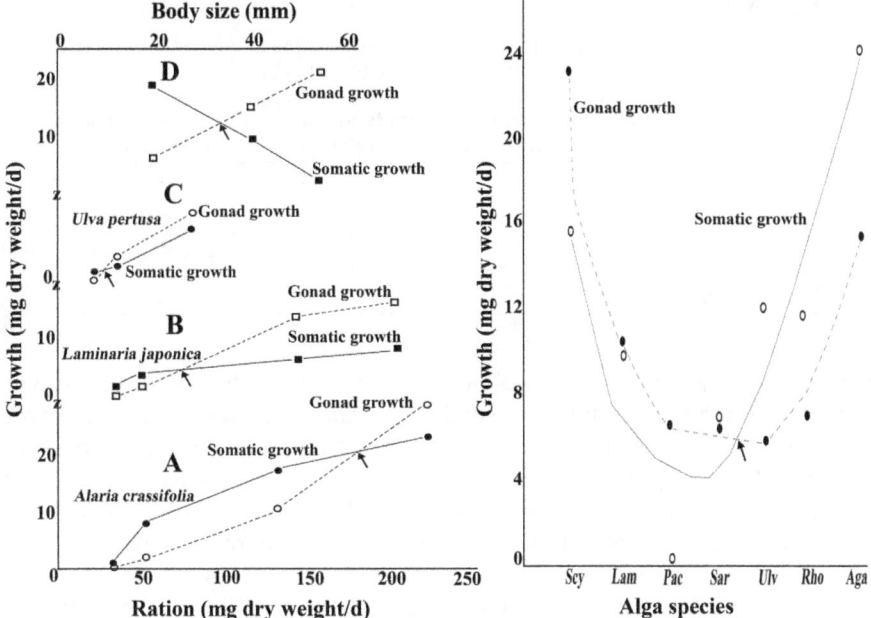

FIGURE 1.7

Left panel: Effects of algal food quality namely A = *Alaria crassifolia*, B = *Laminaria japonica* and C = *Ulva pertusa* and ration on allocation to somatic and gonad growth in *Strongylocentrotus intermedius*. D = Effect of body size on allocation to somatic and gonad growth in *S. intermedius* fed on *L. japonica*. Note the shift (indicated by arrows) of greater allocation to gonad with increasing body size. Right panel: Allocation to somatic and gonad growth of *S. droebachiensis* fed on different algae. *Scy* = *Scytosiphon lomentaria, Lam* = *Laminaria japonica, Pac* = *Pachymeniopsis yendoi, Sar* = *Sargassum tortile, Ulv* = *Ulva pertusa, Rho* = *Rhodymenia palmata, Aga* = *Agarum cribrosum* (drawn from data reported by Lawrence, 1987).

For example, biomass of *S. drobchiensis* decreases from 1,410 g/m² at 2–3 m depth to 522 g/m² at 12–18 m depth (Table 1.11). As a result, the biomass released as gametes by the urchin also decreases from 101.5 to 18.1 g/m² with corresponding decreases body weight lost on spawning from 8.0 to 2.9% (Keats et al., 1984).

1.5 Gonad Index and Fecundity

1.5.1 Gonad Index

Conventionally, the proportion of gonad to the body size is expressed as Gonado Somatic Index (GSI). However, echinoderm biologists report Gonad Index (GI) values, in view of large quantums of body fluids being maintained

TABLE 1.11

Estimated biomass of the green sea urchin *Strongylocentrotus droebachiensis* and its gametes as a function of depth (compiled from Keats et al., 1984)

Depth (m)	Body Weight Spawned (%)	Body Biomass (g/m²)	Biomass of Gametes Released (g/m²)
0–2	8.0	325	36.6
2–3	7.2	1410	101.5
6–9	3.6	668	24.0
12–18	2.9	522	18.1
Mean for depth range	–	–	22.9

in the unique water-vascular system. GI is calculated by dividing the wet weight of gonads by total live weight or drain (inclusive of gonads) wet or dry weight of the animal and is expressed as a percentage (Hopper et al., 1998). Draining the body fluids from the body reduces wet weight. For example, the reduction ranges from 13% in *Holothuria scabra* to 35% in *H. fuscogilva* (Conand, 1993). Limited available information reveals that 1. In general, planktotrophic holothuroids invest more on female gonads than male gonads (Table 1.12). 2. Surprisingly, lecithotrophic asteroids, which produce larger but smaller number of eggs, invest 4.3 to 8.6% more on male gonads. 3. This is also true of brooding asteroid *Leptasterias hexactis*. 4. It is difficult to comprehend why planktotrophic ophiuroid males invest 5 to 8% more on GI; notably, lecithotrophic ophiuroid males and females make equal investment on GI. 5. In both planktotrophic and lecithotrophic asteroids, nutrient reserves are accumulated in pyloric caeca to enable the ripening ovaries and testes to draw materials required for vitellogenesis and maturation, respectively. Many asteroids exhibit a negative relation between GI and pyloric caeca index (e.g. Crump and Barker, 1985). 6. Amazingly, the planktotrophic *Coscinasterias calamaria* female is able to invest as much as 67% of its body weight on GI. 7. In *Strongylocentrotus dorebachiensis*, quality of algal feeds and habitats induce great changes in GI (Table 1.13). Food preference, possibly induced by chemicals and protein content of algae may be an important factor in absorption efficiency (see also Pandian, 1985a, b), which ultimately provides more nitrogen for growth and reproduction.

1.5.2 Fecundity

Intraspecific comparisons: The total number of oocytes contributing to fecundity is assured by waves of oogonial proliferation and subsequent oocyte recruitment (see Pandian, 2013). Fecundity is a decisively important factor in recruitment at population level. With regard to fecundity, the terms used by fishery biologists have to be introduced to echinoderm reproductive biology: 1. Batch Fecundity (BF) is the number of eggs released per spawning. Because BF is related to the volume of space available to accommodate the

TABLE 1.12

Gonad index (%) of some echinoderms. Pl = Planktotroph, Lc = Lecithotroph

Species	Gonad Index		Remarks	Reference
	Female	Male		
Holothuroids				
			Oocyte (µm)	
Thelenota ananas	1.03	1.12	200 Pl	Conand (1981)
Microthele fuscogilva	2.18	1.74	170 Pl	Conand (1981)
Stichopus variegatus	2.7	2.8	180 Pl	Conand (1991)
Holothuria atra	3.3	3.3	150 Pl	Conand (1991)
M. nobilis	5.04	2.89	150 Pl	Conand (1981)
H. scabra versicolor	5.8	4.0	210 Pl	Conand (1991)
H. scabra	7.2	5.4	190 Pl	Conand (1991)
H. spinifera	8.7	1.44	148 Pl	Asha and Muthiah (2008)
Actinopyga echinites	9.0	6.9	165 Pl	Conand (1991)
A. mauritiana	14.3	13.8	135 Pl	Hopper et al. (1998)
H. fuscocinerea	20.5	11.11	121 Pl	Benitez-Villalobos et al. (2013)
Echinoids				
Strongylocentrotus droebachiensis*	50	50?	Pl	Thompson (1979)
Ophiuroids				
Ophiocoma alexandri**	40	45	70 Pl	Benitez-Villalobos et al. (2013)
O. aethiops**	20	28	80 Pl	Delroisse et al. (2013)
O. scolopendrina	1.7	1.8		
Asteroids				
Coscinasterias calamaria	67	35	150 Pl,	Crump and Barker (1985)
Storage	*30††*		fissiparous	
Echinaster sp type I	10.7	19.3		Scheibling and Lawrence
Pyloric caeca	*14.6*	*14.2*	840 Lc	(1982)
Echinaster sp type II	13.2	17.5		
Pyloric caeca	*12.5*	*10.9*	960 Lc	
Leptasterias hexactis	5.75	10.7		Menge (1975)
Storage	*649†*	*417†*	Brooder	

*g gametes/g soft tissues (in dry weights), **GSI, † = values in calories, Values for storage organ/pyloric caeca are given in italics, †† common to both male and female

ripe ovaries (F = aL^b), geometry suggests that length exponent 'b' would be 3.0. The 'b' value remains at 3, when growth in weight is isometric, i.e. the animal grows without change in shape. In bilaterally symmetrical vertebrates like fishes (e.g. Pandian, 2011), and invertebrates like crustaceans (e.g. Pandian, 2016) and molluscs (e.g. Pandian, 2017), the 'b' values mostly remains negative (< 3) or positive (> 3) (due to change in body shape with growth) resulting in allometry. In radially symmetric echinoderms, growth is measured in diameter of the test (e.g. *Strongylocentrotus droebachiensis*,

TABLE 1.13

Effects of food algae and their combinations, and habitats on gonad index (%) of *Strongylocentrotus droebachiensis*

Algal Species/combinations	Gonad Index
Algal food and combinations (Larson et al., 1980)	
Laminaria longicornis	22.2
Corallina officialis	19.7
Chondras crispus	15.7
Ascophyllum nodosum	9.8
Agarum cribrosum	6.1
Chondras : Laminaria	
25% : 75%	24.7
50% : 50%	24.3
75% : 25%	16.6
Laminaria : Ascophyllum	
25% : 75%	16.0
50% : 50%	18.4
75% : 25%	22.9
Ascophyllum : Chondras	
75% : 25%	14.0
25% : 75%	17.3
50% : 50%	18.6
Habitats (Vadas et al., 2015)	
Boothbay harbor	22.5
Schoodic Point	18.0
Owls Head	12.0

Larson et al., 1980), disk (e.g. ophiuroids, Schoener, 1972), weight (e.g. *S. droebachiensis*, Thompson, 1979) or arm radius (e.g. *Ophidiaster granifer*, Yamaguchi and Lucas, 1984). The 'b' value can be negative (= 2.44), as in *Salmacis sphaeroides* (Rahman et al., 2013). 2. Relative Fecundity (RF) is the number of oocytes/eggs per unit weight or measures like diameter of disk/test. RF is a very useful measure for comparison of spatial and temporal variations in reproductive performance. 3. Potential Fecundity (PF) is the maximum number of oocytes commencing to differentiate and develop. However, due to one or other environmental factors like food supply, level of favorable habitats, a fraction of the developing oocytes may be resorbed. 4. The number of remaining oocytes that develop to maturity is the realized fecundity (RzF). For example, PF of the deep-sea brittle star *Ophiomusium lymani* is 12,000 oocytes. However, 1,800 oocytes develop into eggs, i.e. RzF is 15% of PF (Gage and Tyler, 1982). 5. Incidentally, a distinction must be made between determinate (DF) and indeterminate (IF) fecundity. DF means the presence of number of oocytes undergoing synchronized maturation. Consequently, the eggs are spawned in a continuous pulse within a short time or intermittent pulses during a short period. Conversely, the IF means the presence of a large number of oocytes undergoing asynchronized maturation (e.g. *Holothuria atra*, Pearse, 1968). It results in continuous spawning. In fishes, planktotrophs represent the determinate capital (see Kjesbu et al., 2010) brooders. On the other hand, lecithotrophs may be indeterminate fecund incoming brooders (see Pandian, 2013, 2016). In echinoderms, investigation on these parameters of fecundity may be rewarding.

For fecundity, body size is considered as a critically important factor, as it provides space to accommodate nutrients and oocytes. Expectedly, fecundity increases linearly as a function of body size considered in terms of body weight in crinoids *Promachocrinus kerguelensis* (Fig. 1.8E), viviparous gonochoric asteroid *Leptasterias hexactis* (Fig. 1.8B), oviparous echinoid *S. droebachiensis* dry weight of soft tissues (Fig. 1.8A), drained weight of holothuroid *Leptosynapta clarki* (Fig. 1.8D) and arm radius of parthenogenic *Ophidiaster granifer* (Fig. 1.8C). In all these representative examples, the relationships are linear and positive. This would imply that Oogonial Stem Cells (OSCs) continue asymmetric divisions to generate OSCs and oogonial cells until their life time. The deep sea ophiuroid *Ophiomusium lymani* produce excess oocytes, of which 15% alone are matured (Gage and Taylor, 1982). With enormous ability to regenerate tissues, organs, systems (see Chapter 5) and whole animals (see Chapter 4), the echinoderms are well known to 'enjoy' "indeterminate growth and negligible senescence" (Bodnar, 2015). Conversely, 96% of crustaceans (except 4% broadcast spawners) suffer reproductive senescence (Pandian, 2016).

Interestingly, the following may be inferred from these representative examples: 1. The differences between minimum and maximum fecundity of *S. droebachiensis* remain constant across the entire length of body size and the former is ~ 50% of the latter, indicating the scope for increasing fecundity by providing suitable food in adequate quantities in aquaculture farms. 2. A comparison of survival of offspring at birth in viviparous *Leptasterias hexactis* (8 g body size) and pentactulae in oviparous *Leptosynapta clarki* (at 450 mg drained body weight) reveals that viviparity ensures > 88% survival but oviparity < 30% survival even at the pentactula stage (Fig. 1.8B, D), which may encounter further mortality prior to metamorphosis and recruitment. 3. Manifestation of parthenogenesis in *O. granifer* has reduced the number (mean ~ 1,000 eggs) of lecithotrophic diploid egg (~ 0.63 mm) production, in comparison to that of (mean 20,000) haploid lecithotrophic eggs (mean size ~ 450 μm) in gonochoric asteroids.

Relative fecundity values indicate 10.8 eggs/g eviscerated weight and 362 eggs/g ovary of *H. spinifera* (Asha and Muthiah, 2008) and upto 1000 eggs/g ovary in *H. scabra* (Pitt and Duy, 2004). With increasing body size, RF values of 8 holothuroid species indicate a decrease from 80×10^3 oocytes/g of *Actinopyga mauritiana* to 2×10^3 oocytes per g of *Thelenota ananas* (Fig. 1.8F).

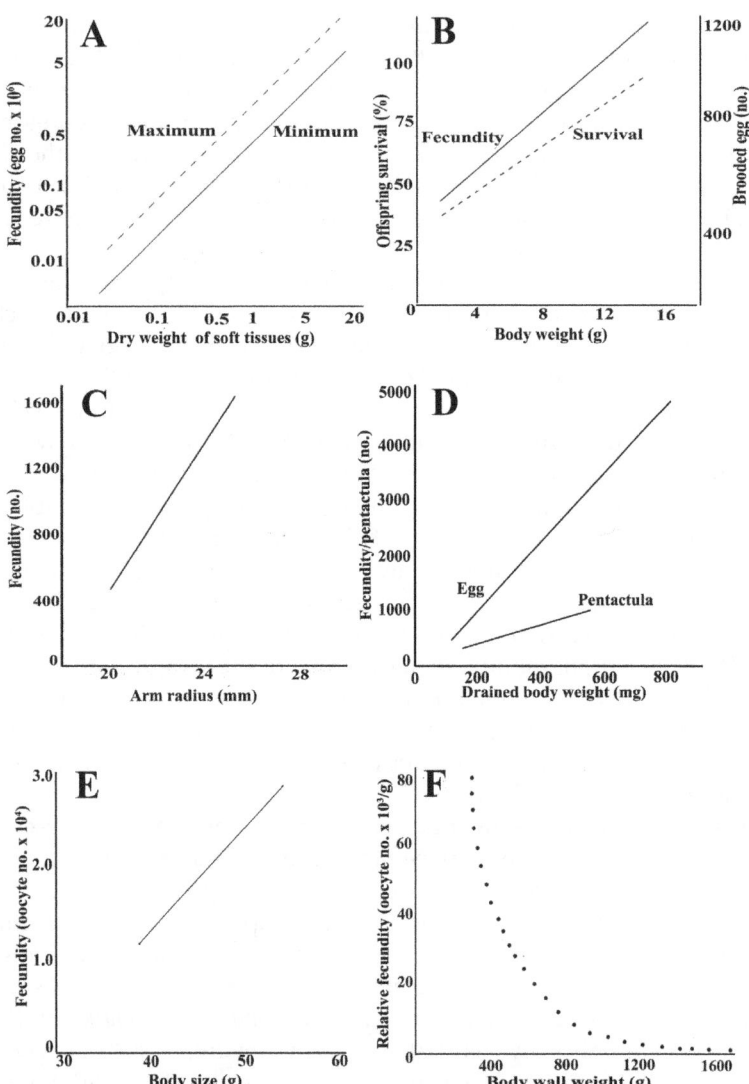

FIGURE 1.8

A. Linear relation between (minimum and maximum) fecundity and soft tissue weight of *Strongylocentrotus droebachiensis* (redrawn from Thompson, 1979). B. Linear relation between fecundity and offspring survival, and body weight in viviparous *Leptasterias hexactis* (drawn from data reported by Menge, 1975). C. Linear relation between fecundity and arm radius in parthenogenic *Ophidiaster granifer* (redrawn from Yamaguchi and Lucas, 1984). D. Linear relation between fecundity and pentactula larvae as a function of body weight in *Leptosynapta clarki* (modified and redrawn from Sewell, 1994). E. Relation between oocyte number and body size in *Promachocrinus kerguelensis* (redrawn from data reported by McClintock and Pearse, 1987). F. Relative fecundity as a function of body wall weight of holothuroids (drawn from data reported by Conand, 1993).

With complicated tubular ovary, 'absolute fecundity' (AF) has been estimated counting all the oocytes in the ovarian tubules of a female (Conand, 1993, Hopper et al., 1998). At best, these values can be considered as potential fecundity; these values range from $0.361 \times 10^3 - 5.195 \times 10^3$ with a mean of 1.739×10^3 in *H. spinifera* (Asha and Muthiah, 2008) to $26.3 \times 10^6 - 41.0 \times 10^6$ in *Actinopyga mauritiana* (Hopper et al., 1998). Values reported for realized fecundity for other holothuroids range from 295×10^3 in *H. fuscopunctata* to $13,281 \times 10^3$ in *H. nobilis* but for potential fecundity, it ranges from $7,861 \times 10^3$ in *T. ananas* to $78,517 \times 10^3$ in *H. nobilis* (Conand, 1993). In some holothuroids, the minimum and maximum values run almost in parallel (Fig. 1.9A), similar to those of *S. droebachiensis* (Fig. 1.8A). Within the geographic range, increasing temperature decreases the size of gonad, from which eggs arise; for example, larger eggs are produced by *S. droebachiensis* inhabiting colder waters.

Field and experimental investigations, especially those of Dr. Sophie George have shown the effects of favorable or unfavorable habitat, high or low ration and food quality on body size, fecundity and egg size (George, 1996). As shown earlier, body size is an important factor to provide nutrients for oocyte development and accommodation for ripening ovary. Consequently, any factor that affects body size may also affect fecundity. But it is not clear whether an echinoderm subjected to less favorable habitat and/or low ration opts to reduce egg size or egg number or both of them.

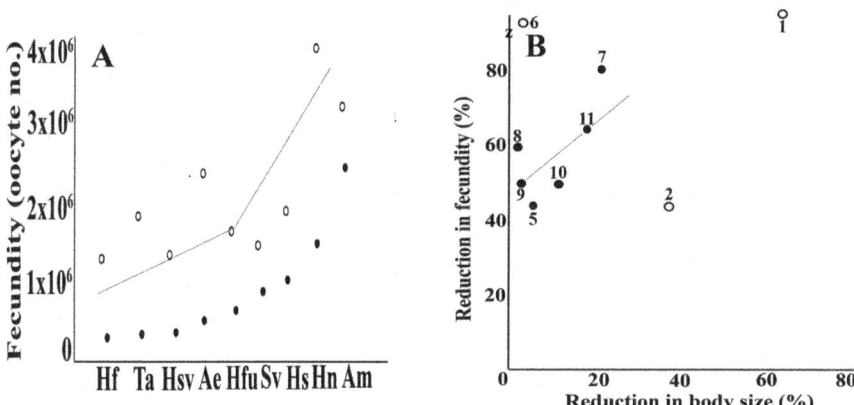

FIGURE 1.9

A. Minimum and maximum values reported for fecundity of holothuroids: Hf = *Holothuria flavomaculata*, Ta = *Thelenota ananas*, Hsv = *H. scabra versicolor*, Ae = *Actinopyca echinites*, Hfu = *H. fuscogilva*, Sv = *Stichopus variegatus*, Hs = *H. scabra*, Hn = *H. nobilis*, Am = *A. mauritiana* (drawn using data reporrted by Conand, 1993). B. Effect of body size reduction on reduced fecundity of echinoids and asteroids from less favorable habitat and fed low ration. For comparison, the values reported in Table 1.14 for body size and fecundity were calculated as percentage of these echinoderms from less favorable habitat and fed low ration.

Figure 1.9B shows the values calculated for the effects of reduction in body size due to reduced ration and/or less favorable habitats. From a critical analysis of values reported in Table 1.14 and Fig. 1.9B, the following may be inferred: 1. Reduction in ration and/or exposure to less favorable habitat do not significantly reduce egg size in both asteroids and echinoids. 2. With an increasing reduction in body size, there is a corresponding reduction in fecundity. Clearly, body size is a decisively important factor in determination of fecundity. 3. In the laboratory, the response of *Leptasterias epichlora* differs; its fecundity increases at the cost of body size (32% reduction) at low ration, but decreases at high ration. This may be a reason for scattering of values in Fig. 1.9B. Therefore, ration seems to be more important than favorable habitat for asteroids and echinoids.

Interspecific comparisons: Using data reported for ophiuroids (Schoener, 1972), asteroids (Villalobos, 2005) and others, attempts have been made to describe the relations between (i) egg size and depth (Fig. 1.10A), (ii) fecundity and depth (Fig. 1.10B), (iii) fecundity and body size (Fig. 1.10C) as well as (iv) direct/indirect development and body size (Fig. 1.10D). Firstly, eggs smaller than 300 µm and larger than 300 µm are considered planktotrophs and lecithotrophs, respectively; this classification holds good for echinoids (see McEdward and Miner, 2001a), and may also hold good for asteroids (however, see also Byrne, 2006) and ophiuroids (however, see Falkner et al., 2015). Secondly, limited data are available on this topic. For example, ~ 21% ophiuroids are reported from abyssal depths of 3,500–6,500 m (see Table 1.3) but no relevant information is available for ophiuroids inhabiting at depths deeper than 2,492 m. With these limitations, the following generalizations seem possible: 1. Planktotrophic eggs of asteroids (110–190 µm) are smaller than those of ophiuroids (250–300 µm). 2. Interestingly, egg size of planktotrophs remains equal at ~ 138 µm for asteroids and ~ 280 µm for ophiuroids throughout the respective ranges of vertical distribution. 3. Conversely, the lecithotrophic egg size increases from 546–1,561 µm for asteroids and 400–600 µm for ophiuroids with increasing depth (Fig. 1.10A). 4. With increasing depth, fecundity decreases in both ophiuroids and asteroids (cf Table 1.11). The decreases seem to be more pronounced in lecithotrophs (Fig. 1.10B). 5. Obviously, planktotrophic asteroids are more fecund than planktotrophic ophiuroids at any depth from 1,000 to 2,000/5,000 m (Fig. 1.10B). The same holds true for lecithotrophs too. 6. Unexpectedly, fecundity also decreases with increasing body size in ophiuroids and asteroids, except in planktotrophic asteroids, in which fecundity shows an almost inverse relation (Fig. 1.10C). 7. The decreases range from 18,000 eggs per female with disk diameter of 6.4 mm to 1,200 eggs per female of 16.3 mm and ~ 500 eggs to ~ 200 eggs/female in planktotrophic and lecithotrophic ophiuroids, respectively. In lecithotrophic asteroids too, the decrease is from 60,000 eggs in the star fish *Brisinga endecacnemos* female of 24 mm

TABLE 1.14

Effects of favorable habitat, ration and food quality on body weight fecundity and egg size in echinoderms from field and laboratory studies

Species	Reference	Body Size (g)		Fecundity (no.)		Egg Size (μm)	
		Favorable	Less favorable	Favorable	Less favorable	Favorable	Less favorable
1. *Echinaster* sp	Scheibling and Lawrence (1982)	5.8	2.0	7609	954	–	–
2. *Leptoasterias epichlora*	George (1994)	11.5	4.5	551	307	953	802
3. *Arbacia lixula*	George et al. (1990)	47.1	34.6	–	–	77	73
4. *Paracentrotus lividus*	George et al. (1990)	71.6	46	–	–	**92**	**90**
5. *L. epichlora**hr	George (1994)	7.8	7.2	583	331	950	957
6. *L. epichlora**lr	George (1994)	4.4†	3.0†	183	255	830	805
7. *Luidia clathrata**	George et al. (1991)	32.1	24.6	100††	20††	**166**	**178**
8. *A. lixula**	George et al. (1990)	43.7	43.4	0.75×10^6	0.3×10^6	77	76
9. *P. lividus**	George et al. (1990)	57.4	56.2	0.80×10^6	0.4×10^6	81	81
10. *Strongylocentrotus droebachiensis**	Thompson (1982)	50**	45**	1.0***	0.5***	91	93
11. *Stylocidaris lineatus**	George and Young (Unpublished)	8.2‡	6.4‡‡	23604	7714	91	93

* Laboratory results for high and low rations, *† Laboratory results also for favorable and less favorable habitats,
** Test diameter (mm), *** Dry weight of gamete released, †† Fertilization success (%), ‡ *Thalassia*, ‡‡ *Sargassum*, Numbers in bold indicate no significant change in egg size, Numbers in italics indicate no significant change in body size, hr = high ration, lr = low ration

size inhabiting at 1,700 m depth to *Pythonaster atlantidis* female collected at 4,700 m depth (not shown in figure). 8. A striking contrast is that ophiuroids larger than 18 mm size do not produce planktotrophic eggs but asteroids larger than 110 mm size alone produce planktotrophic eggs (Fig. 1.10C). 9. In ophiuroids characterized by an indirect life cycle, the female size ranges from 5 mm in *Ophiacantha densispina* to 39 mm in *Ophionotus hexactis* and fecundity from 3 in *Ophiolebella biscutifera* to 550 in *Ophioplocus esmarki* (Schoener, 1972). Briefly, a maximum of 100 offspring is produced by adults not larger than 15 mm size. On the other hand, most of the planktotrophic ophiuroids produce up to 5.5 million eggs (Fig. 1.10D). Notably, the trends reported by Schoener (1972) and Villalobos (2005) for the body size-fecundity relation of lecithotrophic ophiuroids differ, as the former has taken body weight, while the latter disk size into considerations (see Fig. 1.10C, D). These generalizations, of course, are made for the first time more to show the need for further research on egg size and fecundity of ophiuroids and other echinoderms inhabiting greater depths.

Reproductive effort and output: Reproductive Effort (RE) is measured as the relative proportion of substances or energy allocated to reproduction annually. Data on RE is available for the green urchin *S. droebachiensis* (Thompson, 1979) and asteroid *Echinaster* (Scheibling and Lawrence, 1982). For example, RE was almost zero but increased to 0.90–0.95 in urchins of > 5 years old. In *Echinaster* type I and II, temporal and spatial variations are described. Reproductive output is measured as the relative proportion of substances or energy in the gonads at the peak of reproductive cycle (cf Table 1.11). In echinoderms, it is equivalent to GI, for which an account has been provided.

1.6 Egg Size and its Implications

In marine invertebrates, the evolution of egg size has remained a hot topic for life history biologists (e.g. Marshall and Bolton, 2007). The life history trade-off between large numbers of small inexpensive eggs versus a few large yolky eggs has been repeatedly analyzed to understand the adaptive significance of these reproductive strategies. Jaeckle (1995) provides an excellent account on the correlation between egg size, energy content and biochemical composition on one hand and mode of larval development in invertebrates on the other. With the presence of a wide diversity in egg size among closely related species, echinoderms have been extremely useful for the study of developmental evolution (Raff and Byrne, 2006). Egg size is correlated with important reproductive and developmental patterns including fecundity, quality and quantity of maternal investment per egg, Fertilization Success

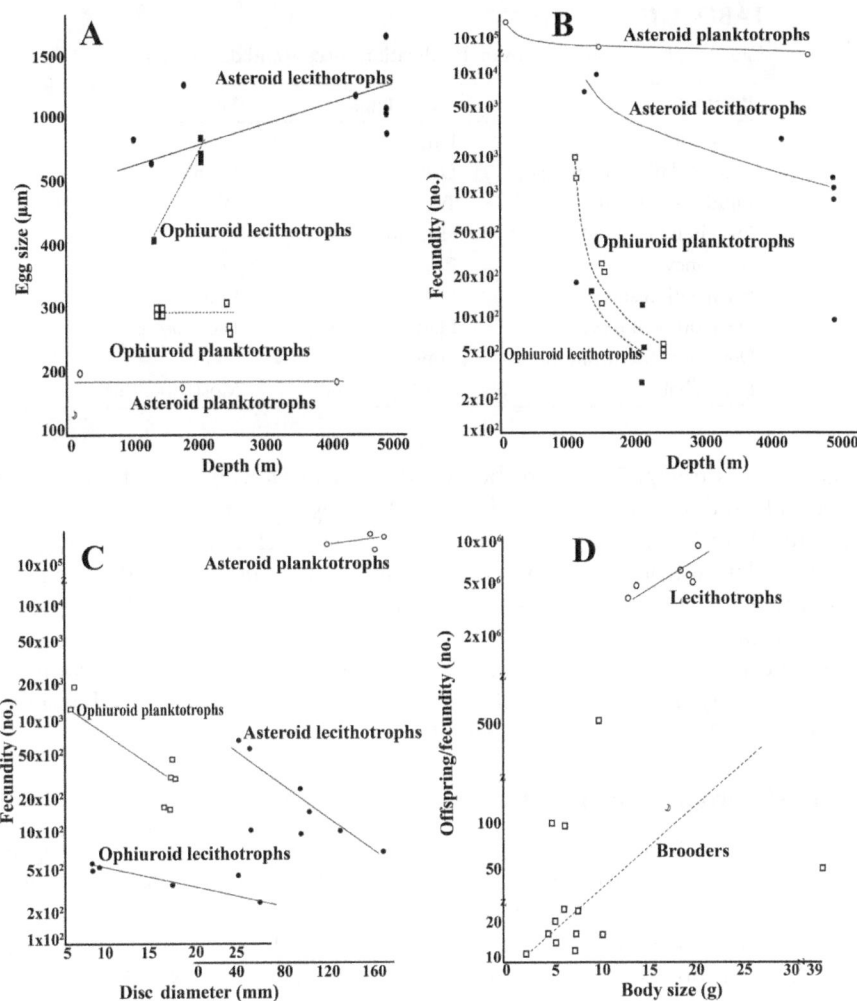

FIGURE 1.10

A. Egg size and B. Fecundity of asteroids and ophiuroids as a function of water depth, C. Fecundity as a function of body size of asteroids and ophiuroids, D. Fecundity as a function of body size in directly and indirectly developing ophiuroids. In Fig. 1A, a single value (ɔ) reported for planktotrophic *Holothuria fuscocinerea* is included (Benitez-Villalobos et al., 2013). In Fig. 1D, a single value (ɔ) reported for the largest viviparous *Leptasterias hexactis* (Menge, 1975) is included. All figures are drawn using data reported by Schoener (1972) and Villalobos (2005).

(FS), durations of embryonic and larval development, larval size and modes of development as well as larval habitat and dispersal (Table 1.15). The range of intraspecific variation in egg size is small (e.g. *Echinometra* type c, 71.83 μm, range 70.8–75.3 μm, Rahman et al., 2002; but, see also *Heliocidaris erythrogramma*, Marshall and Bolton, 2007); however, at interspecific level,

TABLE 1.15

Egg size and its implications on fertilization success and development

Parameter	Small Eggs	Large Eggs
Fecundity (no.)	Large	Small
Maternal investment/egg	Low	High
Lipid investment	Triglycerides	Waxes
Protein investment	High for structure	Low
Bouyancy	Positive	Low
Fertilization success (%)	High	Low
Development mode	Planktotrophs	Lecithotrophs
Development duration	Long	Short
Larval habitat	Mostly pelagic	Mostly benthic

it is very wide (e.g. 50 μm in the apodid *Synaptula reciprocans* to 4,400 μm in the elaspid *Psychropotes longicauda*). Although a continuum of egg size may result in facultative lecithotrophs, egg size dichotomies, reflecting the feeding planktotrophic and non-feeding lecithotrophic strategies have been unequivocally demonstrated within echinoderm classes (Sewell and Byrne, 2008). In fact, the facultative lecithotrophic strategy is evolutionarily short-lived and is rare at a given point of time (Herrera et al., 1996). For example, there are only 2 species out of 175 echinoids, whose life cycle has been analyzed by McEdward and Miner (2001a).

1.6.1 Maternal Investment

Egg size: The selective advantages of planktotrophic and lecithotrophic bimodal egg size distribution have been subjected to mathematic modeling. The early Vance model has viewed that the extremes namely planktotrophy and lecithotrophy confer maximum reproductive efficiency. The bimodal egg size distribution has been demonstrated in invertebrate taxa like molluscs and crustaceans. To know whether the bimodal egg size distribution occurs also in echinoderms, Sewell and Young (1997) have analyzed the species frequency as functions of raw and ln-transformed diameter of eggs of holothuroids (184 species), echinoids (131 species), asteroids (149 species) and ophiuroids (132 species). The ln-transformed data provide an objective means to examine the egg size distribution in a particular taxon and directly test the hypothesis of bimodality of egg size. The raw distributions indicate the presence of four modes in holothuroids, echinoids and asteroids (Fig. 1.11). The first mode comprises mostly species with smaller planktotrophic eggs, the second overlapping mode mostly lecithotrophs and a few brooders with intraovarian development, the third mode lecithotrophs and brooders, and the fourth includes species characterized by unknown developmental mode. On log transformation, only two modes become apparent in three classes of echinoderms. In ophiuroids, however, there are only three modes,

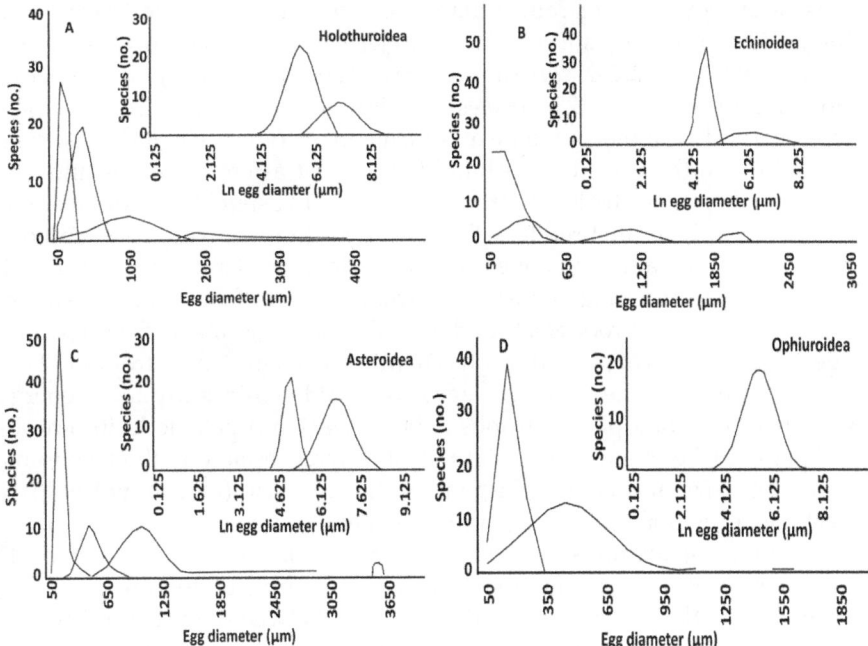

FIGURE 1.11

Species frequency as function of raw and ln-transformed (in the windows) egg size distribution in A. holothuroids, B. echinoids, C. asteroids and D. ophiuroids. Note the bimodal distribution in the first three classes and unimodal distribution in ophiuroids (modified and redrawn from Sewell and Young, 1997).

the first one at 165 μm consisting of mostly planktotrophs, the second at 450 μm consisting of abbreviated, brooded and unknown development and the third at 1,528 μm with a single species *Astrospartus mediterraneus* of unknown developmental mode. But, on natural log transformation, there is only one mode. From this analysis, Sewell and Young have concluded that the ln-transformed distributions are bimodal in holothuroids, echinoids and asteroids but unimodal in ophiuroids. The echinoids and asteroids display a clear demarcation between feeding and non-feeding development but the demarcation is less visible in holothuroids, as lecithotrophic and brooded developmental types are present in both modes.

Energy Investment: The reported values for energy invested on an egg ranges from 3.8 j/egg in *Echinaster* type I to 10.25 j/egg in *Anasterias perrieri* and from 25.9 kj/g eggs (dry weight) in *A. rupicola* to 29.9 kj/g eggs in *A. perrieri* as well as from 17.6 kj/g ovary (dry weight) in *Asterias vulgaris* to 32.6 kj/g ovary in *Acodontaster conspicuous* (Lawrence, 1987). As energy investment in egg(s)/ ovary vary so widely, the need for considering the maternal investment in terms of energy per egg rather than egg size is obvious. McEdward and Morgan (2001) made a survey of echinoderm eggs in relation to their size (in volume) (range: 2.2–23.2 nl) and energy (0.0013–161.3 j/egg) in 49 species belonging to asteroids (22 species), echinoids (18 species), holothuroids (4 species) and a crinoid; for some reason, no ophiuroid was included. Among planktotrophs, there were 15 echinoid species, 4 asteroid species and one holothuroid species; the lecithotrophs included 13 asteroids, 2 holothuroids and one crinoid; within brooders, there were 3 echinoids, 5 asteroids and one holothuroid. With a shift in development mode from planktotrophy through lecithotrophy to brooding, there is significant increase of maternal investment in lecithotrophs and brooders. Brooders make their eggs that are significantly larger than the eggs of planktotrophs (16 nl) and pelagic lecithotrophs (42 nl). In terms of energy, they produce eggs that hold significantly more energy (66.9 j per egg) than planktotrophs (0.009 j/egg) and pelagic lecithotrophs (4.4 j per egg). Yet, there are no significant differences among the taxonomic classes, egg size and energy. In general, the egg size (in volume) holds a direct linear relation with energy more than egg content.

Revisiting the relationship between egg size and energy content of 47 echinoderm species including the 29 taxa and a couple of ophiuroid species, Moran et al. (2013) confirmed the direct relation between egg size and energy content. However, differences in water content among eggs of planktotrophic species alter this linear relationship to some extent. For example, *Echinometra viridis* egg holds a higher percentage of water and measures 1.2 times larger than that of *E. lucunter* but energy content of their eggs is equal. Another factor that may interfere with this linear relation is variations in egg size within a species. Wide range of variations in egg size of *Heliocidaris erythrogramma* has been reported. The variations in egg size are 23% within a brood and 13% among broods. Within a single brood of eggs, the emerging smaller larvae settled within 4 days, while the larger ones may take a longer duration (Marshall and Bolton, 2007). This dichotomy in egg size and larval duration is similar to that of poecilogony in molluscs (see Pandian, 2017).

Biochemical composition: To understand the adaptive strategies of planktotrophic and lecithotrophic echinoderms, data on lipid (inclusive of its classes), protein and carbohydrate contents of an egg provides a better picture and more reliable information (see Fig. 1.12) than gravimetric and/ or calorimetric analyses. Hence, recent studies have focused on lipid and protein contents of a single egg, as they provide ~ 80% of the total energy in an echinoderm egg (Jaeckle, 1995). Energetic lipids (e.g. triglycerides, TG) are utilized during the pre-feeding developmental stage, but structural lipids (acetone-mobile polar lipids, AMPL, cholesterol) and protein remain relatively stable (Table 1.16), as they are used as materials to construct larval body (e.g. Prowse et al., 2008).

In planktotrophs of three echinoderm classes, the proportion of structural lipids contribute 62%, 57% and 63% in eggs of representative ophiuroids

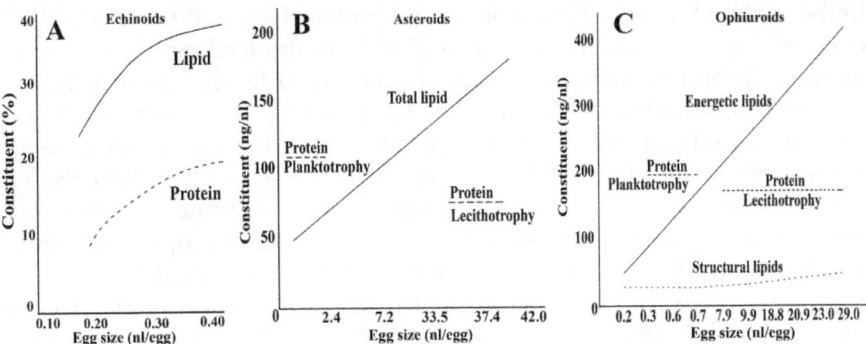

FIGURE 1.12

A. Lipid and protein constituents as a function of egg size in echinoids considering the size of 0.16, 0.17, 0.18, 0.31, 0.38, 0.40 and 0.43 nl/egg of *Diadema maxicanum, Echinometra vanbrunti, D. antillarum, Ec. lucunter, Ec. viridis, Eucidaris thouarsii* and *Eu. tribuloides,* respectively (redrawn from McAlister and Moran, 2012). B. Lipid and protein constituents as a function of egg size of 2.4, 7.2, 33.5, 37.4 and 42.0 nl/egg in asteroids *Patiriella regularis, Meridiastra mortenseni, M. oriens, M. calcar* and *M. gunnii,* respectively (drawn using data reported by Prowse et al., 2008) C. Structural and energetic lipid classes and protein constituent as a function of egg size in ophiuroids considering the size of 0.2, 0.3, 0.6, 0.7, 7.9, 9.9, 18.8, 20.9, 23.0 and 29.0 nl/egg for *Ophiocoma dentata, Ophiactis resiliens, Ophionereis fasciata, O. schayeri, Ophiopteris antipodum, Clarkcoma canaliculata, C. pulchra, Ophiarachnella ramsayi, Ophiocoma endeani* and *Ophiarthrum elegans* (approximate sketch redrawn from compiled figures and table of Falkner et al., 2015).

TABLE 1.16

Structural and energetic lipid classes in eggs of echinoderms. * TG, ** Wax

Lipid Class	Planktotrophs	Pelagic Lecithotrophs
Asteroids (Prowse et al., 2008)		
Structural (%)	63	18
Energetic (%)	37	82
Ophiuroids (Falkner, 2006, Falkner et al., 2015)		
Structural (%)	62	18
Energetic (%)	38	79* + 3**
Structural (%)	57	10
Energetic (%)	43	70* + 12** or 16* + 66**
Echinoids (Sewell, 2005, Byrne et al., 2008)		
Structural (%)	43–55 (49)	–
Energetic (%)	45–58 (51)	–

(4 species, Falkner et al., 2015, 1 species, Falkner, 2006), echinoid (1 species) and asteroids (2 species), respectively (Table 1.16). Conversely, their proportion is limited to 18% in all the lecithotrophs. Clearly, the structural lipids are utilized for the construction of swimming and feeding structures

in the feeding bipinaria/brachiolaria and pluteus larvae. While the levels of energetic lipid classes are limited to 37–43% in the feeding planktotrophic larvae of asteroids, echinoids and ophiuroids, their levels remain as high as 82% in lecithotrophs. Of the energetic lipid classes, TG constitutes 79, 97% in planktotrophic and lecithotrophic echinoderms. Hence, TG serves as the main energy source to meet the developmental metabolic cost. Notably, wax ester, new lipid classes discovered by ophiuroids, constitutes 3–66%. In fact, the role of TG as an energy source is reduced in lecithotrophic ophiuroids and is almost replaced by methyl ester in *Clarkcoma canaliculata, C. pulchra, Ophionereis schayeri* and *Ophiarachnella ramsayi*, and by wax ester in *Ophiocoma endeani* and *Ophiarthrum elegans* (Falkner et al., 2015). The offspring of an asteroid *Patiriella exigua* draw energy from TG, which makes up 95% of the total energetic lipids. Like planktotrophs and lecithotrophs, brooders also do not draw energy from protein, as it remains at a constant level during development.

Notably, the constituents of eggs present different pictures, when expressed in absolute quantitative value, i.e. ng/egg and relative value, i.e. ng/nl egg or %. With increasing egg size from 2.4 to 42.0 ng/egg in asteroids, maternal lipid investment also increases from 218 (range: 122–314) ng/planktotrophic egg to 7,650 (range: 6,117–8,980) ng/lecithotrophic egg (Prowse et al., 2008), i.e. the shift to lecithotrophy requires a 35-fold increase in maternal lipid investment in asteroids. However, the same requires only 13-fold increase in ophiuroids (see Falkner et al., 2015). In asteroids, protein investment also increases from 609 (range: 317–889) ng/planktotrophic egg to 2,731 (range: 2,359–3,103) ng/lecithotrophic egg (Prowse et al., 2008), i.e. the shift to lecithotrophy requires just 4.5-fold maternal protein investment. Corresponding values for the ophiuroid eggs with size range of 0.2 to 29.0 nl/egg indicates an increase from 100 ng/planktotrophic egg to ~ 3,250 (range: 1,000–5,500) ng/lecithotrophic egg (Falkner et al., 2015), i.e. the shift to lecithotrophy requires 33-fold maternal protein investment. Hence, increase in egg size requires accumulation of more lipids in asteroids but proteins in ophiuroids. Clearly, the shift to lecithotrophy has adopted different biochemical pathways in these two classes.

However, a different picture emerges, when relative values are considered. Protein content remains at the respective constant levels in planktotrophic and lecithotrophic asteroids and especially ophiuroids, irrespective of their increasing egg size from 0.6 to 29.0 ng/egg in the latter and 2.2 to 42 nl/egg in the former (Fig. 1.12B, C). In both asteroids and ophiuroids, the lipid level increases linearly with increasing egg size and it is more due to larger accumulation of energetic lipids. But the structural lipid level remains constant in ophiuroids and possibly asteroids. Interestingly, both protein and lipid levels (as percentage) increase with increasing egg size in an echinoid (Fig. 1.12A). Clearly, there is a need to present values either in ng/egg or ng/ng eggs; the former may prove useful for the analysis of maternal investment in a single egg and the latter for comparison.

1.6.2 Egg Size and Fertilizability

In his seminal contribution, Levitan (1993) has proposed an alternate (to fecundity-time hypothesis) hypothesis to explain the evolution of egg size of echinoderms in terms of the probability of Fertilization Success (FS). He has applied the fertilization kinetics model (Vogel et al., 1982) to test the probability of FS in three planktotrophic echinoids *Strongylocentrotus purpuratus* (egg size: 84, range: 70–85 µm), *S. franciscanus* (135, 120–140 µm) and *S. droebrachiensis* (145, 136–160 µm) with respective sperm velocity of 145, 130 and 88 µm/s under varying environmentally relevant egg-sperm densities at three selected egg-sperm contact durations; in these echinoids, given a certain amount of energy, sperm swim more quickly for a shorter duration or slowly for a longer duration. These tests have led him to show an increase in FS with increasing egg size, in spite of the associated decrease in sperm velocity of *S. droebrachiensis*, even when the egg-sperm contact duration is kept at the minimum duration of 3 seconds. Hence, the cost of reduced egg number, due to the production of larger eggs, can be neutralized by an increase in FS. Extending the hypothesis of Levitan by adding an attribute namely larval mortality, Podolsky and Strathmann (1996) have shown that selection favors (i) the extreme large or small egg size and not the intermediary one and (ii) larval mortality-dependent egg size, i.e. small eggs result in more juveniles, when larval mortality is low, while large eggs are favored under high larval mortality risks.

Increase in target size in small eggs of echinoids can be achieved by hydration or addition of external structures like jelly coat and the like. But such structures can also act as a physical barrier for sperm attachment and/or reduced surface area per unit volume. The jelly coat of the echinoid *Arbacia punctulata* consists of several concentric layers of polysaccharide fiber network embedded in a glycoprotein matrix (Bolton et al., 2000). According to Podolsky (2001, 2002), the coat of the sand dollar *Dendraster excentricus* is compressed before spawning but rapidly expands on contact with sea water. It expands to 80% of its maximum volume within 15 minutes of contact with sea water but the expansion becomes stable after 60 minutes (Fig. 1.13A). As it makes up 93% of the absolute volume of an egg and standard deviation > 10 times wider than for ovum, its absolute shift is ~ 14-fold in volume and 6-fold in target area for sperm collision but also increases in the organic cost by 1.2-fold/egg. Incidentally, its energy cost is 7.4% in *A. punctulata* (Bolton et al., 2000). Organic density of the ovum of *D. excentricus* is 67-times greater than that of jelly. Hence, the jelly coat around the ovum is a highly cost-efficient means of substantially increasing the volume from 0.013 to 0.019 µl per egg with increasing an ovum size from 0.001 to 0.0012 µl/ovum (Fig. 1.13B) and the physical target area. Besides, the specific gravity of jelly is ~ 1.023 close to that of sea water (1.0225 at 13°C). Despite increasing the size, the addition of jelly coat over the ovum reduces the sinking rate to 104 µm depth/s, i.e. an ovum without the coat sinks 12-times faster than that with

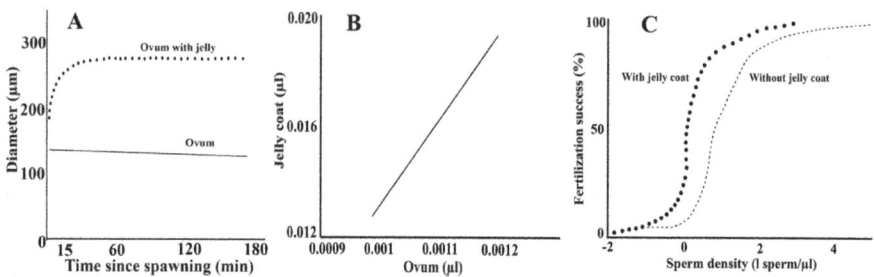

FIGURE 1.13

A. Changes in size of the ovum and jelly coat of *Dendraster excentricus* as a function of post spawning time. B. Jelly coat volume as a function of the ovum volume *D. excentricus* (redrawn sketches from Podolsky, 2001, 2002). C. Fertilization success of eggs with and without jelly coat as a function of sperm density of *D. excentricus* (simplified sigmoid sketches from Podolsky, 2002).

the coat. As a result, an intact egg remains suspended for a relatively longer duration amidst the 'sperm cloud', facilitating higher probabilities of sperm-egg collision and FS. Without harm, the jelly coat can readily be removed by exposing the eggs to acidic waters of pH 5.0–5.5. Using standard fertilization kinetics model (Vogel et al., 1982), Podolsky (2002) has achieved 100% FS at much lower sperm density in eggs with jelly coat than those without the coat (Fig. 1.13C), i.e. with the removal of the coat, FS is predicted to decline by 54–74% due to changes in the rate of sperm-egg collision at different sperm density. Apart from this important physical role in sperm-egg collision, the jelly is implicated with chemical interactive roles in fertilization relation to sperm activation (Suzuki, 1989), chemo-attraction (Ward et al., 1985) and sperm acrosome reaction (Vacquier and Moy, 1977). The quantitative impact of one or more of these chemical factors on the prediction model of Podolsky (2001, 2002) and that of Levitan (1993) in turbulent oceanic waters remains to be tested.

Egg hydration is another important factor to increase the target area of small eggs almost at no cost. Increased hydration of invertebrate eggs is often wrongly interpreted to limit the available ovarian space to store the oocytes. Hydration is an important event that facilitates floatation of pelagic eggs (see Pandian and Fluchter, 1968). Interestingly, there is a remarkable difference in timing of the hydration event. In fishes, it occurs within the ovary/body cavity; for example, the oocyte of pelagic spawning cod is heavily hydrated within the ovary (Pandian, 2013). However, it occurs in the majority of invertebrates, when the spawned eggs come in contact with sea water (e.g. crustaceans, Pandian, 2016, echinoids, Podolsky, 2001). Hence, egg hydration in invertebrates may not limit the ovarian space from storing oocytes/eggs. Research to explain the effect of hydration in small eggs on increasing target area and FS of echinoids and other invertebrates may be rewarding.

1.6.3 Endogenous and Exogenous Nutrients

Planktotrophic asteroids and echinoids develop through an initial pre-feeding period, during which they depend on endogenous nutrient reserves from eggs. This pre-feeding period lasts upto 4-, 6- and 8-arm pluteus stage in *Arbacia punctulata* (egg size: 75–80 µm), *Mellita quinquiesperforata* (110 µm) and *Leodia sexiesperforata* (208 µm), respectively. The facultative planktotrophs like *Clypeaster rosaceus* (280 µm) may bypass the pre-feeding period and metamorphose directly (Herrera et al., 1996). Briefly, the smaller the egg, the earlier is the pluteus arm stage, in which feeding is commenced. Incidentally, the estimates of pre-feeding stage in *Strongylocentrotus* vary with egg size with regard to feeding and ability to grow. For example, twins of *S. purpuratus* (41 µm) are reported not to develop beyond 6-arm stage but they (44 µm) are subsequently shown to successfully complete feeding larval stage, even under the low ration regime. A functional gut is developed in 6-arm pluteus of *S. drobrachiensis*, though the mouth is formed already in the 4-arm stage (McEdward, 1996). Hence, the pluteus is considered to depend on yolk reserves up to the 6-arm stage. But in the presence of the ciliary band, mouth and stomach, both intact and dwarfed 4-arm *S. drobrachiensis* pluteus are reported to filter sea water for food acquisition.

Of available endogenous nutrient reserves, the obligate planktotrophic echinoids like *Tripneustes gratilla* (Fig. 1.14A), *Evechinus chloroticus* (Sewell, 2005) and asteroids like *Patiriella regularis* (Fig. 1.14B) draw energy from energetic lipid TG but not from structural phospholipid. Conversely, TG and protein are scarcely used by pelagic facultative planktotrophic asteroids like *Mediastra calcar* (Fig. 1.14C) and benthic lecithotrophic like *Parvulastra exigua* (Fig. 1.14D). As a consequence, increasingly less TG is available for planktotrophics with decreasing egg size. For example, for egg size of 0.44 nl per egg in *Echinometra viridis* and 0.21 nl/egg in *E. vanbrunti*, the available TG is reduced to 8.3 and 3.4 ng/egg, respectively. Consequently, ration level becomes increasingly important for larval survival and critical size to attain Metamorphic Competence (MC). Within a 25-day period of larval rearing, *E. viridis* attains MC size, when fed low or high ration (Fig. 1.14E). With the smallest egg, *E. vanbrunti* attains the MC size on the 18th day of feeding on high ration (10 cells/µl) (Fig. 1.14F) but those receiving medium-(3 cells/µl) and low (1 cell/µl) ration do not reach Rudiment (R) stage, even after feeding for 25-days.

Egg size can also be manipulated by maternal nutrition to a certain extent but with far reaching consequences on survival and size at metamorphosis. Pre-conditioning the coral-eating seastar *Acanthaster planci* by feeding preferred coral prey *Acropora abrotanoides* or non-preferred coral *Porites rus* or starving for 60-days, Caballes et al. (2016) reported reduction in egg size from 25 to 22 µm and survival of early brachiolaria from 50 to 5%

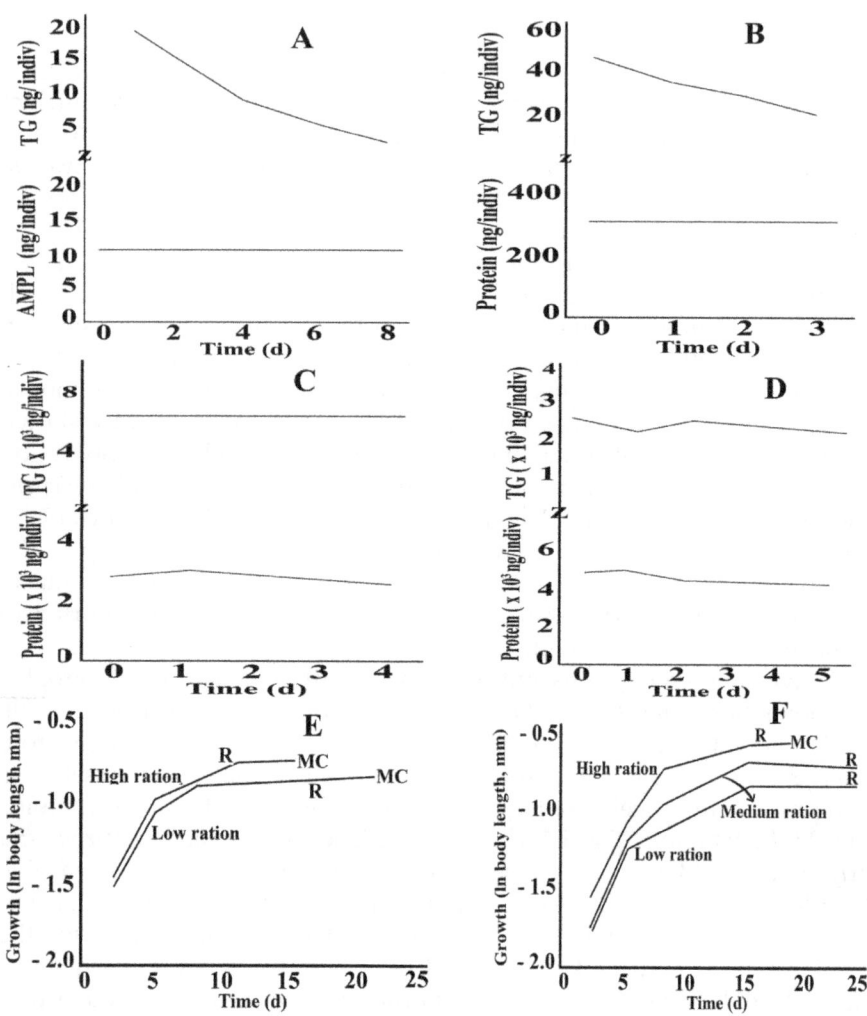

FIGURE 1.14

Sketches to show triglyceride as source of energy for the development of A. pre-feeding echinopluteus larva of *Tripneustes gratilla* (redrawn from Byrne et al., 2008) and B. bipinnaria larva of *Patiriella regularis*. C. Note protein scarcely serves for the development of pelagic non-feeding bipinnaria larva of *Meridiastra calcar* and D. benthic non-feeding bipinnaria larva of *Parvulastra exigua* (simplified sketches drawn from Prowse et al., 2008). E and F. Effect of ration regimes on growth to attain R (plutues with juvenile rudiment) and MC (metamorphic competence) size in *Echinometra viridis* (E) and *E. vanbrunti* (F) (from McAlister and Moran, 2013).

(Table 1.17). Remarkably, FS of these eggs remains at 96–98%, despite the 10% decreasing egg size (cf Levitan, 1993). Egg size can be manipulated to 50 and 25% of its whole size, resulting in twin and quadruplet larvae from the same egg.

TABLE 1.17

Effect of maternal nutrient on egg size and larval size in the coral eating starfish *Acanthaster planci* (compiled from data reported by Caballes et al., 2016)

Parameter	*Acropora*-fed	*Porites*-fed	Starved
Egg size (µm)	25	23	22
Egg volume (µl)	8.3	6.5	5.5
Fertilization success (%)	98	97	96
Normal larva at stage 3	50	45	5
Survival of brachiolaria (%)	50	44	5
Larval size (mm)	0.75	0.72	0.53

1.6.4 Egg Size Manipulations

In many marine invertebrates, correlations between egg size and life history characteristics have been reported at inter-specific level. In echinoids, blastomeres can be totipotent up to 4-cell stage and can be considered as 'eggs' of reduced size (Harvey, 1940). By virtue of their regulative development capacity, embryos can be manipulated to produce eggs of different sizes within a single species. This unique technique of intra-specific egg size manipulation has been used to test egg size-time hypothesis and others as well. These studies have helped to know the effects of intra-specific egg size differences on larval survival and duration as well as size at metamorphosis. Incidentally, quadruplets often display more abnormalities than twins. Unlike the first cleavage cutting the egg in its vertical plane, the second cuts through the blastomeres in the horizontal plane creates more disturbances to the animal/aboral and vegetal/oral axes (Cameron et al., 1996). In many lecithotrophic eggs, quadrupulation may not be possible as the cleavages of the first and second divisions are incomplete (see Schatt and Feral, 1996). However, experimental isolation by cutting embryos of lecithotrophic *Peronella japonica* into equatorial or meridional twins and quadruplets is possible but produces different results; both equatorial twins and quadruplets survive but the meridional twins and quadruplets survive better (Okazaki and Dan, 1954). As on date, eggs of seven echinoid species including an irregular sand dollar *Dendraster excentricus* have been manipulated; of them, *Clypeaster rosaceus* and *Strongylocentrotus droebrachiensis* have been tested for quadruplets, while all others are for twins. Echinoid egg size ranges from 70–90, to 280–350 and to 275–1,200 µm in the obligate planktotrophs, facultative planktotrophs and non-feeding lecithotrophs, respectively (McEdward and Miner, 2001a). However, information available is limited to one non-feeding lecithotroph *Heliocidaris erythrogramma* (400 µm), one facultative planktotroph *C. rosaceus* (270 µm) and five obligate planktotrophs with egg size ranging from 80 to 161 µm (Table 1.18). The technique employed to isolate twins and quadruplets is simple but different authors have used a range of procedures. Essentially, (i) exposure of just fertilized eggs to p-aminobenzene acid prevents hardening of the Fertilization Envelope (FE), (ii) exposure to calcium free or calcium-magnesium free sea water facilitates the removal of hyaline membrane and (iii) vigorous (*Arbacia punctulata*) or gentle shaking or stirring (e.g. *Echinarachnius parma*) and sieving through nytex mesh (species specific size) isolates the twins and quadruplets. Due to exclusion of lipid-rich component, centrifugation of 2-cell blastulae yielded half embryos containing 50% organics in *H. erythrogramma* (Emlet and Hoegh-Guldberg, 1997).

Table 1.18 summarizes relevant information on larval survival (see also Fig. 1.15C) and duration (see also Fig. 1.15D) as well as size at metamorphosis in egg size manipulated echinoids. Firstly, many publications have not reported the required information. Secondly, the reported values for size at metamorphosis on the basis of different structures (test, disk) and units (μm, μm^2, mm^2) render them not immediately comparable. Thirdly, as indicated elsewhere, there are contradictory reports on the functional ability to filter water and acquire food at the 4- or 6-arm pluteus stage (Fig. 1.15A, B). With limited information and other limitations, it is still possible to make a few generalizations: 1. Fairly large egg (270 μm) of the facultative planktotroph *C. rosaceus* enables successful development and metamorphosis of a quadruplet. However, the quadruplet (38 μm) of the obligately planktotrophic *S. droebrachiensis* (egg size 152 μm) is unable to develop beyond the 8R-arm stage, owing to unsuitable shape and inadequate ciliary band to feed enough (McEdward, 1996). Given higher ration, the larvae of *Echinometra vanbrunti* with the smallest egg (0.21 nl/egg) grow to Metamorphic Competence (MC) size, but are unable to pass beyond the R stage, when fed low ration (see Fig. 1.14F). Had Sinervo and McEdward (1988) reared the quadruplet *S. droebrachiensis* on high ration, the quadruplet may have also developed to MC stage. Apparently, the critical egg size, that enables a quarter embryo to successfully complete feeding larval stage and metamorphose, ranges between 152 and 270 μm. Experimental studies on egg size manipulation in planktotrophic echinoids with egg size of ~ 250 μm are urgently required. 2. Similarly, twins of *S. purpuratus* (41 μm) are reported not to develop beyond the 6-arm stage (Fig. 1.15A) but the twins of (each 44 μm) of the same urchin are shown to successfully complete larval stage and metamorphose even under low ration (Allen, 2012). From this analysis, it is proposed for the first time that the regulative developmental capacity of echinoids is limited to egg sizes of ~ 270 and 44–50 μm for the quarter and half embryos, respectively. 3. With decreasing egg size among the obligate planktotrophs, larval survival also decreases. Notably, twins that have been subsequently fed high ration, survive up to 58%, in comparison to 39% of these fed low ration (Fig. 1.15C). Values reported for larval survival and duration are scattered, as quality and quantity of food provided also vary widely. For example, both the intact and twins of *S. droebrachiensis* have been provided *Dunaliella tertiolecta* at 1,000 or 5,000 cells/ml by Allen (2012) but a combination of *D. tertiolecta*, *Isochrysis galbana* and *Rhodomonas* sp at an unknown density (Hart, 1995). Hence

TABLE 1.18

Effects of egg size manipulations on survival, larval duration and size at metamorphosis in echinoids

Egg Size (µm)	Survival to Metamorphosis (%)			Larval Duration (d)			Size at Metamorphosis (test µm)		
	Normal	High	Low	Normal	High	Low	Normal	High	Low
Heliocidaris erythrogramma (Emlet and Hoegh-Guldberg, 1997)									
400	95	–	–	Equal duration			424	–	–
200	95	–	–	~ 3.5 d			370	–	–
Clypeaster rosaceus (Allen et al., 2006)									
270	–	33	31	–	20	18	–	4.75*	4.70*
135	–	25	22	–	18	19	–	4.67*	4.60*
77	–	13	18	–	19	17	–	4.55*	4.46*
S. droebachiensis (Alcorn and Allen, 2009, Sinervo and McEdward, 1988, Hart, 1995)									
161	–	58	39	–	36	54	–	102**	68**
80	–	46	26	–	38	58	–	72**	58**
152	–	–	–	30	–	–	= size	–	–
76	–	–	–	36	–	–	–	–	–
> 125	–	–	–	36	43	57	427	–	–
~ 63	–	–	–	39	–	–	405	–	–
Echinarachnius parma (Alcorn and Allen, 2009)									
142	–	50	38	–	38	54	–	101**	–
71	–	41	36	–	42	62	–	91**	–
Dendraster excentricus (Allen, 2012)									
> 110	–	28	10	–	11.62	11.63	–	14.80	14.50
> 55	–	20	7	–	11.67	11.66	–	14.55	14.50
S. purpuratus (Allen, 2012)									
88	–	–	–	–	30†	65†	–	–	–
44	–	–	–	–	37†	45†	–	–	–
Arbacia punctulata (Allen, 2012)									
80	–	23	25	–	60	65	–	–	–
40	–	4	3	–	68	80	–	–	–

* disc area (µm²), ** disc area (mm²), † values for 50% metamorphics

food availability plays a critical role in their survival. However, there are corresponding decreases in larval duration with increasing manipulated egg size (Fig. 1.15D). Clearly, food availability plays a critical trade off between survival and larval duration with consequent effect on dispersal. 4. Extension of larval duration with decreasing egg size within a species confirms the conclusion arrived by interspecific egg size, and supports the egg size-time hypothesis. 5. In the said relation, alteration by food quality and quantity is limited to a level but not the relationship itself. 6. The relations between size

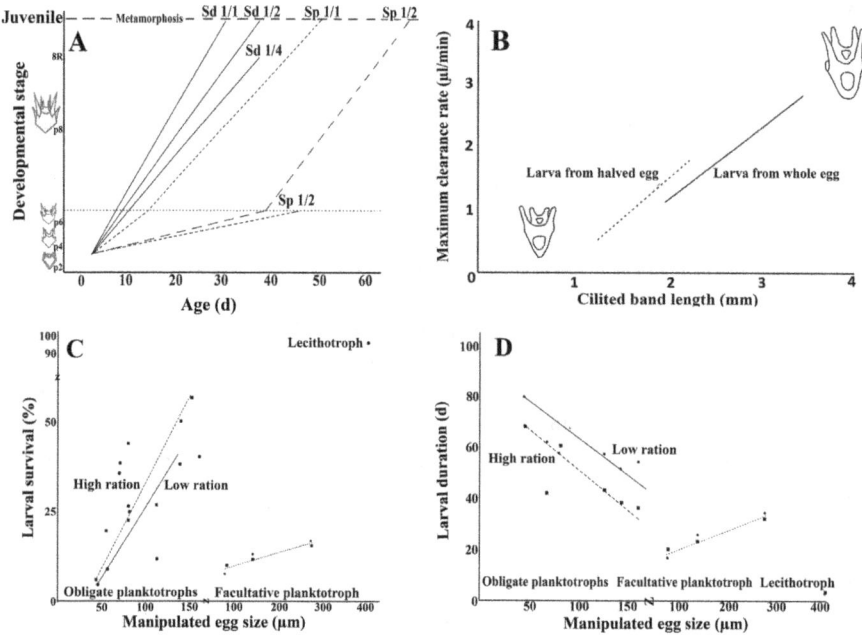

FIGURE 1.15

A. Developmental stage as a function of larval age in Sd = *Strongylocentratus droebachiensis* and Sp = *S. purpuratus*. 1/1 = Full size eggs, ½ = Half size eggs, ¼ = Quarter-size eggs. 2p = 2-arm pluteus, 4p = 4-arm pluteus, 6p = 6-arm pluteus, 8p = 8-arm early pluteus and 8r = 8-arm late pluteus with a juvenile rudiment (compiled and redrawn from Sinervo and McEdward, 1988 and McEdward, 1996). The dashed line labeled with Sp ½ indicates the observation reported by Allen (2012) for the twins of *S. purpuratus* successfully completing larval stage. B. Clearance rate of dwarf larva from halved eggs and normal larva from whole eggs as function of ciliated band length in *S. drobachiensis* (simplified sketch drawn from Hart, 1995). C and D. Effect of manipulated egg size and ration on larval survival and duration in planktotrophic, facultative planktotrophic and lecithotrophic echinoids. The values are taken from Table 1.18.

and age (larval duration) at metamorphosis (see Table 1.18) in *S. droebrachiensis* (125–161 μm) and *E. parma* (142 μm) with larger eggs show significant interaction for both size and age (larval duration) at metamorphosis. However, *D. excentricus* with medium egg size (110 μm) shows no significant interaction for either age or size. *Arbacia punctulata* with smallest egg (80 μm) shows significant interaction between ration and age, i.e. larval duration. 7a. In facultative planktotrophic *C. rosaceus*, survival of the twins and quadruplets is progressively decreased but not the larval duration. 7b. Expectedly, survival of both intact and twins remains high (95%), with equal larval duration (3.5 days) and size at metamorphosis. Briefly, twins in obligate planktotrophs suffer largely, when embryo size and ration are reduced.

1.6.5 Ontogenetic Pathways

Reviewing the ontogenetic developments of 20 species of crinoids, 69 holothuroids, 177 echinoids, 182 asteroids and 67 ophiuroids, McEdward and Miner (2001b) have justifiably considered three features namely morphogenesis, nutrition and habitat to identify ontogenetic pathways. Based on morphogenesis, the life cycle of a vast majority of echinoderms is recognized as indirect, which may be simple or complex. Direct development is limited to a few species of echinoids, asteroids and ophiuroids: for example, brooding occurs in 49 out of 175 echinoid species belonging to 4 orders and 5 families (McEdward and Miner, 2001a). From the nutrition point of view, development can be planktotrophic or lecithotrophic. It may occur in the pelagic realm or benthic habitat. Of 12 possible combinations of these three features, echinoderms have not evolved simple pelagic or benthic planktotrophs. Neither the evolution of pelagic nor benthic planktotrophic echinoderms with a direct life cycle has been possible. Briefly, (i) holothuroids, echinoids, asteroids and ophiuroids but not crinoids have feeding larval stage(s) (Table 1.19), (ii) all the five classes have evolved non-feeding larval stage(s) but (iii) benthic free-living, feeding larvae are not reported in any echinoderm (see also McEdward and Miner, 2001b).

Each class of echinoderms is characterized by a specific larval and/or a number of successive larval stages. Accordingly, doliolaria, auricularia, bipinnaria, echinopluteus and ophiopluteus (Fig. 1.16) are characteristic larvae of crinoids, holothuroids, asteroids, echinoids and ophiuroids, respectively. Based on the description of Hyman (1955) and McEdward and Miner (2001b), the ovoid bipinnaria is a pelagic, feeding larva bearing bilaterally arranged pre- and post-oral ciliated, swimming and feeding bands borne on the arms (e.g. *Asterias vulgaris*). The brachiolaria is a feeding larva but with specialized attachment surfaces (e.g. *A. rubens*). The yolky brachiolaria develops through a non-feeding pelagic brachiolaria, and lacks swimming and feeding ciliary structures, bipinnarian arms and a functional gut (e.g. *Solaster endeca*). The barrel-shaped larva with no arms, ciliated bands, mouth and anus is a non-feeding larva with abbreviated development (e.g. *Astropecten latespinosus*). Yolky non-brachiolaria is brooded, non-feeding larva, which lacks bipinnarian arms, brachiolarian arms and ciliated bands (e.g. *Ctenodiscus australis*). Mesogen is characterized by a complete absence of larval body and it is brooded to develop directly. Pluteus is a free swimming larva of echinoids and ophiuroids, whose larval arms are, however, not homologus. There is less variation and number of larval arms among ophiopluteus than in echinopluteus. For example, echinopluteus is a pelagic larva with eight anteriorly directed arms bearing ciliated feeding structures (e.g. *Strongylocentrotus droebachiensis*). The non-feeding pluteus has only non-ciliated two arms and has no gut (e.g. *Peronella japonica*, McEdward and Miner, 2001a, Fig. 1.16H). The ovoid auricularia is bilaterally symmetrical, planktonic feeding larva (Strathmann, 1971). A single ciliated

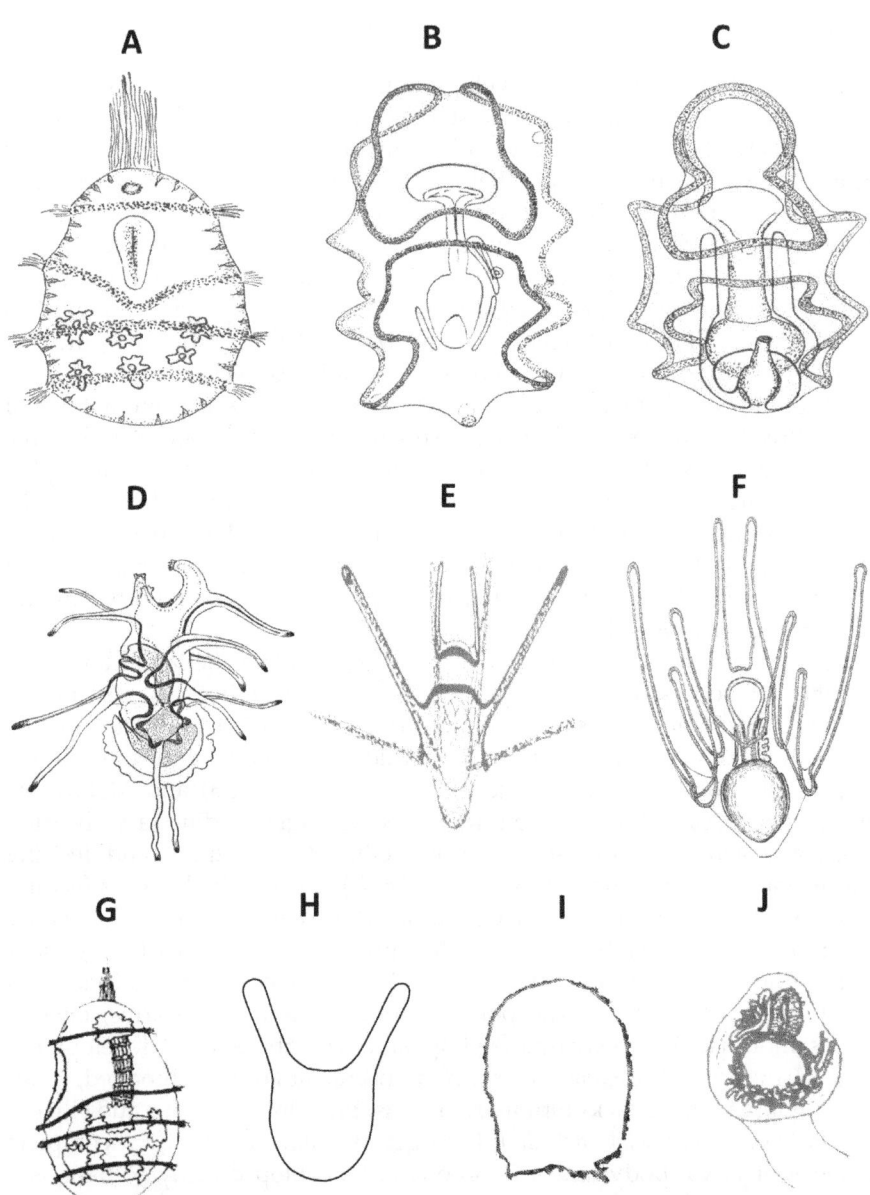

FIGURE 1.16

Larvae of Echinodermata: A. Doliolaria of crinoids, B. Auricularia of holothuroids, C. Bipinnaria and D. Brachiolaria of asteroids, E. Echinopluteus of echinoids and F. Ophiopluteus of ophiuroids. G. Vitellaria larva, H. Non-feeding pluteus of *Peronella japonica*, I. 'schmoo' larva of *Heliocidaris erythrogramma* and J. Vestigial pluteus of *Amphipholis squamata* (All are free hand drawings: A, B, C, D and F are free hand drawings from Hyman, 1955, E modified from McEdward and Miner, 2001a, G from an unknown source, H redrawn from Okazaki, 1975, I redrawn from Emlet, 1995 and J redrawn from Fell, 1946).

band (contrast with bipinnaria) extends along the lobes that project from the body (e.g. *Parastichopus californicus*). But the auricularia is soon transformed into doliolaria (Hyman, 1955). Doliolaria is a ciliated motile but non-feeding larva with no mouth and gut (e.g. *Antedon adriatica*, Hyman, 1955).

Comparing the feeding mechanisms and feeding rates of planktotrophic larvae of 15 species belonging to holothuroids, echinoids, asteroids and ophiuroids (crinoids have only non-feeding doliolaria larva), of which 12 species were reared from fertilization through metamorphosis, Strathmann (1971) has made a valuable contribution on the clearance rate of these larvae. Clearance rate, an approximate indicator of feeding rate, is the volume of water filtered for particle collection per unit time. Hence, these estimates may have a leading effect on the larval duration. (i) The clearance rate increases with increasing body size (Fig. 1.17) and (ii) considering equal body size, the rates are faster for *Ophiopholis aculeata* ophiopluteus representing ophiuroids than those of *Parastichopus californicus* auricularia, representing holothuroids, *Luidia foliolata* bipinnaria and *Pisaster ochracues* brachiolaria representing asteroids. (iii) Hence, the feeding mechanism of holothuroids is far less efficient, while the most efficient are the echinopluteus larvae. Lucas (1982) has reported that the filtration rate of *Acanthaster planci* begins to decrease at *Dunaliella primolecta* concentrations of ~ 4 × 10³ cells per ml and 5 × 10³ cells per ml in bipinnaria and brachiolaria, respectively. However, the ingestion rate of both the larvae progressively increases up to 5 × 10⁴ cells per ml. Not surprisingly, bipinnaria larvae are known to grow and attain very large sizes.

All the 20 reviewed crinoids are characterized by the presence of motile but non-feeding pelagic (Mariametridae, Himerometridae, Tropiometridae) or benthic (Thalassometridae) lecithotrophic doliolaria larva. Comasterid and antedonid species include either pelagic or benthic doliolaria. However, vitellaria larva replaces doliolaria in Notocrinidae (Fig. 1.18).

In holothuroids, the malpadids and psolidids pass through auricularia alone. Within cucumariids, some pass through auricularia alone but others vitellaria alone. Within chiridotids and holothurids, some pass through auricularia + doliolaria, while others pass through doliolaria and vitellaria. Likewise, some synaptids pass through vitellaria alone, but in others auricularia is succeeded by doliolaria (Fig. 1.18).

Echinoids have planktotrophic pluteus alone or non-feeding larva or both of these larval stages. Of 31 families belonging to 11 orders of echinoids, morphogenesis is completed by planktotrophic pluteus alone (Fig. 1.18). All the families namely Echinothuroidea and Phormosomatidae (Echinothurioida) as well as Urethinidae (Holosteroida) have evolved non-feeding larva alone. In Cidaridae (Cidaroida), Fibulariidae and Laganidae (Clypeasteroida), Temnopleuridae (Temnopleuroida), and Echinometridae (Echinoida), the feeding pluteus is succeeded by non-feeding larval stage. Schizasteridae (Spatangoida), another non-feeding mesogen larval stage is added.

TABLE 1.19

Distribution of developmental patterns among the echinoderm classes (condensed from McEdward and Miner, 2001b)

Class	Indirect			Direct	
	Planktotrophy	Lecithotrophy		Lecithotrophy	
		Pelagic	Benthic	Pelagic	Benthic
Crinoidea	–	+ Doliolaria	+ Vitellaria	–	–
Holothuroidea	+ Auricularia	+ Doliolaria	+ Vitellaria	–	–
Echinoidea	+ Pluteus	+ Non-feeding	+	–	+
Asteroidea	+ Bipinnaria	+ Yolky brachiolaria	+ Mesogen	+	+
Ophiuroidea	+ Pluteus	+ Doliolaria	+ Vitellaria, Mesogen	–	+

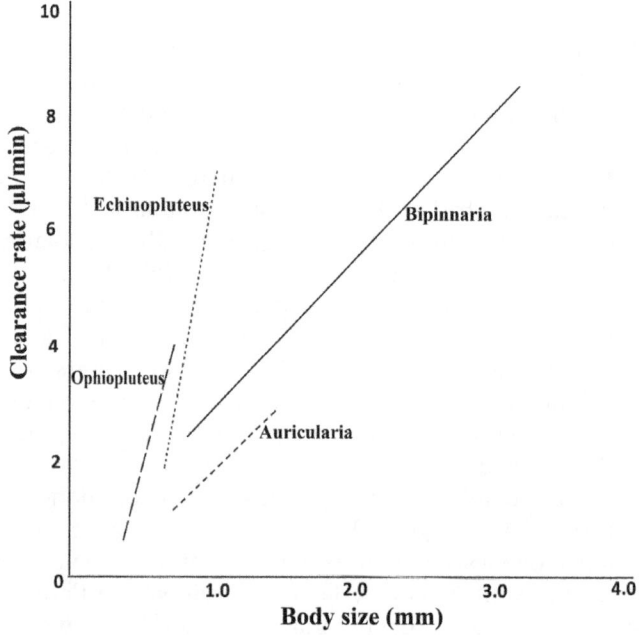

FIGURE 1.17

Clearance rate as a function of body size of echinoderm larvae. Note the trend for bipinnaria includes data reported for brachiolaria of *Pisaster ochraceus* (drawn using data reported by Strathmann, 1971).

Crinoidea

Pelagic or benthic doliolaria

Doliolaria → Vitellaria (Notocrinidae)

Holothuroidea
Doliolaria (Psolidae)

Auricularia or Doliolaria (Holothuriidae)

Vitellaria (some cucumariids)

Echinoidea
Pluteus (31 families in 11 orders)

Non-feeding larva (Echinothuroidae)

Pluteus → Non-feeding larva (Laganidae)

Pluteus →Non-feedig larva → Mesogen (Schizasteridae)

Asteroidea
Bipinnaria (Astropectinidae, Luidiidae, Notomyotida)

Bipinnaria → Brachiolaria (most valvatids)

Yolky brachiolaria (Solasteridae)

Yolky non-brachilaria (Goniopectinidae)

Yolky brachiolaria → Mesogen (Pteriasteridae)

Bipinnaria → Brachiolaria → Yolky brachiolaria (Asterinidae)

Ophiuroidea

Pluteus (Ophioactidae)

Pluteus → Doliolaria (Ophiocomidae)

Pluteus or Vitellaria (Ophiothricidae)

Pluteus or Doliolaria or Vitellaria (Ophiuridae)

Pluteus or Vitellaria or Mesogen (Amphiuridae)

Doliolaria or Vitellaria (Ophiodermatidae)

FIGURE 1.18

Recognized ontogenetic pathways through which larval development occurs in major classes of echinoderms.

Unlike sessile crionoids, sedentary holothuroids and slow motile echinoids (Table 1.1), the relatively fast moving asteroids and ophiuroids have evolved a more complex indirect life cycle involving one or more larval stages. Members of the families Luidiidae (Paxillosida) and Benthopectinidae (Notomyotida) pass through the bipinnaria larval stage alone. Likewise, those of Goniopectinidae (Paxillosida) complete the morphogenesis with yolky non-brachiolaria alone. And so are the valatid Solasteridae with yolky brachiolaria alone. In others like the Archasteridae, Odontasteridae, Mithrodiidae, Oreasteridae and Acanthasteridae, the bipinnaria is succeeded by brachiolaria. In still others (Astropectinidae, Paxillosida), it is succeeded by a barrel-shaped larva. In yet others like the Asteriidae (Forcipulatida), Asterinidae, Poraniidae and Ophidiasteridae (Valvatida), the bipinnaria is followed by a brachiolaria or yolky brachiolaria (Fig. 1.18).

Some ophiuroids pass through pluteus alone (Ophioactidae), pluteus + doliolaria alone (Ophiocomidae), doliolaria or vitellaria alone (Ophiodermatidae), pluteus or vitellaria alone (Ophiothricidae) as well as pluteus, vitellaria or doliolaria alone (Ophiuridae), or pluteus, vitellaria or mesogen only (Amphiuridae) (Fig. 1.18).

Briefly, crinoids lack a feeding larval stage. Doliolaria larva is absent in asteroids and echinoids. Non-feeding vitellaria/mesogen is present in all the classes. Four modes of development namely planktotrophic, facultative planktotrophic, lecithotrophic and egg brooding are common among all the classes (see McEdward and Miner, 2001b). The evolution of small-egg to large egg lecithotrophy requires increased provision of endogenous nutrients reserves. Initial increase may have resulted in facultative planktotrophic. Subsequent increase in provision may have simplified larval morphology with less larval feeding structures. The transition from planktotrophy to lecithotrophy has occurred independently at multiple numbers of times, for example, 14–20 independent times in asterinid asteroids (Byrne, 2006) within echinoids (see McEdward and Miner, 2001a).

1.7 Larval Developments and Thyroid Hormones

Thyroxine, one of the iodinated hormones produced by vertebrate thyroids, was reported to accumulate during larval development in three sea urchins (Chino et al., 1994), an asteroid *Acanthaster planci* (Johnson and Cartwright, 1996) and ascidian *Ascidia malaca* (Patricolo et al., 1984). A gradual accumulation of Thyroid Hormones (THs), drawn from the diatom *Chaetoceros gracilis* during development of *Hemicentrotus pulcherrimus* larva was observed by Chino et al. (1994); the accumulation reached the critical level at the 8-arm stage, when rudiment formation was completed. The exogenous production of thyroxines (T_3, T_4) in the larvae was demonstrated by starving and feeding sand dollar *Dendraster excentricus* on *Rhodomonas* or *Isochrysis galbana* (Hodin et al., 2001).

To T_4 exposure, the larvae of asteroid and echinoid responded differently. The exposure accelerated development of mid-brachiolaria as well as settlement and metamorphosis of *Acanthaster planci* (Fig. 1.19B). Hence, the asteroid brachiolaria was capable of acquiring T_4 from the medium during the exposure. But the acceleration was limited to settlement and metamorphosis alone but not to the pre-feeding 4- and 6-arm plutei of *Evechinus chloroticus* (Fig. 1.19A). In fact, the exposure slowed down the development in the pre-feeding plutei. During the subsequent feeding pluteus stage, the larvae fed on *Dunaliella tertiolecta* acquired adequate T_4 from the diatom and were ready to metamorphose. During this stage alone, exogenous T_4 accelerated settlement and metamorphosis dose-dependently. Clearly the non-feeding pluteus are unable to acquire T_4 from the medium during the exposure.

Experimental ration studies on *D. excentricus* pluteus led Heyland and Hodin (2004) to suggest that the pluteus can synthesize some T_4 or thyroid-like hormone endogenously but not enough to support development to metamorphic competence in the absence of food. That thiurea can only postpone but not prevent settlement and metamorphosis suggests that the larvae can synthesize some thyroid hormone as pre-adaptation for evolution of lecithotrophy in echinoids. Interestingly, independently evolved instances of lecithotrophy are concentrated in Spantangoida, Cidaroida and Clypeasteroidae, to which the planktotrophic *D. excentricus*, the facultative planktotrophy *Clypeaster rosaceus* and lecithotrophic *Peronella japonica* belong. In support of their suggestion, Heyland et al. (2004) found that 12%

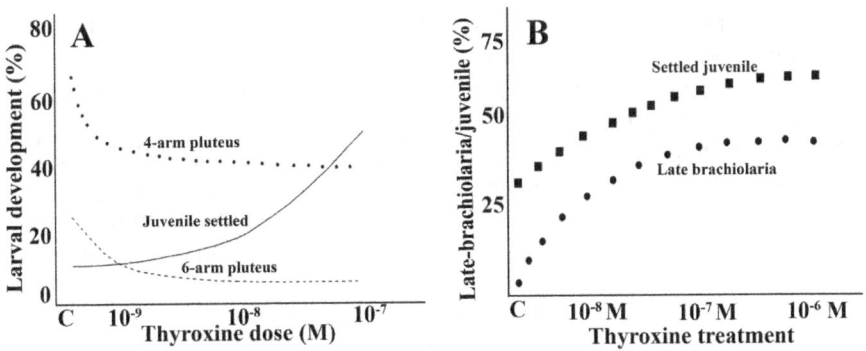

FIGURE 1.19

Rough sketches drawn to show the relationship between larval development/juvenile settlement and thyroxine treatment in A. *Evechinus chloroticus* and B. *Acanthaster planci* (drawn using data reported by Johnson, 1998 and Johnson and Cartwright, 1996).

TABLE 1.20

Effect of food, starvation and T_4 on metamorphosis of echinoids with different egg sizes (compiled from data reported by Heyland et al., 2004)

Species	Egg Size (µm)	Proportion of Metamorphosis (%)		
		Food	Starved	Starved + T_4
Mellita tenuis	100	50	0	0
	121	50	0	57
Leodia sexiesperforata	187	50	0	50
	202	50	12	50

of *Leodia sexiesperforata* larvae can successfully complete metamorphosis in the absence of food and without the exposure to T_4 (Table 1.20). In the obligatorily planktotrophic echinoids *Mellita tenuis* (egg size: 100, 121 µm) and *L. sexiesperforata* (egg size: 187, 202 µm), the authors demonstrated that despite 20% increase in egg size, *M. tenuis* larvae arising from larger eggs are unable to metamorphose in the absence of food and T_4, although the starved larvae are able to undergo delayed (on the 21st day instead of 12th day) metamorphosis in the presence of T_4 alone. However, with < 10% increase in egg size (from 187 to 202 µm), the larger eggs (202 µm) in *L. sexiesperforata* have led at least 12% of the larvae to successfully (delayed only by 2 days) metamorphose in the combined absence of food and T_4. The juveniles arising from the unfed but T_4 induced larvae are small and hold less lipid energy. Still, the fact that the larvae metamorphosed in the absence of food and T_4 led Heyland et al. (2004) to claim the successful transformation of planktotrophy to facultative planktotrophy. Understandably, lecithotrophic sand dollar *P. japonica* larva lacks a mouth and gut. During its embryonic development, T_3 and T_4 contents progressively increase to 30 and ~ 33 mol/10,000 larvae at 96 hours post fertilization (hpf), when the 2-arm pluteus is hatched; the larva metamorphoses subsequently within 7–8 hours. Hence, adequate T_3 and T_4 are synthesized endogenously to support embryonic and larval development as well as metamorphosis (Saito et al., 1998).

Notably, within the range of egg size (70–250 µm), some planktotrophs like *L. sexiesperforata* are capable of synthesizing T_4 endogenously. Within the egg size (280–380 µm) of facultative planktotrophs, some may have the ability to synthesize T_4 endogenously. Incidentally, a single publication by Tsushima et al. (1995) indicated the presence of 60-times higher concentration of carotenoids in lecithotrophic buoyant eggs of *Heliocidaris erythrogramma* than that of negatively buoyant planktotrophic *H. tuberculata*. It is not known whether the carotenoids have any role in determination of the reproductive mode. Hence, egg size, though important, is not the sole determining factor of reproductive mode in echinoids. The development is rather determined by a combination of factors like egg size, ability to endogenously synthesize signaling molecules like thyroids and others like carotenoids.

1.8 Brooding and Viviparity

Internal fertilization is perhaps a pre-requisite for brooding and viviparity. However, echinoderms lack an intromittent organ to facilitate internal fertilization. As a result, more than 44% females of the brooding holothuroid *Leptosynapta clarki*, for example, remain unfertilized (Sewell, 1994). Ventilation current is considered to draw adequate sperm to facilitate internal fertilization simultaneously in all the bursal sacs of *Ophionereis olivacea* (Byrne, 1991). But it is difficult to comprehend it, as ophiuroids do not have madreporite(s).

From his detailed study, Menge (1975) summarized a few striking features in the life history strategies of brooding and broadcasting sea stars. Brooding *Leptasterias hexactis* is small, attains sexual maturity at earlier age and smaller body size and produce fewer but larger eggs, in comparison to broadcasting *Pisaster ochraceus* (Table 1.21) From their extensive analysis of larval and life cycle patterns in 535 species of echinoderms, McEdward and Miner (2001b) concluded that direct development has been documented in asteroids, echinoids and ophiuroids. However, brooding of eggs, embryos and/or larvae also occurs in all the major classes of echinoderms. Table 1.22 lists some examples and external and internal brooding locations of the body. A few crinoids brood eggs and embryos, and then release the fully developed doliolaria (e.g. *Antedon bifida*). Others like *Isometra vivipara* delay release of the early benthic stalked cystidean larva that undergoes prolonged metamorphosis. Still others like the paedomorphic *Comatilia iridometriformis*, the ovarian brooder, release the late cystidean larva, in which metamorphosis is imminent (see McEdward and Miner, 2001b). Some holothuroids brood one or other larval stage for varying periods. Among synaptids, *Trochodota dunediensis* brood doliolaria, *Cucumaria rotifera* auricularia followed by doliolaria and *Synaptula hydriformis* vitellaria.

In brooding asteroids, echinoids and ophiuroids almost all the larval stages are suppressed and the brooding mother usually releases the fully

TABLE 1.21

Comparative accounts on brooding and broadcasting asteroids (compiled from Menge, 1975)

Parameter	Brooding *Leptasterias hexactis*	Broadcast Spawning *Pisaster ochraceus*
Size range (g)	3–8	300–650
Sexual maturity		
Size (g)	2.2	70–90
Age (y)	2	5
Annual fecundity (no.)	44.5	40×10^6
Absolute fecundity (no.)	356	40×10^9

TABLE 1.22

Reported observations on brooding echinoderms (from Hyman, 1955 and others)

Species	Reported Observations
Crinoidea	
Antartic comatulids	Gonads and brood chambers located in the arms at bases of pinnules
Isometra vivipara, Phrixometra nutrix, Notocrinus	Suppressed ciliated free-swimming stage
Holothuroidea	
Psolus antarticus	Young ones adhered to the creeping sole
P. dubiosus	Eggs retained in the tentacular crown
Thyonepsolus nutriens	Depressions/pockets in body surface
P. koehleri	Incubatory pouches on the body
Cucumaria (12 species)	Interradial marsupium
Echinopsolus (2 species)	Interradial marsupium
Leptosynapta minuta	Coelomic incubation
L. clarki	Ovarian incubatory pouch
Asteroidea	
Leptasterias hexactis	Arm 'pouch'
L. mulleri	Disk 'pouch'
L. almus	Aboral body 'pouch'
L. groenlandica	Ray based 'pouch'
Pteraster placenta	Nidamental pouch
P. vivipara	Intra-gonadal brooding
Echinoidea	
Abatus cordatus	Deepened petaloid pouch
Ctenocidaris spinosa	Peristomial brooding
Fibularia nutriens	Crescentric aboral 'pouch'
Ophiuroidea	
Stegophyiura sculpta	Sacci-from bursal expansion
Ophionotus hexactis	Ovarian incubation

developed young ones. Hyman (1955) indicated that among asteroids, ~ 20 species belonging to 8 genera are brooders. From his global survey, Emlet (1995) reported that 15 species out of 149 echinoids are brooders. With the rapid discovery rate of echinoid brooders between 1995 and 2000, McEdward and Miner (2001a) recorded 49 out of 175 species are brooders, i.e. 28% of echinoids are brooders. The ophiuroids present a different picture, i.e. only about 55 out of 2,064 ophiuroid species (< 3%) are brooders (see Stohr et al., 2005). Interestingly, *Ophiolepis paucispina* broods more than one clutch simultaneously but *O. kieri* does it sequentially (Hendler, 1979). A possible cannibalistic ingestion of inmates by *Neoamphicyclus materiae* is suggested (O'Loughlin et al., 2009).

Notably, the asteroids *Pteraster vivipara* and *P. paravivipara* are specialized brooders. Unlike other brooders with eggs up to 2.5 mm size (Hyman, 1955), these asteroids produce smaller eggs (< 150 μm) and juveniles (< 314 μm) and are cannibalistic and thereby attain larger size prior to release (Byrne and Cerra, 1996), a situation recalling nurse embryos of gastropods (see Pandian, 2017). In internally brooding holothuroids like *Leptosynapta clarki*, the pentactula larvae derive substantial nutrients from extra embryonic sources (Sewell and Chia, 1994). In ophiuroids, the extra embryonic nutrients support growth to a large size prior to emergence, as in *Amphipholis squamata* (Fell, 1946).

In brooding ophiuroids like *Ophionereis olivacea*, fertilization may be simultaneous in all the bursae and internal, as sperm are drawn in to the bursae by the ventilation current. In general, the egg measures from 400 μm in *O. olivacea* and *Ophiolepis paucispina* to 760 μm in *Sigsbeia conifer* but 100 μm only in true viviparous *Amphipholis squamata* (Byrne, 1991). With no need to provide nutrient to the embryos, the ophiuroids brood as many as 400–760 embryos, whereas the viviparous *A. squamata* bears only 2.2 embryos.

1.9 Size and Life Span

Echinoderms are small animals with mean life span of ~ 4 years. Holothuroids exhibit the widest size range, the smallest *Parvotrochus belyaevi* attaining a maximum body length of 1.5 mm and the largest *Holothuria scabra versicolor* measuring 48 cm and 2.8 kg (Table 1.23). The star fish *Pisaster ochraceus* attains arm length of 60 cm. These values may be compared with those reported for other aquatic invertebrates like crustaceans and molluscs; the largest body size attained is 4 m breadth and 20 kg for the giant spider crab *Macrocheirus kaempferi* (see Pandian, 2016) and 18 m arm span and 470 kg for the cephalopod *Architeuthis* sp (see Pandian, 2017). Echinoderms do not attain such large body size. Though they are estimated to live long (> 75 years, Ebert, 2008, Ebert et al., 1991), many reports indicate that they do not also display a long Life Span (LS). For example, the longest LS reported is 24 years for the green sea urchin *Strongylocentrotus droebachiensis* at St Andrews harbor, USA (Robinson and MacIntyre, 1997), in comparison to 20 years for freshwater mussels like *Elliptio arca* (North America) and 50 years for *Tindari callistiformis* in the North Atlantic at 3,806 m depth (see Pandian, 2017, Table 1.14). Incidentally, it must be indicated that size is a poor indicator of age in echinoderms. A 20-mm green urchin can be of age between 2 and 10 years and 50-mm urchin can range from 4 to 25 years (Robinson and MacIntyre, 1997). Too little data are available on LS and generation time of echinoderms to make any generalization.

TABLE 1.23

Largest size, life span (LS, y) and generation time (GT, y) of echinoderms

Species	LS	GT	GT/LS	Reference
		Holothuroids		
Holothuria atra	>9	~1	10.0	Chao et al. (1994)
		Asteroids		
Lepasterias hexactis, brooder, temperate	10.2	2.0	19.6	Menge (1975)
Pisaster ochraceus, broadcaster, temperate	10.2	3.4	33.3	Menge (1975)
		Echinoids		
Strongylocentrotus intermedius, temperate, planktotrophic	6–10	2	25.0	Agatsuma (2001a)
S. droebachiensis, temperate, planktotrophic	>8	>2	25.0	Vadas (1977), Scheibling and Hatcher (2001)
Echinus chloroticus, temperate, planktotrophic (predicted)	15	3	20.0	Barker (2001)
Observed at Doubtful Sand	8–9	3–4	41.2	Barker (2001)
Lytechinus variegatus, tropical, planktotrophic	3	1	33.3	Watts et al. (2001)
		Ophiuroids		
Ophiomusium lymani	6	3	50.0	Gage and Tyler (1982),
Acanthaster planci	9			Richard (1994)
		Largest size		
Crinoids				
Synaptids		90 cm long		Hyman (1955, p 198)
		Holothuroid		
Holothuria scabra versicolor		48 cm long, 2,800 g		Hamel et al. (2001)
Cucumaria pseudocurata		3 y 5 y 0.67		Rutherford (1973)
Parvotrochus belyaevi		1.5 mm, smallest cucumber		wikipedia
		Echinoids		
Centrostephanus rodgersii		12 cm test diameter		Andrew and Byrne (2001)
		Asteroids		
Pisaster ochraceus		60 cm arm length from tip to tip, 650 g		Hyman (1955, p 349) Menge (1975)
Thromidia gigas		70 cm arm length, 6 kg		wikipedia

2

Fisheries and Aquaculture

Introduction

The echinoderm fishery is small in scale. But global analyses reveal alarmingly high incidences of overexploitation and depletion of stocks of sea cucumbers (Purcell et al., 2013), urchins (Keesing and Hall, 1998) and other incidental biological resources of the coral reefs. As a result, capture production of sea cucumbers and urchins has begun to decline since 1990's (Fig. 2.1A, B). Unlike others, the capture fisheries of the cucumbers and urchins depend on manual collection. For example, global estimate indicates that the cucumber fisheries alone engage three million participant fishers, especially around the Great Barrier Reefs of Philippines (> 37% of 3 million), Papua New Guinea (> 18%) and Indonesia (15%) (Purcell et al., 2013). Despite a moratorium in India and Indonesia, and regulatory measures in other countries, large scale illegal fishing and trading of cucumbers continues. There is a need for international/bi-national cooperation (e.g. BOBLME, 2015). For example, it is difficult for India to implement a moratorium against the cucumber fishing and trading, especially in the cucumber stock-rich Gulf of Mannar, which adjoins Sri Lanka, where government has neither levied moratorium nor any regulatory measure. There is an urgent need for conservation of the fast declining stocks of sea cucumbers and urchins, and associated biological resources of the corals. There is also a need to minimize the embarrassment of countries implementing a moratorium and regulatory measures. At the same time, echinoderm fishers, especially from the developing countries, have to be engaged in alternate jobs. The scientific and technical intervention for the development of small scale aquaculture of sea cucumbers and urchins seems to be the only option. Thanks to the efforts of FAO, research activity in this area has been accelerated during the last two decades.

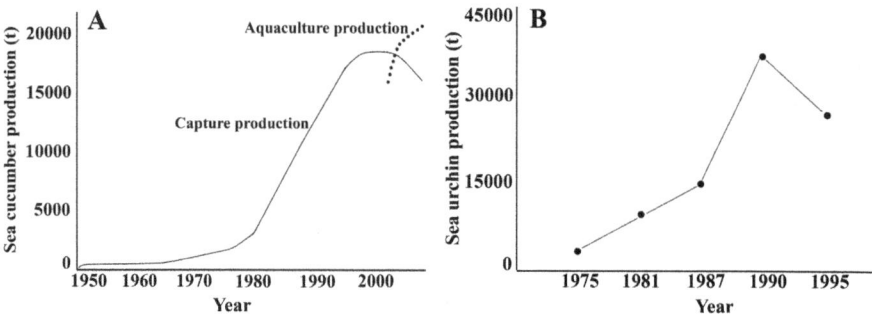

FIGURE 2.1

A. Global capture and aquaculture production of sea cucumbers during the last 55 years (smoothened trends redrawn from Purcell et al., 2013). B. Approximate cumulative production of sea urchins from Japan and USA during the last 20 years (drawn using data reported by Keesing and Hall, 1998).

2.1 Capture Fisheries

Holothuroids have been harvested since ancient times in the Indo-Pacific regions and used as a delicacy, especially during celebrations by the Chinese of South Asian countries. Globally, holothuroid fisheries are based on ~ 30 species. More specifically, only a few deposit-feeding species belonging to two families and five genera are harvested: *Actinopyca* and *Holothuria* (Holothuridae) and *Parastichopus*, *Stichopus* and *Thelenota* (Stichopodidae). The cucumbers are eaten raw, boiled and pickled. The dried body wall is marketed as beche-de-mer (Conand and Byrne, 1993). Processing of the cucumbers reduces to 10–40% of original weight and exportable product fetches 25 US$/kg. Hong Kong and Singapore are the major world markets for the beche-de-mer. In terms of both quantity and value, the trade in Hong Kong consistently have increased from < 1,000 t (2.5 million US$) during 1970's to > 6,000 t fetching 15 million US$. However, the trade has declined to < 4,000 t during 2000's (Anderson, 2010). In India, the decline in capture of *Holothuria scabra* is from 91 t in 1975 to 11 t in 1985 (James, 1989) and from 192,000 specimens in 1991 to 39,000 in 1993 in Papua New Guinea (Lokani, 1996).

Since pre-historic times, humans have been consuming the roe of sea urchins. Remarkably, all the edible sea urchins are found in shallow waters. Some urchins like *Tetrapygus niger* are not palatable, despite its abundance in the Chilean coast. However, bitter gonads of others like *Hemicentrotus pulcherrimus* becomes edible on preservation in brine or alcohol (Lawrence,

TABLE 2.1

Global capture fisheries of sea urchins (condensed from FAO, 1995, see also Keesing and Hall, 1998)

Region	Country/State	Species	Capture (t)
S. America	Chile	*Loxechinus albus*	54,609
Japanese Islands	Japan	*Anthocidaris crassispina*	13,735
	Hokkaido	*Strongylocentrotus intermedius*	5,163
	Pacific	*S. nudus*	3,923
	E. China Sea	*Pseudocentrotus depressus*	3,455
	Japan Sea	*Tripneustes gracilis*	807
	Japan total		27,083
N. Pacific	California	*S. franciscanus*	10,086
	B. Columbia	*S. franciscanus*	6,328
	Mexico	*S. franciscanus*	3,000
	N. Pacific total		19,414
Korea	S. Korea	*Evechinus chloroticus*	3,707
N. Europe	Russia	*S. nudus*	2,344
	Iceland	*S. droebachiensis*	923
Total capture including other countries			117,193

2001b). Available recipes for cuisine of the sea urchin roe are summarized by Lawrence (2001c). At least 16 species of sea urchins are harvested worldwide for food and the harvest totals to 0.12 million metric tons (mt) (Table 2.1). The three major countries, which harvest sea urchins are Chile (*Loxechinus albus*), Japan (*Anthocidaris crassispina, Strongylocentratus intermedius, S. nudus*) and the USA (*S. franciscanus*). Chile is the world's largest producer of sea urchins with landings of 54,609 metric tons (t). Due to overexploitation, however, many countries are encountering a progressive decline in their respective sea urchin fisheries (Keesing and Hall, 1998). Notably, Japan's sea urchin harvest has continued to decline from the peak of 27,528 t in 1969 to 13,735 t in 1995, despite increased sea ranching of sea urchin seedlings from 8 million in 1986 to 60 million in 1992. However, Japan imports 6,303 t at the cost of 243 million US$ to meet the domestic market. In Barbados, the capture rate has also decreased from 1,200 specimens per day in 1980s to 140/day in 1990s. The overexploitation of limited wild stock in Chile has led to opening up of new fishing grounds; but in the newly opened fishing grounds, the harvest has been declining even at the time of Sloan's (1985) review. On the other hand, the harvest of sea urchins in Canada remains stabilized, thanks to the effective implementation of conservation measurements. As sea urchin aquaculture is in its infancy, there is considerable scope for research to overcome bottlenecks and obstacles for the cost effective, commercial production of sea urchins.

2.2 Aquaculture

Since 2005, aquaculture production of sea cucumbers has commenced; ~ 5,000 tons of the cucumbers are being currently produced (Fig. 2.1A). An equal amount of urchins are also produced from aquaculture. At this juncture, the following observations suggest new areas of research to be focused in the years to come:

1. In Japan and the USA, macrophytic algae have been repeatedly tested as food for the urchins in their different forms like fresh, individual species (e.g. Larson et al., 1980) or combination of algae (e.g. Vadas, 1977), frozen (e.g. Bureau et al., 1997), hydrolyzed ingredient (up to 44%), in semi-moist granulated diet (e.g. Motnikar et al., 1997) or as an ingredient (up to 14%) in dry pelleted diet (e.g. Klinger et al., 1997). *Laminaria* spp are recognized as the best or second best food to enhance roe growth (Vadas, 1977, de Jong-Westman et al., 1995, Klinger et al., 1997). However, acceptance and preference of seagrass remain to be tested. In tropical countries like India, benthic macrophytic algae and seagrasses are abundantly available in the coastal zone, rendering their collection at low cost and providing alternate employment for fisher folks. The Palk Bay of India is reported as seemingly inexhaustible 'mines' for a wide range of the seagrasses. As many as 14 species of seagrasses have been recorded from the mid intertidal areas to depth of 50 m along the east and west coasts. Among them, *Enhalus acoroides, Cymodocea rotunda* are high energy containing grasses (Pradheeba et al., 2011). *C. nodosa* holds high lipid content and *Halodule pinifolia* and *H. uninervis* high protein (Thirunavukarassu and Subhashini, 2016). These grasses can be dried and used as an ingredient in the feeds of not only sea cucumbers and urchins but also in the feed industries of chicken and cattle. If they are unacceptable to the urchins and cucumbers, research must be initiated to identify deterrent(s) and suppressant(s) (Table 1.9), and develop inexpensive methods to eliminate them. In fact, the protein rich *Syngodium isoetifolium* is found suitable to feed juvenile cucumbers (James, 2004).

2. The body wall (< 40%) and gonad (10–30%) used for human consumption constitute only a fraction of the cucumbers and urchins. Hence, a major portion of these cucumbers and urchins is discarded as 'waste'. While there is a great demand for marine ingredients for chicken and other animal feed, no publication is yet available on the scope for using these echinoderm 'waste' in the animal feed industry. Interestingly, Radhika-Rajasree et al. (2016) have extracted polyhydroxylated naphthoquinone pigments with high anti-oxidant and anti-cancer properties from the discarded shells and spines of *Salmacis virgulata*. Sea urchins appear to be 'vegetarians' but cucumbers can be 'non-vegetarians'. Diets containing fish (herring) are not acceptable to urchins (Hooper et al., 1996). But fish

feed up to 22% as an ingredient is acceptable to the cucumber *Apostichopus japonicus* and ensures the highest Ae (see also Fig. 1.7C) and growth rate (Sun et al., 2004). Diets containing echinoderm 'waste' as an ingredient to enhance the size and quality of gonad of the urchins and body wall of cucumbers remain to be tested.

3. Besides their high nutritional value, pharmacodynamic studies have shown that sea cucumbers contain mucoitin that retards ageing and acts as an anti-tumor and anti-coagulation agent (Xiyin et al., 2004). Sarhadizadeh et al. (2014) have described cytotoxic, anti-bacterial and anti-fungal activities of *Stichopus herrmanni* against human pathogenic organisms. Rahman (2014) provides a long list of medicinal properties of cucumbers including anti-angiogenic and anti-hypertension factors. Medicinal properties of discarded 'waste' have been just indicated. Future research may show that there is wealth in the discarded 'wastes' of echinoderms.

2.2.1 Sea Urchins

In *Paracentrotus lividus*, year round captive feeding ensures (i) greater survival (45–52%) after KCl injection, in comparison to (20–37%) wild urchins (Table 2.2) and (ii) positive response to KCl injection by larger proportions of males (72%) and females (85%), in comparison to wild males (59%) and females (16%) as well as (iii) release of two-three times more gametes by the responding captive urchins than by wild urchins. Clearly, the urchins are positively amenable for captive rearing and aquaculture for longer durations, especially with reference to reproductive performance and selective breeding for stock improvement.

TABLE 2.2

Survival after KCl injection and year round reproductive performance of *Paracentrotus lividus* (compiled from Luis et al., 2005)

Parameter	Wild		Captive Fed	
	Male	**Female**	**Male**	**Female**
Survival (%)	37	20	52	45
	Wild		Seaweed		Maize		Combination	
	Male	Female	Male	Female	Male	Female	Male	Female
Spawning (%)	–	–	59	16	60	40	72	85
Sperm ($\times 10^6$)	244	–	270	–	311	–	276	–
Oocytes ($\times 10^3$)	–	788	–	863	–	1788	–	1582

Growth and body size: As food, algae and their combinations play a decisive role in regulation of growth and body size. They are important factors, as they provide nutrients for development and space for accommodation of

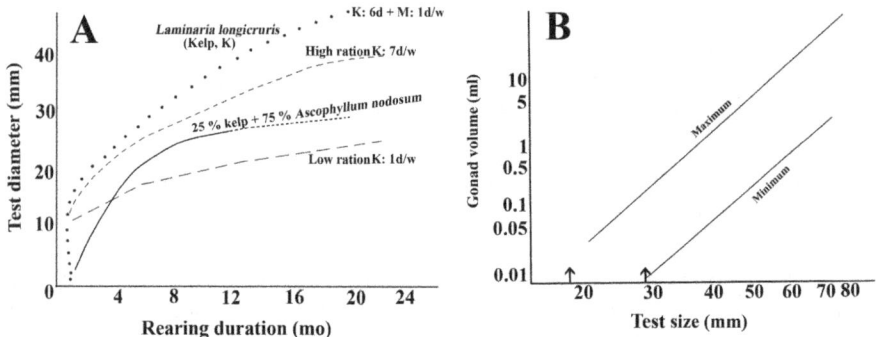

FIGURE 2.2

A. Effects of ration and diet combinations consisting of 25% *Laminaria longicruris* (Kelp) and *Ascophyllum nodosum* or Kelp + Mussel (M) on body growth of *Strongylocentrotus droebachiensis* (compiled and redrawn from Vadas, 1977, Meidel and Scheibling, 1999). B. Relation between body size and minimum and maximum gonad volume of *Diadema antillarum*. Arrows indicate the size at sexual maturity. Arrows indicate the onset of sexual maturity (simplified sketch drawn from Levitan, 1989).

the ripening ovary. In green sea urchin *Strongylocentrotus droebachiensis*, a combination of *Laminaria longicruris* + mussel meal ensures the fastest growth and maximum body size of ~ 40 mm test size on the 12 months (mo) of rearing (Fig. 2.2A). During the corresponding period, an alternate combination of 25% *L. longicruris* + 75% *Ascophyllum nodosum*, however, supports a maximum size of ~ 25 mm test size only (see also Hooper et al., 1996). Short term experiments (6 months) involving reciprocal change from *Nereocystis luetkeana* to *Agarum cribrosum* or *A. cribrosum* to *N. luetkeana* indicate monthly growth of 6.5 and 0.4 g, respectively (Vadas, 1977). Hence, the choice of algal feed and their combination is critically important, as the feed quality introduces 6–16 time differences in growth and body size of urchins. The longest duration of 24-months rearing of the urchin fed on *L. longicruris* once a day and a week ensures ~ 24 mm and ~ 12 mm disk growth/y, respectively (Meidel and Scheibling, 1999). Hence, ration introduces > 2 times difference in growth (Fig. 2.2A). Notably, the same ration regime lasting for 12 weeks during reproductive period does not introduce any significant change in somatic growth (Minor and Scheibling, 1997), as the incoming food energy is mostly allocated for reproductive growth (see also Fig. 1.7D).

Gonad Index (GI): Due to one or other environmental factor like food availability, fecundity, a derivative of GI, of *S. droebachiensis* exhibit minimum and maximum parallel levels throughout its size range (see Fig. 1.8A). Interestingly, field and other observations on GI of *Diadema antillarum* also indicate that (i) gonad volume increases as a function of body size measured in units of test diameter, and (ii) population density and consequent per capita

(0–4 g of *Ulva*) food availability plays a decisive role on body size at maturity and volume of the gonad. Firstly, individuals postpone maturity from ~ 18 mm at low density to 30 mm size at high density (Fig. 2.2B). Secondly, the size range of reproductive urchins is reduced from 18- > 80 mm at low density to 30–75 mm at high density. These observations of Levitan (1989) confirm that per capita food availability determines size at sexual maturity, range of reproductive size and gonad volume, and thereby indicate the importance of ration and stocking density of the urchins in aquaculture farms.

Table 2.3 lists some findings to show that gonad yield of the green urchin can be increased from 5.5–13.8% to 31–38% by manipulations of feed quality and quantity. The following may be inferred from Table 2.3: 1. The longer the experimental duration, greater is the scope to enhance the gonad yield, 2. Ensured feeding by algae and/or synthetic diets almost doubles the total (Bureau et al., 1997) and monthly mean (Walker and Lesser, 1998) gonad yield, 3. Ration level (once a day or week, Meidel and Scheibling, 1999) almost trebles the yield and 4. Supplementation of wheat and corn may (Motnikar et al., 1997) or may not (Klinger et al., 1997) increase the yield. But that with fish (Pearce et al., 2002) or mussel (Meidel and Scheibling, 1999) meal doubles and trebles the yield, respectively. Supplementation of herring meal is neither accepted nor increase body growth (see p 62). But that with other fish meal or

TABLE 2.3

Effects of experimental durations, food quality and quantity and photoperiod on gonad index of *Strongylocentrotus droebachiensis*

Feeding Regime	GI (%) Increase	Reference
Semi-moist diet containing 21% wheat gluten + 42% hydrolyzed algae fed for **30 days**	6 to 15	Motnikar et al. (1997)
Fresh/frozen bull kelp *Nereocystis luetkeana* for **54 days**	5.5 to 10.3	Bureau et al. (1997)
Pelleted diet containing 73% wheat + corn meal with or without kelp fed for **56–84 days**	= at 20 or 23	Klinger et al. (1997)
Fresh *N. luetkeana* fed for **84 days** at 8 or 16 hours photoperiod	13.8 to 17.0	Dumont et al. (2006)
Fed on kelp stable pelleted diet supplemented with corn, wheat, soybean and fishmeal for **84 days**	13 to 28	Pearce et al. (2002)
Fed on *Laminaria longicruris* on low and high ration for **84 days**	13 to 27	Minor and Scheibling (1997)
Mean monthly field and synthetic diet lab-fed for **210 days**	11–13 to 25–30	Walker and Lesser (1998)
Fed on *Laminaria* spp on low and high ration for **660 days**	11 to 31	Meidel and Scheibling (1999)
Supplemented with mussel	to 38	

preferably mussel meal is obligatorily required to enhance the gonad yield. Consequently, the gonad production is costlier and involves more protein-containing feed/diet than that required to increase body growth.

Value addition: Marketing of urchin gonad is dependent on a number of factors including gonad size and quality, species of urchin and time of the year. Important factors that determine the quality include color, texture, firmness and flavor. The urchin market typically prefers bright orange or bright yellow gonads that are firm and sweet in taste (Dumont et al., 2006). A short term experiment undertaken on value addition of the gonad of *S. droebachiensis* suggests that feeding a synthetic diet containing raw carrot and cabbage results in the highest gonad yield and retains its color, texture and taste as 'good' (Robinson and Colborne, 1997). Semi-moist rubbery sinking granules bound by wheat gluten and modified starch are water-resistant up to 48 hours only (Motnikar et al., 1997). However, other binders like alginate, gelatin and guar gum maintain the pellet diet intact for more than 48 hours. With binder constituting 5% of the diet, the consistency decreases in the following order: gelatin < alginate < starch < guar gum. Feed loss incurred with these binders remains at 12% for starch but < 12% for the others. Incidentally, the gonad color is significantly affected by the binders. Starch produces better color than all the others. In terms of gonad yield and ratings for color, texture and taste, 5% starch as binder is proven to be better than all other binders (Pearce et al., 2002). Hence, despite the 12% feed loss, starch is recommended as a binder, as it results in larger and better colored and tastier gonad. Incidentally, photoperiod has no effect on gonad hue (Dumont et al., 2006).

Photoperiod and temperature: The gonad of urchins contains two major types of cells: somatic nutritive phagocytes and germinal cells—oogonia or spermatogonia. In shallow water echinoids inhabiting the temperate and polar region, the cellular mechanisms that stimulate gametogenesis and vitellogenesis are associated with the changes in photoperiod with autumn day length and declining winter temperature, respectively. In *S. droebachiensis*, the gonad grows from May to August with increase in number and size of nutritive phagocytes and gonial cells undergoing mitosis. In the large gonad, vitellogenesis is completed between September to January. Spawning begins in February and lasts up to March. Briefly, the initiation of gametogenesis is associated with autumn day length but winter temperature is essential for completion of vitellogenesis. Hence, the hitherto made photoperiodic manipulations have succeeded in producing out-of-season mature males but not females (Table 2.4). The trigger to initiate vitellogenesis in urchins may be species-specific and each species may have its own environmental or chemical cue. Hence, it is intensely researched. With almost equally long photoperiod prevailing in the tropics, reductions in photoperiod (9 L : 15 D) and water temperature may initiate and accelerate gametogenesis

TABLE 2.4

Effect of photoperiod on gonad index and production of out-of-season mature males and females of *Strongylocentrotus droebachiensis*

Bay-Schmith and Pearse (1987)

Held at fixed short day (8 L : 16 D) for a year, gonads of *S. purpuratus* were active and ripe but in those at long day (16 L : 8 D), the gonads were devoid of gametes

Hagen (1997)

Darkness stimulated out-of-season gonad production of males but not females

McClintock and Watts (1990)

In tropical sea urchin *Eucidaris tribuloides* too, fixed long days (15 L : 9 D) delayed gametogenesis in 20% individuals over the entire year. But 60% individuals held under short days (9 L : 15 D) produced gamete throughout the year

Walker and Lesser (1998)

Exposure to autumnal photoperiod (8 L : 16 D) and abundant food supply accelerated the development of nutritive phagocytes. Consequent to gonial mitosis and gametogenesis, GI of the exposed remained higher at 18% in March than that (11%) of wild urchins

Dumont et al. (2006)

Exposed to 5 different photoperiods for 12 weeks. No significant effect on food consumption and absorption. But the nutritive phagocytes increased from ~ 56 to ~ 82% between the 4th and 12th weeks. Consequently, GI also increased significantly from 8.6 to 16.8% in the 16 days (D) treatment group. Gonad color and its water content did not differ among the treated group

Kirchhoff et al. (2010)

Successfully induced out-of-season spawning in both males and females by the exposure to photoperiod of ~ 14 D : ~ 10 L from May–June to September and to ~ 5°C from October to February. GI increased to a maximum of 25 g during November–December. Out-of-season group spawned 0.14 ml/g, required 10% more sperm to achieve equal (~ 82%) FS with 30% hatchability, in comparison to ~ 0.22 ml spawn/g and 65% hatchability

---

and vitellogenesis, respectively, as has been reported for *Eucidaris tribuloides* (McClintock and Watts, 1990). The manipulations limited to photoperiod have not succeeded to produce out-of-season spawning females. Manipulations of both photoperiod and temperature produce the key to successfully produce out-of-season maturing and spawning females. For technical details of these manipulations, Kirchhoff et al. (2010) may be consulted.

Induction of spawning: Injection of 2 ml of 0.50–0.55 M KCl through peristomial membrane into the coelom of mature adult sea urchins is the proven, readily practicable, cheapest and effective method to induce spawning in many echinoids (Table 2.5). The active substance extracted from the diatom *Phaeodactylum tricornutum* also induces spawning in 96% individuals of green urchin (Starr et al., 1992). The spring phytoplankton bloom is one of the most sharply defined events in temperate and polar seas but is not a defined one in tropical countries like India. Hence, the more complicated procedure of

TABLE 2.5

Induction of spawning in echinoids. For comparison details on ophiuroids are included

Species	Procedure	Reference
Echinoids		
Strongylocentrotus droebachiensis, S. fransiscanus, S. purpuratus	Intracoelomic injection of 0.55 M KCl	Hart (1995) Levitan (1993)
S. droebachiensis, Tripneustes gratilla	2 ml of 0.5 M KCl injected through peristomial membrane into coelom of mature adult	Bottger et al. (2011) Byrne et al. (2008)
S. droebachiensis	NaOH extract of the diatum *Phaedactylum tricornutum* induces 91% of individuals to spawn	Starr et al. (1992)
	Sperm motility can be enhanced by adding 0.1 mM L-histidine to sperm suspension	Saotome and Komatsu (2002)
Ophiuroids		
Ophiactis resiliens, Ophiothrix caeopitosa, O. spongicola	Spontaneous induction of spawning more due to disturbance and stress	Selvakumaraswamy and Byrne (2000)
Many ophiuroids	Repeated thermal shock (from ~ 20 to 34°C) followed by exposure to darkness in aerated water for 2–3 hours induces male to milt	Selvakumaraswamy and Byrne (2000)
Ophionereis fasciata, O. schayeri	(i) Immersion of ovaries in L-methyladenine, (ii) continuous exposure of females to sperm suspension or (iii) continuous exposure of males and females to sunlight fails to induce spawning. Eggs dissected from ripe ovaries are not fertilizable	Selvakumaraswamy and Byrne (2000)
	Injection of PCF remains to be tested	

extracting the active substance from one or other alga may not be useful in tropical aquaculture of urchins.

S. droebachiensis is considered as a representative candidate in urchin aquaculture. Following KCl injection, the volume of semen released by a male is increased with increasing body size, although the relationship is not significant (Fig. 2.3A). Mean volume of sperm released is 30.2 ml (0.25 ml/g body weight, Kirchhoff et al., 2010) equivalent to ~ 3.0×10^{11} sperm/milting. A major fraction of it is released following the injection but a small amount can also be extruded after 2 hours (Levitan et al., 1992). More than 60% of males release ~ 10–30 ml sperm/milting (Fig. 2.3A). Whereas > 90% of eggs remain fertilizable up to 90 minutes after spawning, fertility of sperm rapidly declines to < 10% within 20 minutes after the release

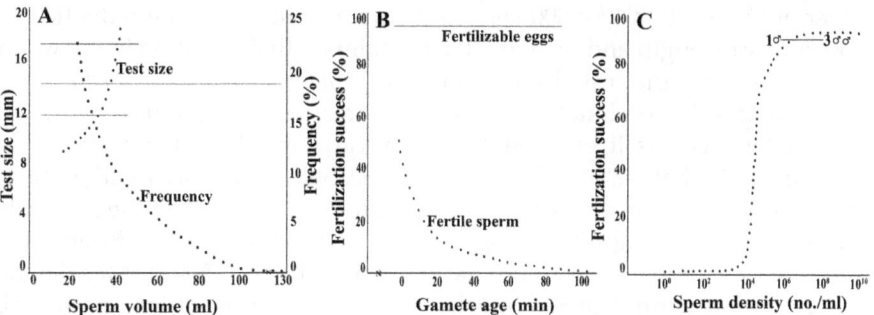

FIGURE 2.3

A. Relationships between body (test) size and frequency as well as sperm volume released by *Strongylocentrotus droebachiensis*. Horizontal lines indicate the range (compiled and redrawn from Levitan et al., 1992. B. Fertilization success as a function of gamete age in *S. droebachiensis*. C. Fertilization success as a function of sperm density and male number of *S. droebachiensis* (compiled and redrawn from Pennington, 1985).

(Fig. 2.3B). Fertilization occurs during the first 20 minutes after mixing the gametes in still water (Pennington, 1985). Over 80% of the eggs are usually fertilized in suspensions containing $> 10^6$ sperm/l. However, FS rapidly declines with dilution of semen (Fig. 2.3C). No fertilization occurs, when sperm density is decreased to 10^4/l. Increasing males from one to three ensures equal FS ($\sim 90\%$) (Fig. 2.3C) but it may increase genetic diversity among offspring.

Larval rearing: In his pioneering contribution, Strathmann (1971) reared the larvae of echinoids *Allocentrotus fragilis*, *S. droebachiensis*, *S. franciscanus*, *S. pallidus* and *S. purpuratus*, clypeasteroid *Dendraster excentricus* and spatangoid *Briaster latifrons*. Larvae of most of them were successfully reared up to metamorphosis. In general, echinopluteus larvae were the most efficient suspension feeders among larvae of the four classes of echinoderms (see Fig. 1.17). Clearance rate of an echinopluteus progressively decreased from 4 µl/min at the density of 10^3 cells per ml to ~ 0.2 µl/min at 10^4/ml. But ingestion rate increased to 7 cells/min at 5×10^3 cells/ml and then decreased to ~ 2.2 cells/min at 10^4 cells/ml. Kelly et al. (2000) investigated larval development and settlement of the echinoid *Psammechinus miliaris* fed different rations of microalga *Pleurocrysis elongata* as well as different algae. The swimming blastulae were decanted and cultured at one larva per ml in bins with 5-µm filtered sea water. On alternate days, the cultures were siphoned through a 40-µm sieve and the culture bins were drained, cleaned and refilled with fresh sea water. Fed at optimal dose of 1,500–4,000 cells/ml, *P. miliaris* larvae displayed more of typical morphology and 17% of them successfully metamorphosed on the 17th day. Those fed at low ration (500 cells per ml) failed to develop and metamorphose. The others fed at super-optimal ration (> 4,000 cells/ml) suffered from extreme reduction in post-oral arm length and were unable to maintain their position in the water column. Among the tested algae namely, *Dunaliella tertiolecta, Pleurocrysis carterae* and 50:50 combination of these two microalgae, *D. tertiolecta* proved as a better feed, as it ensured 66% survival through metamorphosis, in comparison to that (48%) fed on of *P. carterae*. Substrate coated with natural mixed species biofilm was a better inducer for settlement than the substrate coated with monoalga *Tetraselmis suecia* (Kelly et al., 2000). Briefly, the selection of *D. tertiolecta* as food and optimal ration of 1,500–4,000 cells/ml and provision natural mixed species biofilm to induce settlement are the keys for successful larval culture. *D. tertiolecta* is robust and ideally suited to bulk production in the hatchery of other commercially important echinoids like *S. droebachiensis* and *Loxechinus albus*.

Nursery phase field culture studies have revealed the importance of cage depth, bottom type and level of water exchange. Stocking density must be considered more in unit of surface area rather than volume, as urchins are attached to the surface. Cobble bottom (apparently provided larger surface area), weak current (~ 0.1 m/s, providing more protection) and stocking at 2 individuals (no.)/l may be ideal for rearing juvenile (~ 8.0 mm) green urchins 'on-bottom' nursery cage culture (Kirchhoff et al., 2008). During a 9-month long conditioning of brooders in culture system with manipulated photoperiod and temperature (Table 2.4), 47% mortality occurs. Of this, ¾ of mortality has occurred during the first month of stocking. Stocking density at 36% surface coverage in 4 × 100 l capacity tanks ensures maximum survival of brood stock (Kirchhoff et al., 2010).

2.2.2 Sea Cucumbers

The sea cucumbers *Apostichopus japonicus* and *Holothuria scabra* (James et al., 1988) are bred and reared in captivity. Information is also available on induced spawning of *Cucumaria frondosa* (Hamel and Mercier, 2004) and rearing of *H. spinifera* larvae (Asha and Muthiah, 2005). In recent years, remarkable advances in aquaculture of *A. japonicus* and *H. scabra* have been made in China and Vietnam, respectively. Available literature provides more technical information on aquaculture of cucumbers and the same is summarized to encourage other Asian countries to follow the Chinese success.

A. japonicus: Of 134 cucumber species found in China's seas, *A. japonicus* is known for its delicacy and medication properties (Xiyin et al., 2004). In China, there are more than 7,000 ha ponds used for monoculture of the sea cucumber and 2,000 ha ponds for polyculture of the cucumber with shrimp. The abandoned coastal shrimp aquaculture farms due to the emergence of diseases have been successfully restored for sea cucumber culture (Yaqing et al., 2004, Han et al., 2016). This may be noted by other Asian countries like India/who may have to abandon shrimp farms due to the prevalence of many viral diseases. In China, the net income generated by cucumber culture fluctuates between 7,500 and 50,000 US$/ha and is one of the most lucrative aquaculture business in northern China (Yaquing et al., 2004). The hatchery is used for conditioning broodstock and rearing larvae. The Chinese commercial hatchery is capable of producing 5,000 juveniles (1 cm long)/m^3 of water and some hatcheries are capable of even producing 10,000 juveniles per m^3 (Xiyin et al., 2004). Cucumbers weighing 300 g and measuring ~ 20 cm length, collected during June–July (15–17°C), serve as good brooders. They are maintained at 3 no./m^3 for 2–3 days. Temperature is gradually raised to 19°C @ 1°C per day. After 7–10 days, the brooders are ready for induced spawning (Renbo and Yuan, 2004). The conventional thermal shock, water jetting and/or aerial exposure are used to induce spawning (Table 2.6). For more details on construction and operation of a hatchery, Renbo and Yuan (2004) and Xiyin et al. (2004) may be consulted.

TABLE 2.6

Induction of spawning in holothuroids. For comparison details on asteroids are included

Species	Procedure	Reference
Holothuroids		
Holothuria scabra	Temperature shock of up to 3–5°C for several weeks induces milting in males earlier. The presence of sperm in water induces female to spawn	James (1994), Renbo and Yuan (2004)
Cucumaria frondosa	Manipulation of temperature in combination with light induces maximum success of 10%	Hamel and Mercier (1996a)
H. fuscogilva	Immerse in water containing dried alga of *Schizocytrium* sp induces maximum success of 10%	Battaglene et al. (2002)
C. frondosa	Injection of Perivisceral Coelamic Fluid (PCF) drawn from spawning/milting (duration: between prior to spawning and release of gametes) into yet to spawn mature individuals initiate spawning behavior (aggregation) and gamete release with 100% success. 2 ml of PCF from spawning male or female is adequate to induce milting in 89–94% males but induction of 97% females requires 3 ml PCF from milting male	Hamel and Mercier (1999)
Asteroids		
Acanthaster planci	Injection of 10^{-5} M L-methyladenine (MA) to induce ovulation	Caballes et al. (2016)
5 species belonging to 4 asteroid families	By immersing ovaries in the ovulatory hormone MA 1–2 µM for 30 minutes or until eggs are released	Saotome and Komatsu (2002)
	Ovary is immersed in 10^{-5} MA until eggs are released	Byrne (2006)
Mediastra oriens, M. calcar, Parvulastra exigua	Females are immersed in 3×10^{-5} MA until spawning occurs	Prowse et al. (2008)

In this context, the remarkable discovery of Perivisceral Coelomic Fluid (PCF) acting as chemical inducer and ensuring successful spawning in > 90% males and females of *Cucumaria frondosa* is an important milestone in aquaculture of sea cucumbers (Table 2.7). This discovery indicates that the lunar cycle photoperiod and other stresses like thermal shock, water jetting and/or aerial exposure may not synchronize reproduction and spawning in majority of holothuroid individuals. From the findings of Hamel and Mercier (2004), the following may be inferred: 1. Synchronized development is related to mucus synthesis, enabling the transfer of sexual information between conspecifics. As a result, maximum aggregation occurs just before spawning events, which minimize sperm dilution and increase FS. 2. The injection of PCF, drawn from mature cucumber displaying spawning activity, into a mature yet-to spawn individuals triggers spawning behavior and subsequent gamete release in up to 100% individuals in captivity and in the field (Table 2.6). It is likely that PCF present in the semen serves as a trigger, as well. PCF collected from post-spawning (1 hour after spawning) cucumber has already lost its effectiveness. Apparently, PCF acts a carrier of one or more short-lived molecule(s) of sexual nature during spawning. Research on isolation of the molecule and preservation of its effectiveness may be of immense importance. For PCF is neither sex-specific nor species-specific (Table 2.7). Briefly, holothuroids seem to employ species specific mucus to induce aggregation behavior and non-species specific PCF to ensure high FS.

In China, a small quantity of semen is added to freshly spawned eggs to ensure egg to sperm ratio of 1:3–5 and water is stirred constantly to increase FS (Xiyin et al., 2004). Development proceeds through hatching of blastula, auricularia, dolilolaria, pentactula and juvenile stages. Typically, the ball shaped body is the early sign of auricularia passing into doliolaria, moving to the bottom is a sign of doliolaria passing into pentactula and indicates the need for placement of settlement plates. According to Renbo and Yuan (2004), auricularia are stocked at 0.5 indiv/ml (or) $3–4 \times 10^5$ indiv/m^3 (Xiyin et al., 2004), and fed on microalgae at 2.5×10^4 cells/ml. Juveniles are stocked at the densities shown in Table 2.8 and fed on a diet consisting of *Sargassum thunbergii*, fish meal and others. Besides ~ 22% protein required in the diet, the rich presence of threonine, valine, leucine, phenylalanine, lysine, histidine and arginine is required to promote rapid growth (Sun et al., 2004).

The credit for pioneering researches on breeding and rearing of *Holothuria scabra* goes to Central Marine Fisheries Research Institute, India (James et al., 1988). That individual cucumbers (240–415 g) spawn on a regular basis from August, 2001 and once or twice every month from March, 2002 to August, 2003 in the farms of Vietnam (Pitt and Duy, 2004) clearly shows that the sandfish is comfortably amenable for captive breeding. In India, the cucumber was stocked in one ton tank with bottom spreaded with mud to

TABLE 2.7

Injection of 2 ml PCF from spawning donors as an inducer of spawning in holothuroids (compiled from Mercier and Hamel, 2002)

Sex	Proportion of Spawning Individuals (%)	GI Before Spawning	GI After Spawning	Oocytes Spawned ($\times 10^6$)
	Bohadschia marmorata			
Male	69	20	7.6	0.7
Female	84	20	6.5	0.9
	Holothuria atra			
Male	78	19	5.2	1.2
Female	87	13	3.1	1.1
	H. leucospilota			
Male	67	20	7.7	0.9
Female	55	22	7.2	0.7

TABLE 2.8

Recommended stocking density of *Apostichopus japonicus* juvenile (modified and compiled from Liu et al., 2004)

Juveniles (no./kg)	Density (no./m³)
> 8000	7500
7000	4500
5000	3500
3000	2500
1500	1500
600	650
< 200	200

15 cm height to enable them to bury. Thermal stress was used to induce spawning during the peak spawning period from March to May and minor peak spawning during November–December. About 0.3 million auricularia were reared in 750 l water and were fed on *Isochrysis galbana* at 2–3×10^4 cells/ml at 28°C. Juveniles were fed on protein rich *Sargassum* spp and the seagrass *Syngodium isoetifolium* (James, 2004).

In Vietnam, brooders (150–200 g) were stocked. Aerial exposure or alternate cold and heat shocks (5°C above or below ambient temperature) induced spawning. Spawning occurred regularly every month but the frequency of spawning individuals ranged from 20% in September to 70 and 75% in April and December, respectively. Auricularia were fed on *Chaetoceros* spp and *Rhodomonas salina*, the best alga for auricularia culture. Mean survival up to juvenile stage was 25, 5 and 0.4% for the larvae fed on *Chaetoceros*, *Spirulina* and *Nanochloropsis*, respectively. Growth of juveniles decreased in the following order, when fed on: shrimp feed > *Spirulina* > 5 powder (corn, rice, soybeans, black and green beans) > shrimp starter. Grow-outs grew rapidly @ 1–3 g/day and the best growth was achieved with the growth of 30 to 300 g in 3 months (Pitt and Duy, 2004).

Many studies on hatching success and larval rearing were successfully undertaken (Hair et al., 2016). Hatching success of *H. scabra* increased from 15 to 66%, when eggs were incubated at the density of 8.0 and 0.5 no./ml (Asha et al., 2011). An interesting study was undertaken by Asha and Diwakar (2013) to rear the relatively less efficiently filter-feeding auricularia (see Fig. 1.17) and non-feeding (p 28) doliolaria stocking them at different densities and feeding them on *Chaetoceras* sp. With increasing size and corresponding ciliary band-body size ratio, the auricularia larva presumably began to depend on its yolk resource. Understandably, the survival of the auricularia also progressively increased at 0.5 no./ml but the large late auricularia and doliolaria required more volume of water. Hence, they survived better at the stocking density of 1.0 no./ml (Fig. 2.4). Even at this stocking density, survival decreased from 70% in the late auricularia to < 10% in doliolaria. Hence, doliolaria is a critical point switching from feeding auricularia to non-

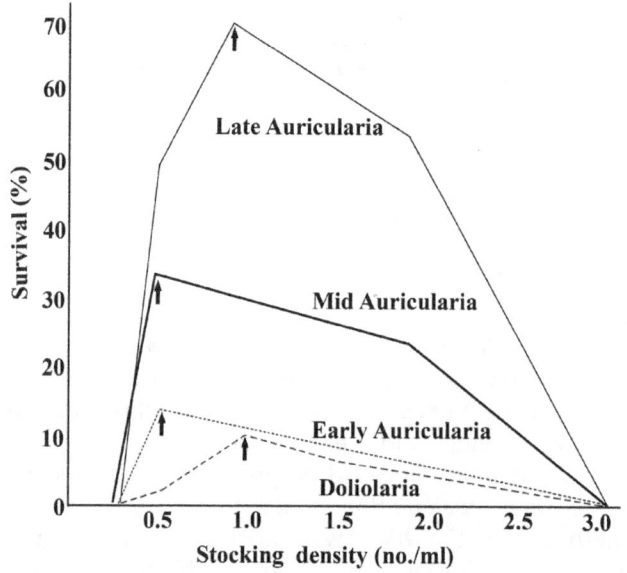

FIGURE 2.4

Survival of *Holothuria scabra* larvae as a function of stocking density. Arrows indicate maximum survival of the larvae at different stages. Note the highest survival of early auricularia and mid auricularia at stocking density of 0.5 no./ml as well as late auricularia and doliolaria at 1.0 no./ml (drawn from data provided by Dr. P.S. Asha).

TABLE 2.9

Development time required for the completion of larval stages and to attain juvenile stage of Chinese and Indian holothuroids

Stage	A. japonicus ~ 20°C Renbo and Yuan (2004)	Holothuria scabra ~ 30°C Asha et al. (2011)	H. spinifera ~ 30–32°C Asha and Muthiah (2003)
Blastula	7 hours	–	3 hours
Auricularia	2 days	1 day	2 days
Doliolaria	8–9 days	–	11 days
Pentactula	10–11 days	9 days	14 days
Juvenile	12–17 days	20 days	15 days

feeding doliolaria. In *H. spinifera* too, Asha and Muthiah (2002) noted high mortality on the 9th day.

Notably, hatching success of *A. japonicus* is > 90% (Xiyin, 2004), in comparison to 66% reported for *H. scabra* (Asha and Diwakar, 2013). Less than 3.6% of *H. spinifera* auriculariae survive to settlement (Asha and Muthiah, 2002, 2007), whereas > 10% of *H. scabra* auriculariae successfully settle in Vietnam (Pitt and Duy, 2004). Clearly, China and Vietnam seem to have perfected the techniques to select better brooders to ensure highest hatching success and larval survival. Surprisingly, the durations required for successful completion of embryonic and larval development, and attain juvenile stage are almost equal for *H. scabra* (14–15 days) in India and Vietnam as well as *H. spinifera* in India (~ 16 days) and *A. japonicus* in China, despite *A. japonicus* is reared at ~ 20°C, while others at ~ 30°C (Table 2.9). Yet neither Chinese nor Indians have estimated the time and production costs of juvenile *A. japonicus*, *H. scabra* and *H. spinifera*. Researches in this area shall be important to estimate the cost and profit of aquaculture of sea cucumbers and urchins.

3

Sexual Reproduction

Introduction

In animals, echinoderms especially the echinoids have served as a model for the study of fertilization and early embryonic development for the last 100 years. The interspersion of asexual reproduction within sexual reproduction in many holothuroids, asteroids and ophiuroids has attracted much attention. Briefly, reproductive biology of echinoderms remains a fascinating field of biology.

3.1 Sexuality

A vast majority of echinoderms are gonochores. Although no precise estimate is yet made, it is likely that > 99% of them are gonochores. This may be compared with < 75% for Mollusca (Pandian, 2017), 92% for Crustacea (Pandian, 2016) and 99.5% for teleostean fishes (Pandian, 2013). Notably, the occurrence of hermaphroditism is not uncommon among echinoderms. It occurs in a few holothuroids (e.g. *Cucumaria laevigata, Mesothuria intestinalis,* Hyman, 1955), asteroids (e.g. *Asterina gibbosa, Fromia ghardaqana,* Hyman, 1955) and in many ophiuroids (e.g. *Amphipholis squamata, Ophiura meridionalis,* Hyman, 1955). There are a couple of records on anomalous occurrence of hermaphroditism in echinoids (see Moore, 1932, Byrne, 1990). But there are no records even for the anomalous occurrence of hermaphroditism in crinoids. Notably, asexual reproduction is also not known to occur in adults of these two classes of echinoderms. In echinoderms, the occurrence of natural parthenogenesis is very rare and is reported to occur in two asteroids *Asterina miniata* (Newman, 1921) and *Ophdiaster granifer* (Yamaguchi and Lucas, 1984). It can also be induced by drastic changes in population density (e.g. *Diadema antillarum,* Bak et al., 1984) or experimentally in *Lytechinus pictus* (Brandriff et al., 1975) and others (Ishikawa, 1975).

3.2 Gonochorism

In gonochores, the benefits arising from recombination during gametogenesis and fusion of two gametes from two sexually different individuals during fertilization increase genetic diversity, the raw material for evolution. Not surprisingly, gonochorism has been successful and is maintained in plants and animals, and echinoderms are not exception to this dictum.

3.2.1 Sex Ratio

Within gonochoric echinoderms, sex ratio seems to be strictly maintained at and around 0.5 ♀ : 0.5 ♂. Spatial and temporal changes in the ratio due to sex-dependent mortality (e.g. *Heliocidaris erythrogramma*, Dix, 1977) and movement skew the ratio slightly. For example, the female ratio ranges from 0.43 in *Holothuria scabra versicolor* to 0.57 in *Holothuria nobilis* (Table 3.1). These values may be compared with those of crustaceans, in which the female ratio widely oscillates from 0.32 in the shrimp *Potimirim brasiliana* to 0.68 in *Exopalaemon cariniculata* (Pandian, 2016). That the strict maintenance of sex ratio around 0.5 ♀ : 0.5 ♂ clearly indicates that sex in echinoderms is determined genetically and role played by environment may be marginal. In many gonochoric ophiuroids, 1–2% individuals are, however, hermaphrodites (e.g. *Ophiomusium lymani*, Gage and Tyler, 1982, *Ophioderma januarii*, Borges et al., 2009, *Ophiocoma scolopendrina*, Deloroisse et al., 2013). It must be noted that in protandrics, the female ratio increases from 0.0 to ~ 0.50 in *Leptosynapta clarki* (Sewell, 1994) and to 0.89 in *H. atra* (Harriot, 1985). However, sex cannot be distinguished morphologically in echinoderms, except in brooding females. In the Antarctic dendrochirotid brooding holothuria *Psolidiella mollis,* males have long genital papilla and females have up to five interradial internal marsupia (see O'Loughlin et al., 2009). In general, size between sexes is equal. However, the presence of 'mini male' clinging to the female is reported in ophiuroids *Ophiodaphne formata*, *O. scripta*, *Ophiosphaera insignis* and *Astrochlamys bruneus* (see Stohr et al., 2012). To these, *Ophiodaphne materna*, *Ophiosphaera insignis* and *Amphilycus androphorus* have to be added (Hyman, 1955). At best, sex can be distinguished in some echinoderms anatomically from the gonad color (Table 3.2). But, not many authors have reported the gonad color. In gonochoric ophiuroids like *Ophiactis resiliens*, *Ophionereis fasciata*, *Ophiothrix caespitosa* and *O. spongicola*, sex can be recognized through the transparent body wall (Selvakumaraswamy and Byrne, 2000). In protogynic holothuroid *L. clarki* too, color of the gonad also changes from white testes to yellow ovaries during sex change, which can be recognized through semitransparent body wall (Sewell, 1994). Available information indicates that the gonad color changes on maturation and after spawning. For example, the testis color of *H. spinifera* changes from white to creamy white on maturation but the white

TABLE 3.1

Sex ratio in some echinoderms

Species	Female	Male	Reference
	Crinoids		
	Holothuroids		
Holothuria scabra	0.45	0.55	Hamel et al. (2001)
H. scabra versicolor	0.43	0.57	Hamel et al. (2001)
H. spinifera	0.47	0.53	Asha and Muthiah (2008)
H. santori	0.53	0.47	Navarro et al. (2012)
H. nobilis	0.57	0.43	
H. atra*† < 100 g	0.11	0.89	Harriot (1982)
> 100 g	0.41	0.51	
	Asteroids		
Coscinasterias calamaria†	0.50	0.50	Crump and Barker (1985)
Echinaster guyanensis†	0.57	0.43	Mariante et al. (2013)
Stephanasterias albula†	1.00	0.00	Mladenov et al. (1986)
Asterina burtoni††	0.00	0.00	Achutuv and Sher (1991)
Ophidiaster granifer**	1.00	0.00	Yamaguchi and Lucas (1984)
	Echinoids		
Salmacis sphaeroides	0.56	0.44	Rahman et al. (2013)
	Ophiuroids		
Ophiocoma aethiops	0.50	0.50	Benitz-Villalobos et al. (2013)
O. alexandri	0.50	0.50	Benitz-Villalobos et al. (2013)
Ophiolepis kieri < 4 mm disc	0.04	0.96	Hendler (1979)
~ 6 mm disc	1.00	0.00	
Ophioderma januarii‡	0.51	0.48	Borges et al. (2009)‡
Ophiomuscium lymani‡‡	0.44	0.40	Gage and Tyler (1982)

* protandric, ** parthenogenic, † sexual and asexual, †† obligatorily asexual, ‡ = 0.01 ♀, ‡‡ = 0.05 ♂

color is restored after milting. Similarly, the ovary color also changes from white to dark yellow but returns to white color after spawning (Asha and Muthiah, 2008). However, sex cannot be distinguished even from histology of the undifferentiated gonads remaining between successive reproductive cycles. Interestingly, Navarro et al. (2012) have noted the color change of *H. santori* gonads during specific gametogenic stages; accordingly, the ovary color is changed from translucent-light pink to intense orange-red, light orange-pink and brown green-translucent with orange blotches during stage II, III, IV and V, respectively. The testis too changes its color from translucent-salmon white to white-light beige, beige-translucent white and brown-green and translucent with orange blotches during stage II, III, IV and V, respectively. Gonad color is also altered by the diet quality and binder used to pelletize the feed (see p 66).

TABLE 3.2

Gonad color of echinoderms

Species	Maturing Gonad		Reference
	Testis	Ovary	
Crinoids			
Promachocrinus kerguensis	Cream colored pinnules	Bright orange colored pinnules	McClintock and Pearse (1987)
Holothuroids			
Stichopus chloronotus	Whitish	Whitish	see Conand et al. (1998)
Holothuria scabra	–	Brown with visible white spotted oocytes	Mary Bai (1980)
H. spinifera	Creamy white	Dark yellow	Asha and Muthiah (2008)
Isostichopus fuscus	Whitish to off white	Whitish to orange	Mercier et al. (2004)
I. badionatus	Creamy white	Pink or red	Guzman et al. (2013)
Leptosynapta clarki	White	Yellow	Sewell (1994)
Asteroids			
Echinaster type I	Creamy white to orange	Creamy white to dark green	Scheibling and Lawrence (1982)
Echinaster type II	Creamy white to orange	Creamy white to bright orange	
Nepanthia belcheri	Creamy white	Olive green	Ottesen and Lucas (1982)
Asteroids	Pale	Pink or Orange	Hyman (1955, p 288)
Echinoids			
Strongylocentrotus droebachiensis	Bright yellow	Bright orange	Motnikar et al. (1997)
Ophiuroids			
Ophiactis resiliens	Orange	Pink red	Selvakumaraswamy and Byrne (2000)
Ophionereis fasciata	White	Pale pink	
O. schayeri	White	Beige	
Ophiothrix caespitosa	White	Yellow	
O. spongicola	Orange-red	Red	

3.2.2 Reproductive Systems

Unlike the most complicated reproductive system in mollusc, the system in echinoderms ranges from no discrete organ/system in crinoids to simpler ones in other four classes. In crinoids, the gonads are usually located in the genital pinnules or in the arm and become evident as swollen areas on sexual maturity. They are simple masses of sex cells filling the genital cavity in the pinnules and lead to genital canal, an extension of coelom. The gametes escape by rupturing the pinnular wall. In *Promachocrinus kerguelensis*, gonadal bodies are located in the pinnules that are borne on the rachis or arm. Sex of pinnules is distinguishable by color in *P. kerguelensis* (Fig. 3.1A, B, C). The number of eggs carried in each pinnule decreases with increasing pinnule number and location of the genital pinnule. But it increases with increasing arm length (McClintock and Pearse, 1987).

In holothuroids, the gonad is composed of a multitude of elongated or branched filamentous tubules joined basally into a tuft attached to the left side of the dorsal mesentery and hangs freely into the coelomic cavity (Fig. 3.1D, E, F, see also Fig. 4.12). From the gonad base, the gonoduct proceeds through the mesentery into the gonopore mounted (or not mounted) on a genital papilla (Hyman, 1955). In the gonads of some temperate holothuroids like *Strongylocentrotus californicus*, the fecund tubules are elongated but sacculonated in *Thelenota ananas*. The gonads consist of one tuft in the family Holothuroidae, but two tufts in Stichopodidae (Conand, 1993). In *Stichopus*

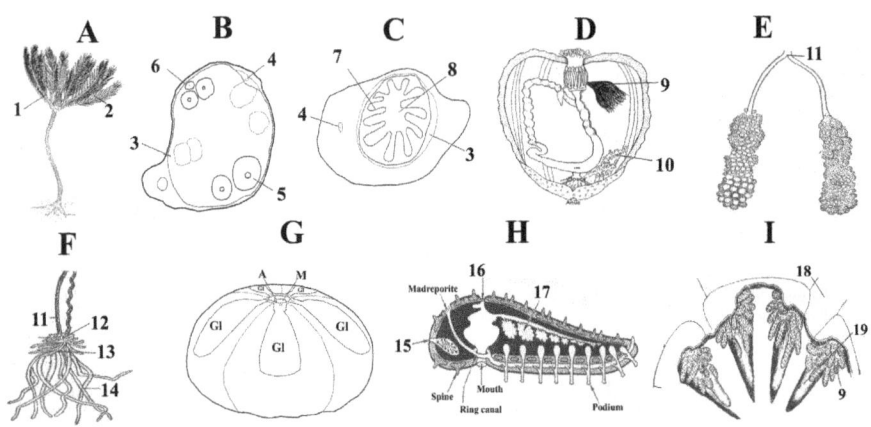

FIGURE 3.1

A. Stalked crinoid (from an unknown Google source). Sketches of cross sections through genital pinnule of B. female and C. male of *Promachocrinus kerguelensis* (from McClintock and Pearse, 1987), D. Sketch of anatomy of a holothuroid (original), E. Sketch of gonads of Elasipoda. Hermaphroditic gonad of F. *Cucumaria laevigata* and G. Reproductive system in a representative sea urchin, Gl = Gonadal lumen, M = Madriporite, A = Anus (redrawn from Strenger, 1973), H. Sketch of longitudinal section through disk and arm of a star fish (from an unknown Google source), I. Reproductive system of *Ophioglypha* after removing is disk wall, Gonad arrangement in J. *Stegophiura vivipara*, K. *Amphipholis squamata* (testis on radial, ovary on inter-radial site of the bursal slit), L. *A. constricta* (both gonads on the inter-radial sites) and M. *Ophiomitrella clavigera* (with two hermaphroditic gonads per bursa) (the remaining figures are all hand drawn from Hyman, 1955). 1 = Arm, 2 = Pinnule, 3 = Water-vascular system, 4 = Genital coelom, 5 = Previtellogenic oocytes, 6 = Completely developed oocytes, 7 = Spermatogenic column, 8 = Testicular lumen, 9 = Gonad, 10 = Respiratory tree, 11 = Gonoduct, 12 = Undifferentiated tubules, 13 = Female tubules, 14 = Male tubules, 15 = Gonad, 16 = Anus, 17 = Pyloric coeca, 18 = Arm base, 19 = Bursal sac (the remaining are free hand drawings from Hyman, 1955).

japonicus, the prodigious tubules grow from 2 cm length in October to 25 cm in July (see Hyman, 1955).

Typically, asteroids have 10 gonads, two in each arm lying freely laterally in the proximal part of the arm (Fig. 3.1H). Each gonad opens into a gonopore at the point of attachment to the interbrachial septum near the proximal end of the gonad. Some asteroids have ventrally located gonads that open orally. In many others like *Mediaster*, *Brisingaster*, the gonads are numerous and located in a row along the side of each arm. With a united single duct, this gonadal tuft, called 'serial', opens also in the usual site through gonopore. In many species like *Leptychaster*, the 'serial' state of the gonad is limited to males but the female have the usual five pairs, two in each inter-radius (Hyman, 1955).

In regular echinoids, the reproductive system consists of five gonads suspended by mesenterial strands along the inner surface of interambulacra (Fig. 3.1G). On ripening, they become voluminous bilobed bodies. Branches of the genital and hemal sinuses interconnect all the five gonads. Consequently, a single short gonad duct emerges from all the five gonads (see Walker et al., 2001). At the aboral end, the single gonoduct opens through a gonopore. Due to the progressive disappearance of inter-radial structures, the number of gonads is reduced to four in most irregular echinoids, to three in spatangoids (e.g. *Abatus*, *Schizaster*) and to two in *S. canaliferus* (Hyman, 1955).

In ophiuroids, the gonads consist of small coelomic sacs attached to the coelomic wall of the bursae (Fig. 3.1I), mostly near the bursal slits or directly to the slit walls. The number and arrangement of the gonads vary. There can be a "single large gonad appended to each bursa or two per bursa, one on its radial and the other on its inter-radial side or a number of small gonads variously disposed sometimes in cluster, sometimes in rows" (Hyman, 1955). In *Gorgonocephalus*, the gonads are numerous. With a very small disk devoid of bursae in some brittle stars, the gonads are located in the proximal arms or one pair per arm joint, as in *Ophiocanops fugiens*. The sex cells are released through the bursal slits. In *Ophiothrix gracilis*, the 10 gonads narrow to a definite ciliated gonoduct opening onto the side of the bursal slit (Hyman, 1955).

Reproductive barriers between species are classified into prezygotic and postzygotic categories. The former reduces the ability or the opportunity of two species to mate with each other and the latter reduces hybrids fitness (Lessios, 2007). At fertilization, there are species specific mechanisms that facilitate intra-specific fertilization but reduce or eliminate inter-specific fertilization. Bindin is a protein that codes the acrosome process of the sperm after its contact with the egg (Lessios, 2007). It is an adhesive responsible for the attachment of sperm to the vitelline layer of conspecific eggs. It is an unsoluble granular material in the acrosome vesicle of sea urchin *Strongylocentrotus purpuratus* (Vacquier and Moy, 1977). For example, cross fertilization between *Echinometra mathaei* and *E. oblongata* is minimized or eliminated at two steps: They are (i) the reduced attachment of the acrosomal process of heterospecific sperm to the vitelline layer of the egg and (ii) the inability of attached heterospecific sperm to develop continuity with the egg plasma membrane. At these steps, incompatibilities are reciprocal. Thus, a barrier to heterospecific gene flow is mediated by molecular interactions during a specific part of the fertilization process, as sperm acrosome surface and the egg vitelline layer contact each other (Metz et al., 1994).

The ensuing statements by Hyman (1955) clearly indicate the coelomic origin of the gonadal primordium. During the development of crinoids, "a primary gonad appears…as an elongated strand of compact cells located in the aboral vertical mesentery. The strand extends into the stalk….later migrates into the arm" (Hyman, 1955). In holothuroids, "the gonads originate in the dorsal mesentery near the stone canal from mesenterial cells" (Hyman, 1955). In the asteroid *Asterias rubens*, "the gonad arises as a group of primitive germ cells in or close to the wall of the left somatocoel….this group of cells grows into a compact mass to become the genital sinus, which elongates into genital rachis (Hyman, 1955). In echinoids, "the gonads develop from the genital stolen, a strand of cells budded off from the left somatocoel" (Hyman, 1955). In ophiuroids, "Primordial Germ Cells (PGCs) become distinguishable by the large nuclei and as a group of cells in the wall of left somatocoel adjacent to the stone canal" (Hyman, 1955). Briefly, the genital primordium is developed from the left somatocoel in asteroids, echinoids and ophiuroids as well as from the vertical and dorsal mesentery in crinoids and holothuroids, respectively. With radial symmetry, the PGCs do not undertake the dorso-lateral migration to coalesce with gonadal primordium, as has been described for bilaterally symmetrical invertebrates like crustaceans (Pandian, 2016) and molluscs (Pandian, 2017), and vertebrates like fishes (Pandian, 2011).

3.2.3 Gametogenesis

Among aquatic metazoan invertebrates, echinoderms display the simplest reproductive system. No report is yet available on the presence of an intromittent organ in echinoderms. The system does not include sperm storage organs like spermato sac, as in some males of crustaceans (Pandian, 2016), and spermatheca, as in some crustaceans (Pandian, 2016), copulatory bursa, seminal receptacle, as in molluscs (Pandian, 2017, Table 3.12) and the like in females. Echinoderms display also the usual simplest process of gametogenesis. They do not produce either spermatophore, which is ubiquitous among aquatic invertebrate groups (Pandian, 2016) or spermatozeugma, as in molluscs (Pandian, 2017). Female echinoderms do not also produce nurse eggs and embryos, as in molluscs (Pandian, 2017), polychaetes (Rasmussen, 1973) and crustaceans (e.g. Munuswamy and Subramoniam, 1973). Nor do they encapsulate their eggs, as in crustaceans (Pandian, 2016) and molluscs (Pandian, 2017). In echinoderms, the presence of dimorphic sperm is reported from the abyssal echinoid *Phrissocystis multispina* (Eckelbarger et al., 1989a). Two types of sperm namely the typical euspermatozoa and bipolar-tailed paraspermatozoa are produced. Developments of both sperm types are identical until the spermatid stage, when nucleus of the paraspermatozoan undergoes chromatin reduction. Both sperm types have acrosome typical of other echinoid sperm. Mixtures of both sperm types tend to clump due to entanglement of sperm axonemes in paraspermatozoan flagellum. This may enhance fertilization success by reducing diffusion of euspermatozoa during milting. Incidentally, the presence of lipid-like bodies in the middle piece of sperm in bathyal deep sea soft-bodied echinoderms *Phormosoma placenta*, *Sperosoma antillense*, *Araeosoma fenestratum* and *A. belli* suggest that the sperm are long-lived and required additional energy stores, not found in sperm of other echinoderms (Eckelbarger et al., 1989b).

Except for the pyloric caeca of asteroids, no internal storage organ responsible for nutrient provision to gonad maturation especially vitellogenesis is described for the other four classes of echinoderms. However, information is available for the role played by the unusual Nutritive Phagocytes (NPs) within the gonad during gamete maturation in echinoids (e.g. Walker et al., 2001). In gonochoric animals, oogenesis is completed following fertilization and the diploid oocyte undergoes second meiotic division and releases the second polar body (e.g. fishes, Pandian, 2013). Uniquely, the entire process of oocyte maturation is completed within the ovary and haploid oocytes are released in echinoids. Hence, the following description of gametogenesis is limited to the contrasting features displayed by echinoids and asteroids.

A brief account on energy metabolism and gonad development is necessary for a better understanding of the process of gametogenesis in echinoids, as an example (Walker et al., 2001). Glycogen is the dominant carbohydrate stored by the NPs, as they grow. During the early gonad growth phase, glycogen content ranges from 18 to 25% of the gonad dry weight in females and 13 to 19% in males. Once gametogenesis is commenced, glycogen content of the gonad declines to < 5%. However, oxidation of 1 mg of glucose requires ~ 32.5 m mole O_2. Due to its poor perfusion in the coelomic cavity, it may be virtually impossible for any internal tissue of an echinoid to acquire adequate oxygen. As a result, anaerobiosis dominates the metabolic process. In *Strongylocentrotus droebachiensis*, 76–92% of cellular energy can be produced from anaerobic metabolism (Bookbinder and Shick, 1986). Not surprisingly, the ovaries of *S. droebachiensis* produce lactate to maintain cellular ATP levels under anaerobic conditions. In echinoderms, the dominant pathway for pyruvate metabolism under hypoxic condition produces lactate via lactate dehydrogenases. The limited musculatures have restricted the scope for diversifying the anaerobic pathways in echinoids. This may also account for the low respiration rates generally found in echinoderms (Lawrence and Lane, 1982).

The gonad begins to grow following the increase in number and size of NPs in echinoids. Expectedly, cumulative gonad growth rate is dependent on daily gonad growth rate, as in *Paracentrotus lividus* (Fig. 3.2C). Hence, the increase in GI as an index of gonad growth is justified. A survey of feeding experiments with synthetic diets containing ~ 20% proteins in *Loxechinus albus*, *Evechinus chloroticus*, *S. droebachiensis* and *S. franciscanus* indicates that the relationship between daily gonad growth rate (GI/d) and food intake (G/d) rapidly increases initially but is saturated with a maximal gonad growth rate of ~ 0.17 GI/d (Fig. 3.2A). Protein composition of a diet influences greatly the protein content of the developing NPs. The correlation between the gonad growth rate (g/d) and respiration rate (O_2 uptake/g/d) in *Parechinus angulosus* (Fig. 3.2D) and related calculations indicate that the high synthetic demands for the production of Major Yolk Protein (MYP) may account for > 50% of the total cellular energy utilized by the developing NPs.

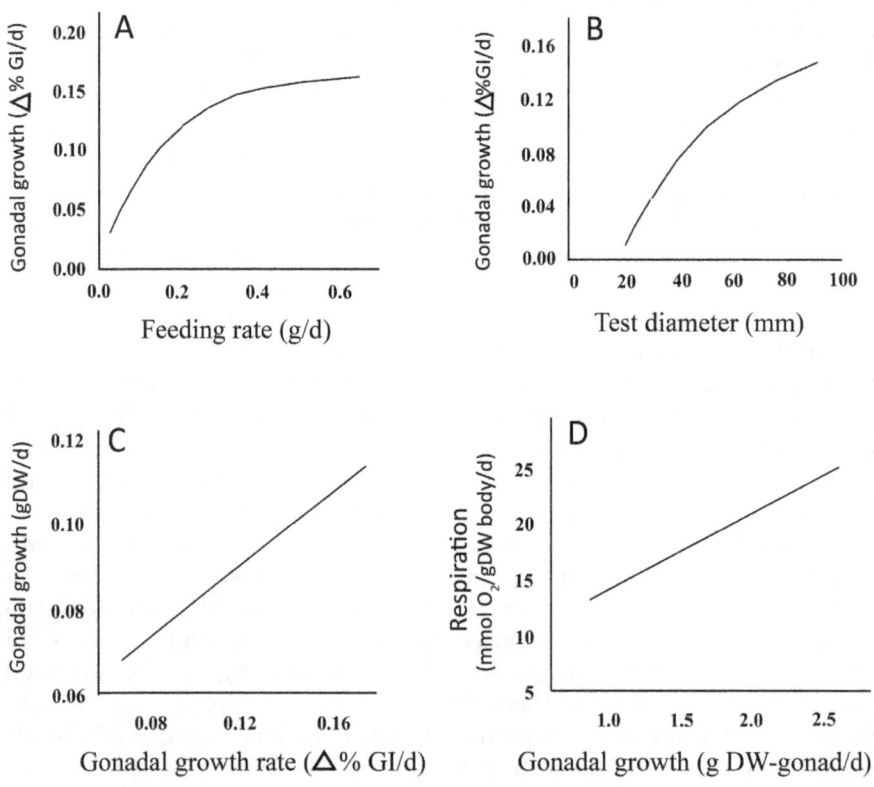

FIGURE 3.2

A. Gonad growth rate as a function of feeding rate in four sea urchin species (simplified sketch drawn from Marsh and Watts, 2001). B. Gonad growth rate as a function of body size in seven sea urchin species (redrawn from Marsh and Watts, 2001). C. Gonad growth as a function of gonad growth rate of *Paracentrotus lividus* (drawn using data of Spirlet et al., 2000). D. Respiration as a function of gonad growth rate of *P. angulosus* (drawn using data of Stuart and Field, 1981).

Figure 3.2B shows that the relationship arrived at for the gonad growth rate (GI/d) as a function of body size in seven echinoids. This would mean that during the 3-months (~ 100 days) period of gonad growth, the gonad may grow @ ~ 8%/day (Walker et al., 2001).

In echinoids, the duration of gametogenesis can broadly be divided into three phases (Fig. 3.3). During the first NPs phase, clusters of residual ameiotic gonial cells of previous generation and gonial cells of ensuing generation are present in the gonad. NPs begin to increase in number and size due to accumulation of (glycoprotein) MYP. The process is responsible for 'bulking' the gonad to a GI of 20%. Feeding regime regulates the abundance and accumulation of NPs during the usual as well as out-of-season (Walker and Lesser, 1988). The second phase corresponds to the simultaneous gonial mitosis and utilization of NPs. This period is regulated by 'autumnal' photoperiod. In *S. purpuratus*, the vitellogenin or *MYP* precursor gene has an estrogen responsive element in its promoter. The promoter is that portion of the gene, which interacts with transcription factors resulting in expression of the *MYP* and production of new MYP (Shyu et al., 1986, 1987). During the third phase, gonad maturation especially vitellogenesis occurs, when temperature drops. Spermatogenesis occurs in the lumen of testis, where immotile fully differentiated sperm are stored. In *S. purpuratus*, the time required from spermatogonic mitosis to fully differentiated spermatozoa is ~ 12 days, which is comparable to that (12–15 days) in fishes (see Pandian,

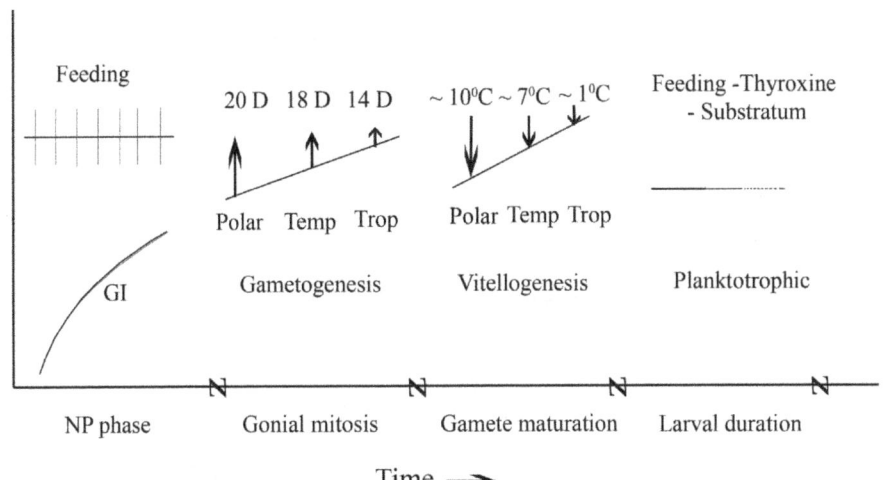

FIGURE 3.3

In echinoids, gametogenesis includes (i) NP phase, (ii) gonial mitosis and (iii) gamete maturation (vitellogenesis), which are regulated by feeding, photoperiod and temperature, respectively in polar, temperate (Temp) and tropical (Trop) regions. For completion, planktotrophic larval duration and factors that regulates the duration are also included.

2013). Briefly, the duration of gametogenesis includes an intensive feeding phase, during which the number and size of NPs increase, a second phase involving gonial mitosis and utilization of NPs as regulated by photoperiod and the third maturation phase including gamete maturation especially vitellogeneis initiated by declining temperature. It has been shown that feeding regulates the gonad growth, short photoperiods (20 D, 18 D and ~ 14 D in polar, temperate and tropical echinoids, respectively) the gametogenesis and declining temperature (~ 10°C, ~ 7°C and ~ 1°C [more due to rain, e.g. *Holothuria fuscocinerea*, Benitez-Villalobos et al., 2013] in polar, temperate and tropics, respectively) the vitellogenesis (Fig. 3.3). Optimal feeding, thyroxine and availability of suitable substrate regulate the duration in planktotrophic larvae. Hence, there is considerable scope for manipulations of gonad size (for 'broiler' production) and time of its maturation (for 'layer' production) as well as shortening the larval duration in aquaculture farms.

In many asteroids, the pyloric caeca serve as a storage organ to provide the required nutrients and ensure gonad growth and gamete maturation. Negative correlation between the indices of gonad and pyloric caeca is apparent in many asteroids. For example, the gonad index of *Cosinasterias calamaria* increases from May, reaches the peak in October–November and declines during spawning season from January to March (Fig. 3.4). Conversely, pyloric caeca index rises to the peak in July and declines thereafter. In *Psammechinus miliaris*, the gut serves as a storage organ and a

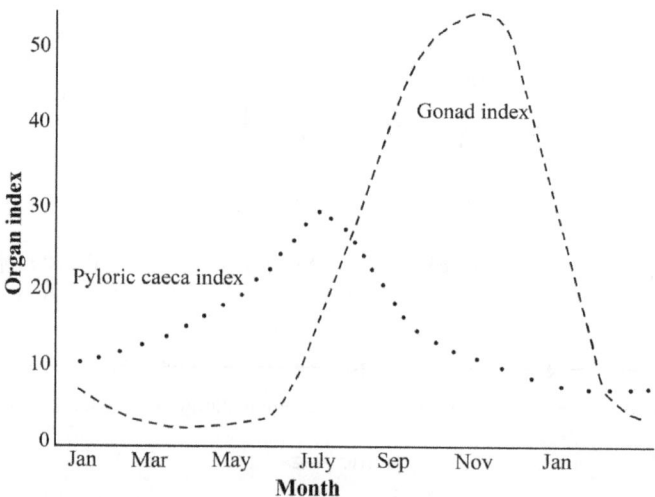

FIGURE 3.4

Indices of gonad and pyloric caeca of *Cosinasterias calamaria* as a function of the calendar month (modified and redrawn from Crump and Barker, 1985).

negative correlation between alimentary index and gonad index has been reported (see Kelly and Cook, 2001). In *Pisaster ochraceus*, photoperiod is reported to regulate the seasonal cycles of both gonad and pyloric caeca (Pearse and Eernissi, 1982). Although these negative correlations do suggest the mobilization and transportation of nutrients from the pyloric caeca or gut to the gonad, no publication is yet available on the route and chemical form of nutrients transported to the gonads.

3.3 Parthenogenesis

In echinoderms, parthenogenesis is rare. Occurrence of natural parthenogenesis is reported from an asteroid, an ophiuroid and an echinoid species and in the latter heavy mortality is suggested to induce parthenogenesis. Of course, efforts have also been made to chemically induce parthenogenesis in a few echinoids and an asteroid. All populations of the coral-reef inhabiting asteroid *Ophidiaster granifer* and all the collected specimens of them from the west Indo-Pacific were females. Of 322 specimens examined by Yamaguchi and Lucas (1984), 64 were either without the gonads or gonads without gametes. The females attained sexual maturity at the size of 19.2 mm arm radius and at around the age of 1+. Spawning occurred in November. The oocytes measured ~ 0.63 mm. Fecundity increased linearly as a function of body size (Fig. 1.9C). On injection of L-methyladenine, females collected during October–November spawned but those injected during April–November did not. The non-feeding larva metamorphosed within 10 days but extended the duration to six weeks in the absence of a suitable substrate. Reporting the sudden and an extreme decrease in population density (to 0.5%) of the echinoid *Diadema antillarum* from ~ 0.024 to 0.0 no./m^2 at the Atlantic mouth of Panama canal at 3 m depth due to pollutants, Bak et al. (1984) noted that gonads of two individuals out of 25 specimens were packed with blastulae. They suggested that the extreme decrease in population density might have induced parthenogenesis.

Garret et al. (1977) have brought to light an interesting case of simultaneous presence of sexual and parthenogenic female morphotypes in brittle star *Ophiomyxa brevirima*. Multi-locus assessment of 35 females and their brooded offspring has revealed that 14 females, i.e. 40% of the examined females are ameiotic/amictic parthenogenic morphotype and 17 females, i.e. 49% are sexual morphotypes. Not surprisingly, sex ratio of *O. brevirima* is 0.76 ♀ : 0.24 ♂.

In *Asterina miniata*, only a few eggs matured at a time and are exuded over a long period. Among these eggs, some of them developed parthenogenically. Following the formation of fertilization membrane, the cleavage commenced. Even the most nearly parthenogenetic eggs took twice as long to reach the

TABLE 3.3

Parthenogenetic induction in *Lytechinus pictus* eggs (modified from Brandriff et al., 1975)

Inducer	Fertilization Membrane	Completion of Nuclear Migration (min)	First Cleavage (min)	First Swimmers (h)	Blastulae (%)
Fertilized	+	25	87	12.5	92.5
Ionophores	+	50	155	18.5	15.3
NH$_4$OH	–	52	210	19.5	7.5

cleavage stage, as did fertilized eggs. However, Newman (1921) has not reported the successful completion of development in these parthenogenetic eggs. Since 1887, efforts were made to induce parthenogenesis in echinoids. Exposure to weak acids or bases or shaking eggs in water containing chloroform induced the formation of fertilization membrane. Elevation of pH by addition of NH$_4$OH initiated successive cycles of DNA synthesis and chromosome condensation. Eggs of *Lytechinus pictus,* on exposure to 0.5 μM ionophore or 1.7 mM NH$_4$OH solution, induced hatching of parthenogenic blastulae, albeit the duration of development was delayed at every stage (Table 3.3). Briefly, exposure to NH$_4$OH and ionophore A23187 in conjuction with hypertonic secondary treatment led to successful production of parthenogenic diploid echinoplutei, unlike the parthenogenic haploid, diploid and triploid plutei induced by earlier authors in *Arbacia punctulata, Paracentrotus lividus* and *Strongylocentrotus purpuratus* (Brandriff et al., 1975).

3.4 Hermaphroditism

In echinoderms, hermaphroditism is not uncommon but it is not as common as in crustaceans (Pandian, 2016) and molluscs (Pandian, 2017). It occurs in a few holothuroids and asteroids and in many ophiuroids but not in crinoids and echinoids. Stray incidences of anomalous hermaphroditic specimens are listed by Hyman (1955) in an ophiuroid *Ophionotus hexactis,* six asteroids *Asterias batheri, A. rubens, Asterina gibbosa, Fromia ghardaqana, Leptasterias groenlandica* and *Marthasterias gracilis,* and 10 echinoids *Arbacia lixula, A. punctulata, Dendraster excentricus, Echinus esculentus, Echinocardium cordatum, Paracentrotus lividus, Psammechinus microtuberculatus, Sphaerechinus granularis, Strongylocentrotus droebachiensis* and *S. pulcherrimus.* For more records and description, Boolootian and Moore (1956) may be consulted. There are others, in which a small proportion of individuals is hermaphrodite; for example, sex ratio of *Ophiomusium lymani* is 0.44 ♀ : 0.15 ⚥ : 0.40 ♂ (Gage and Tyler, 1982).

In some of these hermaphroditic echinoderms, sexuality ranges from self-fertilizing simultaneous hermaphroditism to sequential hermaphroditism. In sequentials, sex change naturally occurs once in a life time in a single direction. Accordingly, protandrics change sex from male to female and protogynics does it in the reverse direction (see Pandian, 2013). Almost all these types of hermaphroditism occur in echinoderms. The simultaneous presence of one or more ovaries and testes or ovotestes (e.g. *Amphiura monorima, Ophiolebella biscuticera*) in as many as eight ophiuroid species reveals that they are 'potential' simultaneous hermaphrodites. In fact, self-fertilization is reported to occur in one of these hermaphrodites namely *Amphipholis squamata* (see Stohr et al., 2012). Hence, it is likely that all these ophiuroids are functional simultaneous hermaphrodites. There are reports for the presence of protandric echinoderms (see Sewell, 1994). A few holothuroids (*Holothuria atra, Leptosynapta inhaerens, L. clarki, Labidoplax media*), ophiuroids (*Ophiolepis kieri, Amphiura stepanovii, Ophiocantha bidentata, Ophionereis olivacea, Ophiothrix* sp) and many species belonging to the three asteroid genera (*Asterina, Asterias* and *Fromia*) are protandrics. Protogyny is reported from holothuroids *Cucumaria crocea* and *C. leavigata* and possibly in *Mesothuria intestinalis* (Table 3.4).

Beyond these reported incidences of hermaphroditism, not many publications are available on the functional aspects of reproduction in them. In general, the life span of holothuroids is longer (> 5 years) than ophiuroids (< 5 years). Similarly, it is longer in oviparous aquatic invertebrates than viviparous ones. During the respective life span of these protandrics, male gonads mature first. As a result, male ratio progressively increases with increasing body size (Fig. 3.5A) up to a point of size/age, when sex change is initiated. At this point, some (2–3%) sex changing individuals display simultaneous hermaphroditism. But then, female ratio increases with concomitant decrease in male ratio. *Ophiolepis kieri* has a shorter life span during the female phase and undertakes fewer ovarian cycle than oviparous *Holothuria atra*. Whereas male ratio declines to almost zero level with increasing body size in viviparous *O. kieri* (Fig. 3.5A), the ratio in gonochoric oviparous *H. atra* declines to ~ 0.3 level only (Fig. 3.5B). In another protandric brooding holothuroid *Leptosynapta clarki*, the ratio remains as high as 0.5 even among the largest size group (Fig. 3.5C). Secondly, sex change occurs, when the cucumber attains body weight of ~ 570, ~ 750 and ~ 950 mg. An important consequence of the absence of the intromittent organ is that 22 females out of 50 remain totally unfertilized and FS in the others too averages to 48% only, despite the presence of > 50% males (Sewell, 1994).

When these observations on hermaphrodites were made, no information was yet available on Primordial Germ Cells (PGCs) and their derivatives Oogonial Stem Cells (OSCs) and Spermatogonial Stem Cells (SSCs). It is now possible to infer some information on this aspect. In holothuroids, PGCs are stored at the base of the gonoduct (Fig. 3.1 E–F), from which

TABLE 3.4

Hermaphroditism in echinoderms

Species, References	Reported Observation
Simultaneous hermaphroditism, Ophiuroids (Hyman, 1955)	
Amphiura monorima	One ovotestis per interradius
Ophiolebella biscutifera	Two ovotestes on the interradial side and one on the radial side per bursa
Amphipholis squamata	One ovary on the interradial side and one testis on the radial side per bursa
Amphiura constricta	Ovary and testis are on the interradial side per bursa
A. megallanica	2–4 interradial and 1–2 radial gonads
Ophiura meridionalis	1 or 2 testes radially and 1 or 2 ovaries interradially per bursa
Stegophiura nodosa	Two testes and one ovary on interradial side per bursa
S. vivipara	1 testis and 1 ovary or 2 ovaries on radial side per bursa
Protandric hermaphroditism	
Leptosynapta clarki (Sewell, 1994)	Male as young and female as older ones
Holothuria atra (Harriott, 1982)	Sex ratio shifts from 1 female : 7 males at 100 g size to 2 females : 1 male at 1000 g size (Fig. 3.5A)
Ophiolepis kieri (Hendler, 1979)	Sex ratio shift from peak of 1 female : 22 males at 4 mm size to 20 females : 1 male at 5 mm size (Fig. 3.5B)
Nepanthia belcheri (Ottesen and Lucas, 1982)	
Asterina gibbosa, A. panceri, Fromia ghardaqana (Delavault, 1975)	
Protogynic hermaphroditism	
Cucumaria crocea, C. laevigata (Hyman, 1955, p 173)	Gonadal tubules at the base of the gonoduct remain undifferentiated at first but with elongation of the tubule, they become ovarian and release eggs. With further elongation, the female elements are phagocytized by coelomocytes and the same tubules then produce sperm
Mesothuria intestinalis? (Hyman, 1955, p 173)	The gonadal tubules branch distally forming tufts, of which some are testicular and others are ovarian. Not known whether they function sequentially or simultaneously

undifferentiated PGCs, SSCs and OSCs are derived sequentially in protandrics and protogynics. The fact that old spent tubules are resorbed or phagocytosized by coelomocytes, new testicular and ovarian tubules are generated clearly indicates that the PGCs are retained at the base of the gonoduct until the life time. Incidentally, the temporal splitting of the ovarian and testicular maturation may have driven *Nepanthia belcheri* to manifest asexual reproduction. In simultaneous hermaphroditic ophiuroids

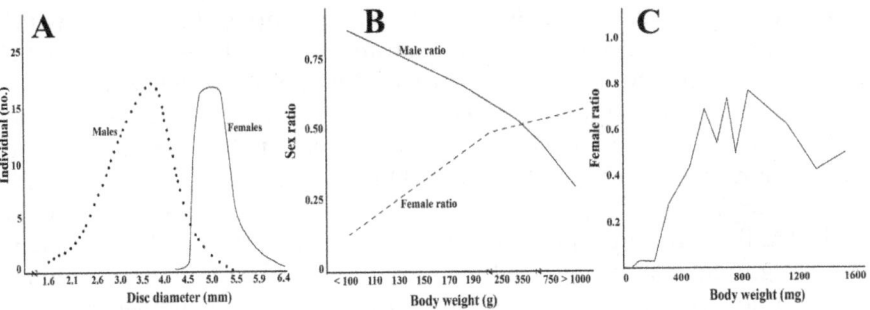

FIGURE 3.5

Sex ratio/number as function of body size in protandric A. Viviparous *Ophiolepis kieri* and B. Oviparous *Holothuria atra* (drawn using data reported by Harriott, 1982 and Hendler, 1979, respectively). C. Protandric brooding *Leptosynapta clarki* (simplified sketch redrawn from Sewell, 1994).

too, the genital plates are the sites, where the PGCs are retained for the life time; it is from these sites ovaries and testes as well as ovotestes arise. Notably, NPs play also another important role in phagocytosis and elimination of relict gametes and gametogenic cells not only in echinoids (see Walker et al., 2001) but also in many asteroids like *Echinaster sepositus* and *Leptasterias hexactis* and possibly in *Odontaster validus* (see Pearse, 1969b). Clearly, a new generation of oogonial and spermatogonia are derived from the gonads through OSCs in females and SSCs in males. Briefly, echinoderms retain PGCs or OSCs and SSCs in the gonads (asteroids, echinoids) and gonoducts (holothuroids) until their life time. As it is possible to identify the sex of some holothuroids by picking the gamete sample using a needle (Mercier et al., 2000), it may also be possible to castrate or ovariectomized the amenable holothuroids. However, such a surgery may be lethal to asteroids and ophiuroids. Chemical means of castration and ovariectomization or testicular or ovarian part of ovotestis in ophiuroids like *Amphiura monorima* and *Ophiolebella biscuticera* may help us to know more about the role played by PGCs and their derivatives in retention or reversion of sex in these echinoderms.

3.5 Reproductive Cycle

A survey of reproduction in echinoderms reveals that 1. Almost all echinoderms are iteroparous. Semelparity is not reported for any echinoderm. 2. Sexual maturity is determined more by environmental factors like food availability, population density and others. Hence, age/size at sexual maturity vary spatially and temporally. For example, the size, at which *Evechinus chloroticus* become sexually mature, ranges between 35 and 45 mm at Kaiteriteri but between 50 and 60 mm in Dusky Sound, New Zealand (see Barker, 2001). 3. The largest gonad size at which spawning occurs also varies widely. For example, *E. chloroticus* spawns, when GI attains 8–15% of body weight during December–February in Dusky Sound but 19–27% of body weight during November–December in Marlborough Sounds (see Barker, 2001). 4A. To ensure higher recruitment, milting in males is synchronized with spawning in females. In fact, milting induces spawning in females of many holothuroids. An active principle present in the previsceral coelomic fluid of holothuroids is reported to induce spawning. B. Spawning aggregation is not uncommon in echinoids. Boudouresque and Verlaque (2001) report spawning aggregation in *Paracentrotus lividus*.

In many temperate (e.g. *Comanthus japonica*, Dan and Kubota, 1960) and circum Antarctic (e.g. *Promachocrinus kerguelensis*, McClintock and Pearse, 1987) crinoids, gametogenesis is synchronized between pinnules. Full grown oocytes are released to the ovarian lumen only a few hours (*C. japonica*, Holland and Dan, 1975) or days (*Florometra serratissima*, Mladenov, 1986) before spawning. Expectedly, *C. japonica* and *P. kerguelensis* are synchronized annual spawners. However, others like *Lamprometra kluzingeri* (Rutman and Fishelson, 1985) and *F. serratissima* (Mladenov, 1986) are continuous spawners.

With geographic range limited to tropics and sub-tropics (Hamel et al., 2001), gametogenesis occurs almost throughout the year in most holothuroids (e.g. *Holothuria scabra*, Krishnan, 1968). With no storage organ (like pyloric caeca) or cells (NP) but vitellogenesis limited to short duration, a single spawning (e.g. *Actinopyga mauritiana*, Hopper et al., 1998, *Bohadschia vitiensis*, Omar et al., 2013) or bi-annual with major peak from December to February and minor peak during August–October (e.g. *H. scabra*, Krishnaswamy and Krishnan, 1967) spawning occurs. However, the duration of gametogenesis in Mexican *H. fuscocinererrea* is restricted from January to March and spawning occurs during July in 79% of females and 88% of males (Benitez-Villalobos et al., 2013). Interestingly, *H. arenacava* (in Kenyan marine protected area) displays an annual cycle with gametogenesis commencing in July during the south-west monsoon, when temperature and light intensity are the lowest and peaks in February and March at the end of north-east monsoon, when temperature and light reach their annual maxima. The correlation between light intensity and gonad growth is higher ($r = 0.93$) than that ($r = 0.7$) between gonad growth and temperature (Muthiga, 2006). Apparently, photoperiod plays a more important role than temperature in regulation of gametogenesis in tropics also.

In most echinoids with storage NP cells, initiation of gametogenesis is associated with 'autumnal' short period and completion of vitellogenesis with dropping 'winter' temperature, resulting in synchronized annual reproductive cycle. In them, spawning occurs during a short duration of

3–4 months from January to April. Expectedly, spawning period of *Loxechinus albus* is shifted from June–July at 25° S to November–December at 40–45° S but to August–November at 52° S in the west coast of South America (Vasquez, 2001). In many annual spawners, fertilization success is ensured by spawning aggregation, as in *P. lividus* (Boudouresque and Verlaque, 2001) and *Evechinus chloroticus* (Barker, 2001), besides spawning season being limited to ~ 3 months (*P. lividus* in France from December–March, *E. chloroticus* in New Zealand from November–February). However, *P. lividus* spawns twice a year, once during early spring and second time during autumn in Ireland, Marseilles, France, Corsica and Algeria (see Boudouresque and Verlaque, 2001). *Strongylocentrotus intermedius* is also an annual spawner (during autumn) in Japan Sea but bi-annual (spring and autumn) in Funka Bay and the southern Pacific Ocean (Agatsuma, 2001a). Likewise, *Echinometra mathaei* is an annual spanner in Japan, Kenya and South Africa but a continuous spawner in Hawaii, Red Sea and Australia. *E. lucunter* is an annual spawner in south Florida and Puerto Rica but a continuous spawner in Panama (Kelly and Cook, 2001). Hence, regional differences in photoperiod and temperature may induce a second spawning in some of these annual spawners.

With the presence of pyloric caeca as a storage organ, a definite annual reproductive cycle with spawning during autumn is reported for many asteroids (e.g. *Coscinasterias calamaria*, Crump and Barker, 1985, *Echinaster guyanensis*, Mariante et al., 2010).

In the Mexican ophiuroids *Ophiocoma aethiops* and *O. alexandri*, gametogenesis is synchronized between males and females, and occurs almost throughout the year with peaks in February in 80% of *O. aethiops* and May–October in 70% of *O. alexandri*. But their GI cycle peaks during September–October, although spawning occurs from May to November/December in both brittle stars. Hence, GI cycle may not precisely represent the spawning cycle (Benitez-Villalobos et al., 2013). In Australian waters (33° 5′S–35° 5′S), spawning occurs from January to April, March to June and May to November in *Ophionereis schayeri*, *Ophiothrix spongicola* and *Ophiactis resiliens*, respectively (Selvakumaraswamy and Byrne, 2000). In the absence of a definite storage organ and/or cells in ophiuroids, availability of food seems to regulate the reproductive cycle. For example, the annual and biannual spawning periodicities in *Ophiura albida* and *O. texurata*, respectively are related to their feeding habits. The former relies on seasonal microalgae but the latter is more an omnivore (Tyler, 1977).

4

Asexual Reproduction

Introduction

Echinoderms have an amazing ability to regenerate not only the lost tissues, organs (e.g. madriporite) and systems (e.g. digestive system) but also some of them can agametically clone and reproduce asexually. Agametic clonal reproduction is bidirectional, when the bisected animal develops into two fully functional individuals, as in echinoderms but unidirectional, when only one half regenerates, as in solitary ascidians (see Rychel and Swalla, 2009). Asexual or fissiparous clonal reproduction provides a mechanism for potential rapid amplification of individual genotypes (Mladenov, 1996). Further, cloning saves time and resources, and avoids risks that are involved in sexual reproduction (Skold et al., 2009). In these echinoderms, the population structure is clonal and each genet constitutes many discrete genetically almost identical individuals or ramets (e.g. holothuroids: *Stichopus chloronotus*, Uthicke et al., 1999, asteroids: *Coscinasterias calamaria*, Johnson and Threlfall, 1987, *Asterina burtoni*, Karako, 2002 ophiuroids: *Ophiocomella ophiactoides*, Mladenov and Emson, 1990). The ramets can readily be recognized by the presence of an asymmetrical body in asteroids and ophiuroids, which may also be differently colored, as in *Ophiactis savignyi*; unpigmented genets and ramets of asexuals can also be distinguished from gray-green colored sexuals (McGovern, 2003). In holothuroids, the ramets differ from genets by color and body profile; for example, genets of *Holothuria hilla* are bright brown with white or yellow papillae but ramets are lighter in color, and smaller in body diameter and length (Lee et al., 2008a). Due to their amenability and commercial value, the number of publications for the relatively less speciose holothuroids is more, in comparison to more speciose ophiuroids. Understandably, the available publications have been reviewed from time to time. But these reviews are justifiably more a summary (e.g. Mladenov and Burke, 1994). In his subsequent review, Mladenov (1996) explored various factors that trigger fission. The excellent review by Dolmatov (2014) is limited to holothuroids. From a more broad-based, comprehensive and incisive analysis, the ensuing account has brought to light for the first time the following: (i) it indicates the limitation of clonal autotomy to a half dozen asteroid species, (ii) distinguishes fission at the antero-posterior body profile in holothuroids from oral-aboral body profile in carnivorous asteroids with pyloric caeca and ophiuroids with no special storage organ, (iii) highlights the incidence of larval cloning in four classes of echinoderms and (iv) approaches the subject of clonal reproduction more from the angle of stem cells (cf Skold et al., 2009), as an objective of this book is to bridge the gap between conventional zoologists and molecular biologists.

4.1 Types and Characteristics

Barring crinoids, a few echinoderms of the other four classes are capable of cloning during embryonic (e.g. *Staurothyone inconspicus*?, O'Loughlin et al., 2009), larval (*Luidia* sp, Bosch et al., 1989), juvenile (e.g. *Holothuria atra*, Chao et al., 1993) and immature (e.g. *Stichopus chloronotus*, Conand et al., 1998) stages, and from juvenile to life time (e.g. *Coscinasterias tenuispina*, Alves et al., 2002). Cloning involves fission, budding or clonal autotomy. Of these, fission is ubiquitous in holothuroids (Table 4.1), asteroids (Table 4.2) and ophiuroids (Table 4.3). Budding is limited to larval cloning in holothuroids, asteroids and echinoids but not in ophiuroids. Clonal autotomy is limited to six asteroid species. As indicated, regeneration of missing organ(s)/system(s) may demand different potencies of stem cells in antero-posterior body profiled holothuroids and oral-aborally flattened asteroids and ophiuroids. In the holothuroids, each ramet 'inherits' different sets of organs/systems (for details see Conand et al., 1987, Purwati, 2004). Conversely, the ramets of radially symmetrical asteroids and ophiuroids 'inherit' from the genet almost equally (except for, say, madriporite, see Mladenov et al., 1983) or unequally, when fission results, say, in 3 + 2 ramets. Clearly, antero-posterior profile seems to facilitate fission in relatively more number of holothurian species. Yet, even in the unequally fissioned asteroids and ophiuroids, the smaller and large fragments do inherit a fraction of the oral disk. Notably, the incidence of larval cloning is reported in increasing number of species. With globose or flattened body profile with no arms, echinoids do not clone as a juvenile and adult, but their larvae undergo larval cloning in *Strongylocentrotus purpuratus* (Eaves and Palmer, 2003), *Dendraster excentricus* and *Lytechinus variegatus* (Vickery and McClintock, 2000). Incidence of clonal reproduction is being discovered in more and more echinoderm species, but it must be noted that the incidence is not more than 3.7, 2.6, 0.3 and 1.9% for holothuroids, asteroids, echinoids, and ophiuroids, respectively (Table 4.4).

TABLE 4.1

Facultative and induced fission, and larval cloning in holothuroids (compiled from Crozier, 1917, Deichmann, 1922, Mladenov and Burke, 1994, Eaves and Palmer, 2003, Razak et al., 2007, Dolmatov, 2012, 2014, Hartati et al., 2016)

Facultative natural

Holothuroidae: *Actinopyca mauritiana, A. miliaris, Holothuria atra, H. difficilis, H. edulis, H. hilla, H. impatiens, H. leucospilota, H. parvula, H. portovallartensis, H. rowei, H. surinamensis, H. theeli*

Stichopodidae: *Stichopus chloronotus, S. monotuberculatus, S. naso*

Cucumaridae: *Ocnus loctea, O. planci, Squamocnus aureoruber, Staurothyone inconspicua, Colochirus robustus*

Schlerodactylidae: *Cladolabes schmeltzi*

Phyllophoridae: *Havelockia versicolor*

Induced

Holothuroidae: *Actinopyca miliaris, Holothuria arenicola, Bohadschia marmorata*

Sclerodactylidae: *Oshimella ehrenbergi*

Stichopodidae: *Apostichopus japonicus, H. scabra, Stichopus herrmanni, Thelenota ananas*

Cucumaridae: *Colochirus quadrangularis, Cucumaria grubei, C. syracusana, C. versicolor*

Larval cloning by budding

Stichopodidae: *Parastichopus californicus*

TABLE 4.2

Obligate (?) and facultative fission, clonal autotomy and larval cloning in asteroids (from Hyman, 1955, Emson and Wilkie, 1980, Mladenov and Burke, 1994 and others)

Obligately natural?

Asterinidae: 5-rayed *Asterina burtoni* (Res Sea)

Asteriidae: *Stephanasterias albula* (Noth Lubec, Maine, USA)

Facultative natural

Asterinidae: *Asterina anomala*, multi-radiated *A. burtoni* (Mediterrnean Sea), *A. exiqua, A. heteractis, A. regularis, A. wega, Asterias rubens, A. vulgaris, Aquilonastra coralicola, Nepanthia belcheri, N. brevis, N. briareus, N. fisheri, N. variablis*

Asteriidae: *Allostichaster inaequalos, A. insignis, A. polyplax, Coscinasterias acutispina, C. calamaria, C. muricata, C. tenuispina, Patiriella regularis, Sclerasterias alexandri, S. euplecta, S. heteropes, S. richardi, Marthasterias gracilis*

Solasteridae: *Pisaster brevispinus, Echinaster sepositus, Stichaster australis*

Oreasteridae: *Oreaster reticulatus*

Clonal autotomy

Linckiidae: *Linckia columbiae, L. diplax, L. guldingi, L. multifora, Ophidiaster cribrareus, O. robillardi*

Larval cloning by budding and fission

Luidia sp, *L. foliolata, Oreaster, Pisaster ochraceus, Distolasterias brucei*

TABLE 4.3

Facultative fission and larval cloning in ophiuroids (from Hyman, 1955, Mladenov, 1996 and others)

Facultative natural

Astroschematidae: *Astrogymnotes catasticta, Astrocharis ijimai, A. virgo*

Euryalidae: *Astroceras annulatum, A. nodosum, Asteromorpha perplexum*

Gorgonocephalidae: *Schizostella bifurcata, S. bayeri*

Hemieuryalidae: *Ophioholcus sexradiata*

Ophiomyxidae: *Ophiostiba hidekii, Ophiovesta granulata*

Amphiuridae: *Amphiocantha dividua, Amphiodia dividua, Amphipholis torelli, Amphiura sexradiata, Ophiostigma isocanthum*

Ophiactidae: *Ophiactis acosmeta, O. arenosa, O. cyanosticta, O. hirta, O. inaculosa, O. lymani, O. modesta, O. mulleri, O. nidarosiensis, O. parva, O. plana, O. profundi, O. rubrapoda, O. savignyi, O. seminuda, O. simplex, O. versicolor, O. virens, O.* sp

Ophiocanthidae: *Ophiologimus hexactis*

Ophiocomidae: *Ophiocoma pumila, O. valenciae, Ophiocomella ophiactoides, O. schmitti*

Ophionereididae: *Ophionereis dictydisca, O. dubia, O. sexradia*

Ophiotrichidae: *Ophiothela danae, O. hadra, O. mirabilis*

Larval cloning by budding

Ophiopholis aculeata (Balser, 1998)

Larval cloning in echinoids

Dendraster excentricus, Lytechinus variegatus (Vickery and McClintock, 2000)

TABLE 4.4

Reported incidences of asexual reproduction in echinoderms

Modes	Crinoidea	Holothuroidea	Asteroidea	Echinoidea	Ophiuroidea
Obligate	–	–	2?	–	–
Facultative	–	23	34	–	47
Autotomy	–	–	6	–	–
Larval cloning	–	1	5+	3+	3+
Induced fission	–	13+	–	–	–
Species (no.)	~ 700	> 1,000	~ 1,800	> 900	2,604
Asexual (no.)	–	37+	47+	3+	50+
Asexual (%)	–	3.7	2.6	0.3	1.9

Briefly, ~ 2% of echinoderms alone are capable of cloning, besides sexual reproduction. Nevertheless, clonal reproduction renders the echinoderm a fascinating group of animals.

In general, clonally reproducing echinoderms display a few characteristic features: 1. Cloning occurs in planktotrophic (e.g. *H. atra*) and lecithotrophic (e.g. *H. impatiens*) gonochorics and protandrics (e.g. *Nepanthia belcheri*,

Ottesen and Lucas, 1982) but not in brooding and viviparous echinoderms. As protandrics act either as male or female at a given time, clonal reproduction is restricted to oviparous gonochorics alone. 2. Clonal species are less fecund. For example, fecundity of the hexamerous fissiparous *Ophiocomella ophiactoides* is 7,400 lecithotrophic eggs, in comparison to one million planktotrophic eggs of the non-fissiparous pentamerous *Ophiocoma pumila*. In terms of fecundity, a single *O. pumila* genet may be equivalent to ~ 135 offspring of *O. ophiactoides* (Mladenov and Emson, 1990). In holothuroids, fecundity is in the range of 6,600 planktotrophic- and 800-lecithotrophic eggs in fissiparous *H. atra* and *H. impatiens*, respectively (Harriott, 1985); these values are far less than that of non-fissiparous *H. scabra* spawning one million eggs (Hamel et al., 2001). 3. Autotomy to regenerate one or more arms is ubiquitous in ophiuroids (Wilkie et al., 1984, Yokoyama and Amaral, 2010) as well as in other classes of echinoderms (see Chapter 4.1). In them, regeneration is limited to repair and replacing the missing body parts alone. However, it is extended to clonal reproduction in only six asteroid species. In order to distinguish them, the former is continued to be called autotomy but the latter is named clonal autotomy. Clearly, autotomic clonal reproduction is too costly in terms of both stem cells and nutritive resource. 4. Larval cloning species do not continue clonal reproduction either as a juvenile or adult (Table 4.4). Clearly, species characterized by autotomic cloning or larval cloning do not clonally reproduce by fission; clonal autotomic species do not undergo larval cloning; likewise, larval cloning species do not undergo clonal autotomy. Hence, these three types of clonal reproduction mutually eliminate each other. Nevertheless, clonal reproduction neither eliminates evisceration in at least 17 holothuroid species (e.g. *H. atra, S. chloronotus*, see also p 141) nor autotomic loss of arms in *Allostichaster capensis, A. polyplax, Coscinasterias calamaria* and *C. tenuispina*.

4.2 Fission and Reproduction

Incidence: In echinoderms, fission, clonal autotomy and budding are the methods of asexual reproduction. Of 137+ species, 87% of the echinoderms have chosen fission as the favored and prevalent method of clonal propagation; only 9 and 4% have opted for larval cloning and clonal autotomy, respectively. Fissionability, i.e. the percentage of population undergoing fission ranges from 8.3% in *Holothuria leucospilota* (Purwati et al., 2004) to 72% in *H. atra* (Chao et al., 1993, 1994) among holothuroids, 21% in *Allostichaster polyplax* (Overdyk et al., 2015) to 100% in *Stephanasterias albula* (Mladenov et al., 1986) among asteroids and from 52% in *Ophiactis savignyi* inhabiting the sponge *Haliclona* sp (Chao and Tsai, 1995) to 100% in those of

Discovery Bay (Emson and Wilkie, 1984). For *H. atra* alone from Taiwan (21 ° 57′ N) down to 23°30′ S on the One Tree Island (OTI), Great Barrier Reef (GBR), descriptive accounts on heterogony, i.e. co-existence of clonal and sexual reproduction are available (Table 4.5). Reports on the natural incidence (e.g. *H. atra*, Asha and Diwakar, 2015) and induced of fission are available for many holothuroid species (e.g. *H. arenicola*, Egypt, Razek et al., 2007, *H. atra* and *Bohadschia marmorata*, Mauritutius, Laxminarayana, 2006). Likewise, detailed descriptions are reported for *S. albula* from Maine, USA down to New Zealand for *Coscinasterias calamaria*, *C. muricata* and *C. tenuispina* (see Table 4.6). However, availability of similar accounts is

TABLE 4.5

Fissiparity (%) and size-wise and temporal separation in clonally and sexually reproducing holothuroids. d = density, Sex ratios (Sr) are shown as ♀ : ♂ : no gonad, F_1 = Fission

Species, Location, Reference	Fissiparity [F] (%), Density (no./m²)	Size-wise and Temporal Separation
Holothuria atra		
23°, 25′ S Herron Island Harriott (1982)	F varies from 10% at Crest to 58% at Mid-Reef	Sr switches from 0.13 : 0.87 at < 100 g to 0.68 : 0.32 in > 1000 g. F peaks in Aug–Feb but spawning in May–June
21°, 57′ N, Taiwan, (Chao et al., 1993, 1994)	F = 72% (10–70 g), d = 1.0/m², fissiparous in Wanlitung with 92% in > 100 g, but not in Nanwan	F peaks in September, Spawning in May–June
One Tree Reef (OTR), GBR, 23°, 30′ S, Lee et al. (2008b)	F = 13% at d = 0.25/m² but 10% at 1.0/m²	F only up to 175 g, sexuals between 180 and 575 g
Reunion Island Conand (1996, 2004)	F = 19% upto 75 g, 16% upto 105 g at d = 4.14/m²	Anterior 8.5%, posterior 10.6%
23° 25′ S, Thorne et al. (2013), 5 years study	F = 27%, d = 0.77/m²	F peaks in Aug. Spawning in May and Sep–Nov
H. leucospilota, Darwin, Australia, Purwati (2004)	F = 8.33%	In 140 g, 2 modes at 25 and 65 g but 15 g only one mode
H. parvula, Fort St Catherine, Bermuda (Emson and Mladenov, 1987)	F = 64% (range: 40–80%)	F peaks > 25°C, Aug to Nov Sexual July to Sep
H. hilla, OTR 23°, 30′ S Lee et al. (2008a)	F = 59%, growth @ ~ 4 g/ month	F in < 100 g size during Mar–Oct, Sexual in > 100 g
H. difficilis, OTR 23°, 30′ S Lee et al. (2009)	F = 37%, double body weight in 6 months, Sr 0.3 : 0.0 : 0.7	F from May–Aug, cooler months; sexual Oct–Nov, warmer months
Stichopus chloronotus, St Gilles, La Saline, Conand et al. (1998).	F = upto 70 g only, Sexual maturity at 70 g and age 1 year, d = 0.02–1.2/m²	F_1 produces > ♂ than ♀, Gonad of 10 and 30 mg for anterior and posterior ramets, against 340 mg for intact

limited to Jamaica for *Ophiocomella ophiactoides* from Florida and *Ophiactis savignyi* to Jamaica (Table 4.7).

Methods: In antero-posterior body profile of holothuroids, two methods of fissions are described: (i) a constriction approximately at the body center leads to fission, as in stichopodids (e.g. *Stichopus chloronotus*, Uthicke, 2001b), (ii) twisting and stretching method, in which anterior and posterior fragments slowly revolve upto 360° in opposite direction, resulting in a constriction around the center of the body. Subsequently, the two fragments move in an opposite direction until the body is teared at the constriction and then they

TABLE 4.6

Fissiparity (%) in population, altered sex ratio (♀ : ♂ : indeterminate) and interval between successive fission in some asteroids. ng = no gonad, F = fissionability, Ff = fission frequency, T = temperature, P = photoperiod

Species, Location, Reference	Fissionability (%), Sex Ratio: ♀ : ♂ : ng	Arm No, Interval between Successive Fission
Stephanasterias albula Maine, USA, Mladenov et al. (1986)	F = 100% 0.0 : 0.0 : 1.0	2–8 arms; 97% with 5 arms fission interval 1 or 2 years Ff : T, r = 0.89, Ff: P, r = 0.42
Asterina burtoni Achituv and Sher (1991) Karako et al. (2002)	F = 80% 0.0 : 0.125 : 0.988 Spermatogenesis upto 8 arms only but not in 9 and 10 arms	Multiradiate fissiparous in Mediterrean Sea, Pentamerous non-fissiparous in Red Sea, Non-fissiparous in Gulf of Suez
Allostichaster acutispina Japan, Seto et al. (2000, 2013), Shibata et al. (2011)	Uozu F = 100%, 0.013 : 0.8 : 0.18 Kursako 0.64 : 0.28 : 0.08 Lab 0.31 : 0.31 : 0.37	Fission interval > 5 months, in lab > 1 y, 6–8 arms
Allostichaster capensis 46°, 45′ S, Argentina Rubilar et al. (2005, 2015)	F = ~ 60%, occurs in all sizes and peaks in 30 mm R 0.002 : 0.033 : 0.964	3–8 arms, > 65% with 6 arms, fission interval once/> ~ 6 months Ff : T, r = 0.988, Ff: P, r = 0.991
Nepanthia belcheri 22–31°C, Australia Ottesen and Lucas (1982)	F = 50%, protandric, oocyte 500 μm but not spawned, spermatogenesis occurs	2–9 arms, 45% with 7 arms Fission interval = ~ 1 year
A. polyplax, New Zealand Overdyk et al. (2016)	F = 21% 0.3 : 0.4 : 0.3	
Coscinasterias tenuispina 20°, 57′, Brazil Alves et al. (2002)	F = ~ 40%, fission peaks at 35–45 mm R	
C. calamaria, New Zealand, Crump and Barker (1985)	Aquarium Point: F = 30%, 0.0 : 0.57 : 0.41 Maori Bay: F = 44%, 0.51 : 0.11 : 0.38 Wellers Rock: F = 17%, 0.44 : 0..57 : 0.0 Echinoderm Reef: F = 59%, 0.37 : 0.37 : 0.24	
C. muricata, New Zealand, Skold et al. (2002)	Aquarium Point: 0.0 : 1.0 : 0.0 Maori Bay 0.71 : 029 : 0.0	

TABLE 4.7

Effect of microhabitat (sponge, epiphytic turf, coralline surface) and their effect on sex ratio in two ophiuroid species. Sex ratios are shown as ♀ : ♂ : no gonad, d = density, Specimen numbers are indicated by square brackets

Species, Location, Reference	Reported Observations
Ophiactis savignyi	
Jamaica, Emson and Wilkie (1984)	Discovery Bay: 100% clonal, 1–3 mm R Bogue Island: 78% fissiparous: 21% sexual
Jamaica, Mladenov and Emson (1988)	Belize, Bermuda, Jamaica: 100% fissiparous
Taiwan, Chao and Tsai (1995), [3559]	52% in sponge *Haliclona* sp 71% clonal and < 3.9 mm R sex ratio: 0.04 : 0.96
Florida, McGovern (2003), [333]	Genets switches to sexual at 3.8 mm R, d = 77/1 sponge, ranges from 3 to 529/1 sex ratio: 0.47 : 53
Jamaica, Mladenov and Emson (1988) Fallmouth mangrove swamp	
Haliclona mobitila [536]	0.00 : 0.10 : 0.887
Tedania ignis [380]	0.03 : 0.03 : 0.94
Lissodendoryx isodictyalis [359]	0.002 : 0.06 : 0.94
Sub total	**0.01 : 0.06 : 0.932**
Twin Cays mangrove	
Haliclona mobitila [53]	0.08 : 0.0 : 0.920
Tedania ignis [1133]	0.02 : 0.096 : 0.902
Lissodendoryx isodictyalis [1118]	0.003 : 0.230 : 0.767
Sub total	**0.034 : 0.109 : 0.863**
Mean	**0.017 : 0.084 : 0.863**
Haliclona mobitila [53]	0.008 : 0.005 : 0.987
Tedania ignis [1133]	0.003 : 0.046 : 0.995
Lissodendoryx isodictyalis [1118]	0.001 : 0.080 : 0.920
Sub total	**0.04 : 0.4 : 0.956**
Ophiocomella ophiactoides	
Mladenov and Emson (1988), Jamaica Maze Cove	
Halimeda opuntia	0.23 : 0.10 : 0.695
Amphiphora fragilissima	0.14 : 0.10 : 0.755
Mean	**0.185 : 0.10 : 0.725**
Epiphytic	0.19 : 0.20 : 0.61

are completely separated, as in holothurioids. However, no information is yet available on the method of fission in cucumarids (see Table 4.1).

In the radially symmetrical asteroids, three distinct methods of fission occur. In 83% of pentamerous *Stephanasterias albula*, for example, a single furrow is formed in one side of the disk, which then progresses across the disk until the sea star is bisected into two fragments. In the remaining 17%

sea stars, splitting of the disk begins with furrows formed on two more or less opposing interradii. Further stretching leads to separation of the two fragments (see Mladenov et al., 1986). In the predominately seven-armed *Nepanthia belcheri*, a shallow furrow forms across the center of the disk, defining the fission plane and the sea star pulls itself into four- and three-armed ramets; the four-armed genet divides into a pair of two-armed ramets (see Ottesen and Lucas, 1982). In other radially symmetrical ophiuroids, fission commences with softening and furrowing of the disk through two opposing interradii, resulting in splitting of the two ramets (e.g. *Ophiocomella ophiactoides*, Wilkie et al., 1984).

Mechanism: Holothuroids cease feeding prior to fission (Purwati, 2004) and egest unusual fecal mass (Chao et al., 1993) and thereby clean their digestive system. Obviously, the behavioral fission process is neurally mediated and is a complex one involving every organ and organ system (e.g. Uthicke et al., 2001b). According to Dolmatov (2014), the dermis of holothuroids consists of almost exclusively connective tissues, which possess a unique ability to alter its mechanical property from a 'stiff' to 'soft' state. Two proteins tensilin and stiparin are involved in alteration of the mechanical property of the dermis. These proteins bind the collagen fibrils. The interaction of matrix metalloproteinases, their inhibitors and enzymes form cross-link complexes between fibril and collagen at one site (in natural fission) and two sites in *H. atra* (Purwati et al., 2009) and three sites in *Stichopus herrmanni* (Hartati et al., 2016) induced by rubber band ligation. The process may be accompanied by contraction of radial muscles. As a result, one or more constrictions are formed in the body. The twisting and stretching in holothurioids may accelerate the entire process.

4.3 Clonal and Sexual Reproduction

Unlike sexual reproduction, in which growth and reproduction competes for resources, there is an intense competition for nutritive resources between growth, clonal and sexual reproduction in asexually reproducing echinoderms. Besides being protected within sponges, the clonal ophiuroids, for which information is available, draw sediment particles, algae and sometimes small larvae brought by the continuous inflowing water. They have no specific storage organ. Holothuroids incessantly ingest sediments; with the transformation of their body from a sauage to stout profile, their dermis begins to serve as a sort of storage organ. Notably, the holothuroids and ophiuroids are herbivores and ingest continuously. Contrastingly, asteroids are carnivores and feed once a day or a few days, and store excess resource in the pyloric caeca. It is from this angle, the available information on competition for resource is analyzed. Strikingly, the analysis has for the first time brought to light the contrasting strategies adopted by 'herbivorous' ophiuroids and holothuroids, on one hand and carnivorous asteroids, on the other.

In Fig. 4.1, the size ranges of clonal species are indicated by a thin line at the bottom for each species; the ranges, in which clonal and sexual reproduction occurs, are indicated by dotted and continuous lines. Peak seasons of clonal and sexual reproduction are also shown by a dotted bow-like curve and inverted ∧- or ⌐-shaped continuous lines, respectively. Notably, the peaks of cloning and spawning durations are distinctly separated by both range of body size and season in the two clonal ophiuroid species and six clonal holothuroid species, for which adequate information is available. For example, clonal size is limited to 60 and 70 g but sexual reproduction occurs in sizes from > 60 to 300 g in 1 year old *Stichopus chloronotus* and > 70 to 1,400 g in *H. atra*. In *H. hilla*, *H. difficils* and *H. parvula*, both size and season effectively separate clonal and sexual reproduction (Fig. 4.1, Table 4.5). Durations of cloning and spawning over-lap only during April in *H. leucospilota*, in which fissionability is the lowest at 8.3%. Contrastingly, clonal and sexual reproduction co-exists in asteroids almost throughout the size range. They are either separated by season or by the almost absence of sexual reproduction as in *Stephanoasterias albula* in Maine, USA, *Asterina burtoni* in the Mediterranean Sea, and Uozu (Japan) population of *Coscinasterias acutispina*. In *C. tenuispina*, for example, spawning occurs during September–October but cloning peaks during November–January (Fig. 4.1, Table 4.6).

Further analysis reveals the presence of two groups in these three classes of echinoderms. In Group 1 asteroids like *Nepanthia belcheri*, *C. tenuispina* and *A. capensis*, fission extends up to 92–94% of their respective terminal size (Fig. 4.2A). In Group 2, it ceases already at 67% of terminal size, as in *A. insignis* (Barker and Scheibling, 2008) and *A. polyplax* (Overdyk et al., 2015). In them, fissionability ranges from 50 to 60% in Group 1 and 21% in Group 2 (Table 4.6). Correspondingly, sexual reproduction is likely to have a higher share in Group 2 than in Group 1. In holothuroids, the clonal potential is retained till 71–80% of terminal size in *Stichopus variegatus* and *H. fuscogilva* (Table 4.12). These holothuroids represent Group 1, in which clonal and sexual reproduction can coexist almost throughout the life time (Fig. 4.2B). In Group 2 holothuroids, clonal reproduction in Group 2a is limited to 5 and 20% of total body size of *H. atra* and *S. chloronotus*, respectively (Fig. 4.1A, B). For the Group 2b, these values are 47, 55 and 60% for *H. parvula*, *H. hilla* and *H. difficilis*, respectively. Fissionability is ~ 70% in Group 1 but is reduced to 50–61% only in Groups 2a and b. Hence the differences in the clonal size range of these two groups and their corresponding decreases in proportions of fissionability are reflected in those of clonal and sexual reproduction. Evidently, the size range of holothuroids, in which clonal reproduction occurs, is progressively decreased from 70–80% in Group

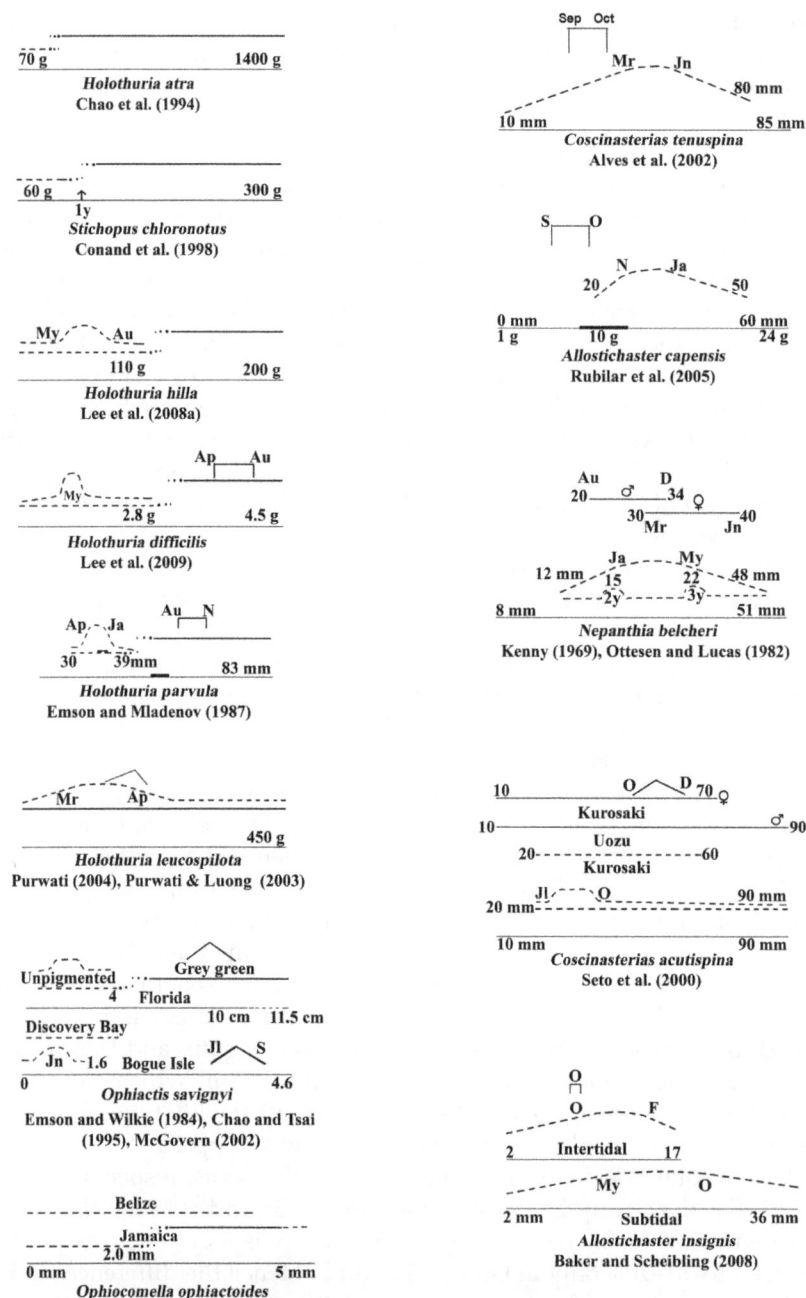

FIGURE 4.1

Sexual and asexual reproduction as functions of body size and time in (left panel) holothuroids and ophiuroids as well as (right panel) asteroids. For details see text.

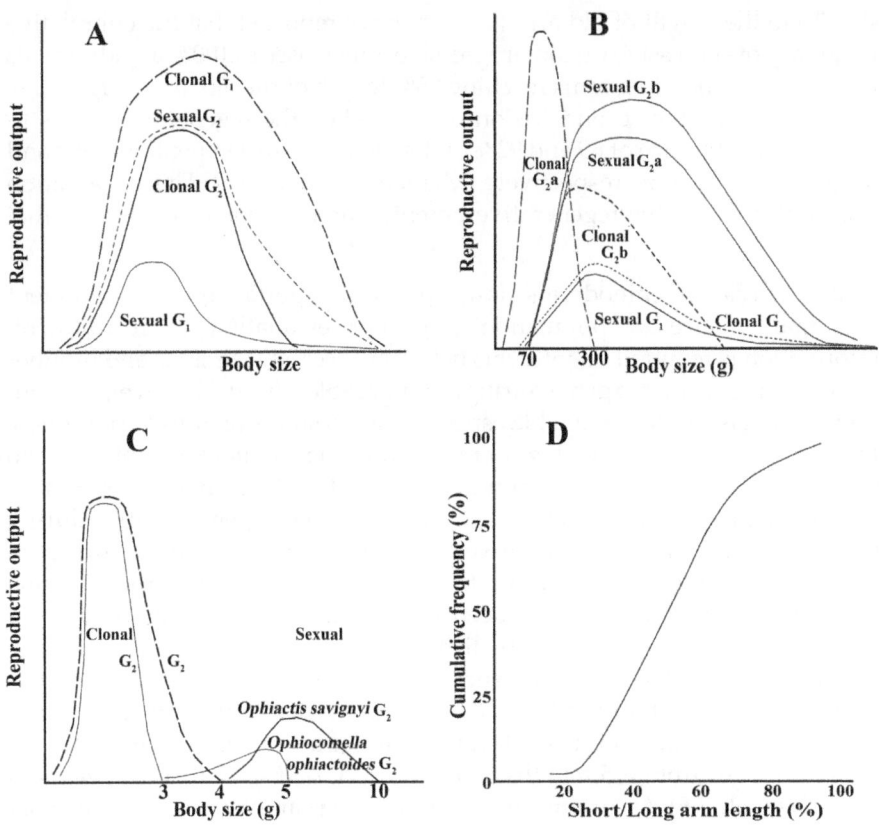

FIGURE 4.2

Sexual and clonal reproduction in Group 1 and 2 in (A). Asteroids, Group 1 and 2a and 2b in (B). Holothuroids (C). Ophiuroids Group 2 only. (D). Cumulative fission frequency as a function of short/long arm length in *Ophiocomella ophiactoides* (drawn from data reported by Mladenov et al., 1983).

1 to < 20% in Group 2. In ophiuroids, sexual reproduction is limited between > 2 and 5 mm size in *Ophiocomella ophiactoides* in Jamaica but as much as between 3.8 and 10 mm in *Ophiactis savignyi* in Florida (Fig. 4.2C, Table 4.7). Notably, sexual reproduction succeeds clonal propagation at a species specific body size in known ophiuroid species, indicating the presence of only Group 2 in ophiuroids.

Interestingly, the second fission occurs in clonal *Ophiocomella ophiactoides* and *S. albula*, even while arm regeneration of the first fission is ongoing. Mladenov et al. (1983) have found that almost all individuals of *O. ophiactoides* commence fission, even before the arms regenerate to a length equal to old arms. Considering the percentage of regenerating arm (RE) size (S) to the length of old arm (L), they have reported that the cumulative frequency of successive fission progressively increases to 100% in individuals with their regenerating arms reaching 95% length of the old arms (Fig. 4.2D). In *S. albula* too, the second fission occurs, when the regenerating arms RE have reached the size of 69 and 92% of the old arms (RE) in pentamerous and hexamerous sea stars, respectively (Mlanedov et al., 1986). This observation also confirms that the regenerative potential of asteroids is lower than that of ophiuroids.

Products: In clonal echinoderms, fission plays an important role in recruitment and maintenance of population size. For estimation of recruitment, information is required on intervals between successive fissions and number of ramets arising from a genet during a comparable period. However, relevant information is scarcely available especially for holothuroids and ophiuroids. From a survey of transferring animals from one to another habitat, Chao et al. (1993, 1994) estimated growth rate as 20 to 23 g/month for *H. atra*. Let us assume the ramet size as 35 g arising from a genet of 70 g during the first month of peak cloning season. The ramet could have grown to ~ 70 g and split by the third month of the peak season, i.e. from a single genet, four ramets could have been generated. There are reports hinting that 65 g weighing *H. leucospilota* displays two modals, one at 25 g and the second at 65 g (Table 4.5). From such bits and pieces of information, it was estimated that a holothuroid genet generates at least four offsprings within a year (Fig. 4.3). For asteroids, relevant available information reveals that the interval ranges from > 5 months/> 1 year in *A. acutispina* to 1 or 2 years in *S. albula* (Table 4.6). Assuming an average of one year interval and each genet generates three ramets, as in *N. belcheri* (Fig. 4.3); hence asteroids may recruit a maximum of three offsprings from a genet per year. A single publication available on the interval for *O. ophiactoides* clearly indicates that it divides once every 89 days, and a genet can undergo annually four successive fissions and generate 16 ramets/genet/year (Fig. 4.3). Not surprisingly, ophiuroids are named as brittle stars. Briefly, the fission rate decreases in the following order: ophiuroids < holothuroids < asteroids.

Potential: With differences in inheritance potential, ramet survival differs widely. For example, anterior ramet of the holothurian *S. chloronotus* inherits a mouth, oral tentacles and others, while posterior the anus, cloaca and others (Table 4.8). In natural populations of a few holothurians, the frequency of regenerating posterior ramets is higher than the anteriors. Hence, the posterior ramet has greater potential for regeneration. However, the anterior ramet can also regenerate, even when receiving 52, 45 and 22% of the total body length in *S. chloronotus*, *H. atra* and *H. leucospilota*, respectively (see Conand et al., 1998).

Available information on the time required for successful clonal reproduction in 12 holothuroid species, five asteroids species and a single ophiuroid species is summarized in Table 4.9, which also includes a couple of

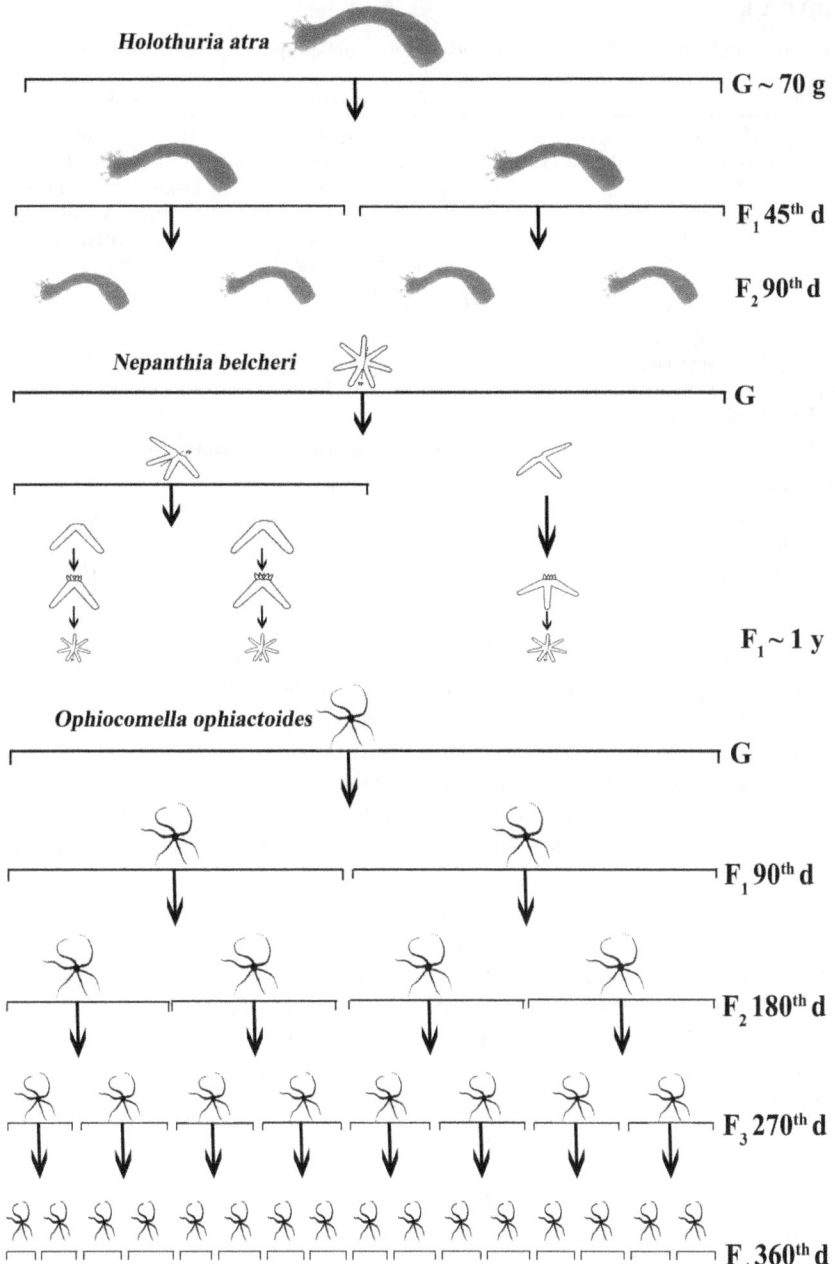

FIGURE 4.3

Approximate estimated intervals between successive fissions and number of ramets arising from a single genet in representative holothuroid, asteroid and ophiuroid during a period of one year.

TABLE 4.8

Inheritance and survival (%) of anterior and posterior presumptive ramets of holothuroids

Species	Anterior	Posterior
Stichopus chloronotus	Mouth	Anus
(condensed from Conand et al., 1987)	Oral tentacles	Cloaca
	Intestine	Disintegrated intestine
	Polian vesicle	Cuverian tubules
	Integument	Muscular band
	Right respiratory tree	Right respiratory tree
Natural fission		
Holothuria atra, Conand (1996)	7	8
H. atra, Conand (1996)	9	11
S. chloronotus, Conand et al. (1996)	7	9
H. leucospilota, Purwati (2004)	13	29
Induced fission (from Reichenbach et al., 1996)		
H. nobilis, 300 g	43	86
H. fuscogilva, 3165 g	0	20
Actinopyca mauritiana, 170 g	0	38
S. variegatus, 3650 g	0	80

values for autotomy in *Antedon bifida* and *Luidia clathrata*. All these values are plotted against body weight or arm number. Irrespective of the size ranging from 70 g in *H. atra* to 3,650 g in *Stichopus variegatus*, the time required for successful clonal reproduction varies from 90 days in *H. atra* to 180 days in *Thelenota ananas* or at the maximum calculated duration of 212 days in *Bohadschia marmorata* (Fig. 4.4) The single value available for ophiuroid *Ophiocomella ophiactoides* is 83 days and falls within this range. Incidentally, the arm regeneration in *Amphiura filiformis* is likely to be completed in ~ 60 days (Dupont and Thorndyke, 2006). In *A. bifida*, visceral regeneration is completed within 31 days (Mozzi et al., 2006). Briefly, either autotomized arm regeneration in *A. filiformis* or autotomized visceral regeneration in *A. bifida* or clonal reproduction in holothuroids and ophiuroids are all successfully completed within a period of 21 and 170 days. On an average, the holothuroids, ophiuroids and possibly crinoids require ~ 100 days to undergo successful clonal reproduction. Conversely, the available information for a half a dozen asteroid species indicates that irrespective of the number of arms varying from two to nine, clonal reproduction requires a long period of > 6 months in *Allostichaster capensis* to 1–2 years in *Stephanasteriasis albula* (Table 4.9). On an average, the asteroids require not less than 1 year for successful completion of clonal reproduction. Surprisingly, arm regeneration in an autotomized pentamerous *L. clathrata* also requires an equally long one year period. Hence, regenerative potential following either clonal reproduction or autotomy decreases in the following order: crinoids < ophiuroids < holothuroids < asteroids.

TABLE 4.9

Size at which fission occurs/is induced and time required for completion of clonal reproduction in holothuroids, asteroids and other echinoderms. *completion of regenerating arm

Species		Size (g)	Time (d)	Reference
Holothuroids				
1. *Holothuria atra*		< 70	90	Chao et al. (1993)
2. *H. atra*		96	150	Laxminarayana (2006)
3. *H. atra*		59		Purwati et al. (2009)
	Anterior		77	
	Posterior		70	
	Middle		90	
4. *H. nobilis*		300		Reichenbach et al. (1996)
	Anterior		83	
	Posterior		97	
5. *H. fuscogilva*				Reichenbach et al. (1996)
	Anterior	350	83	
	Posterior	3,165	104	
6. *Actinopyga miliaris*		–	120	see Reichenbach et al. (1996)
7. *A. mauritiana*		170		Reichenbach et al. (1996)
	Anterior		0	
	Posterior		65	
8. *Bohadschia marmorata*		110	212	Laxminarayana (2006)
9. *Stichopus herrmanni*		600–1300	40–80	Lambeth (2000)
10. *S. chloronotus*			90	see Reichenbach et al. (1996)
11. *S. variegatus*	Anterior	1300	83	Reichenbach et al. (1996)
	Posterior	3650	104	
12. *Thelenota ananas*			180	see Reichenbach et al. (1996)
Crinoid				
13. *Antedon bifida*	Visceral		21	Mozzi et al. (2006)
Ophiuroid				
14. *Ophiocomella ophiactoides*			83	Mladenov et al. (1983)
Asteroids				
15. *Stephanasterias albula*	5 arms	360		Mladenov et al. (1986)
16. *Allostichaster acutispina*		150–360		Shibata et al. (2011)
17. *A. capensis*	3–8 arms		> 180	Rubilar et al. (2005, 2015)
18. *Nepanthia belcheri*	2–9 arms		~ 360	Ottesen and Lucas (1982)
19. *Luidia clathrata**	5 arms		~ 360	Pomory and Lares (2000)

In his impressive review, Hotchkiss (2000) has grouped living asteroid families. Of them, 20 families are strictly pentamerous and 14 familes multiradiates. Of the latter, nine families are both pentamerous and multiradiate. They have evolved independently many times. The heterogonic (clonal and sexual) families (i) Asterinidae, (ii) Asteriidae, (iii) Solasteridae, (iv) Ophiasteridae (or Linckiidae, see Hyman, 1955 autotomic) and (v) Luidiidae (larval cloning)

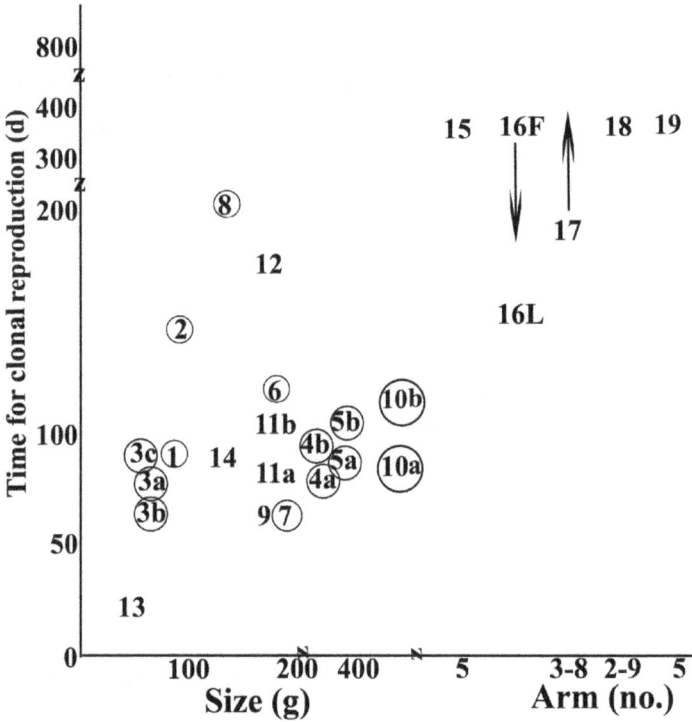

FIGURE 4.4

Effect of body size/arm number on the time required for completion of clonal reproduction in holothuroids and asteroids (data drawn from Table 4.9 are plotted. Values, for which size is not known, are not circled). L and F are values from laboratory and field, respectively. A single value on regeneration of arm in *Luidia clathrata* reported by Pomory and Lares (2000) is also included. L = Value from laboratory and F = Value from field.

are included within the nine (pentamerous cum multiradiate) families. In a non-heterogonic but pentamerous cum multiradiate *Labidiaster radiosus*, "new rays are added until far into adult life", "as many as six of these rays may bud from the disc simultaneously", "they may be distributed at rather regular intervals and the entire circumference of the disc" (see Hotchkiss, 2000). Apparently, the excess resource in the body induces increasing number of rays in member species of these plastic multiradiate families. It appears that member species of five multiradiate cum pentamerous heterogonic families have also opted to increase the number of arms but beyond a point, begun to split and switched to clonal reproduction.

Regenerating arms are henceforth indicated by smaller alphabets and numerals in text and figures. The pentamerous genets of *A. capensis* regenerate 1, 2, 3 or 4 arms (Rubilar et al., 2005). This would imply that the demand for

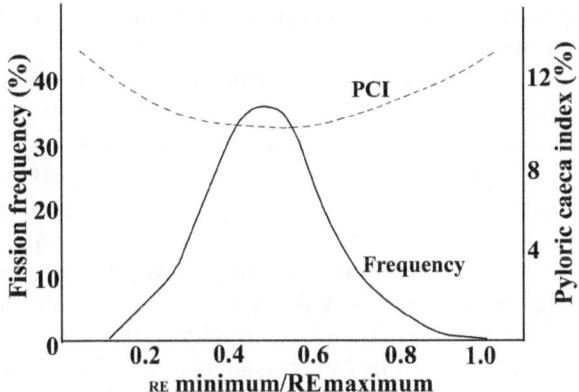

FIGURE 4.5

Fission frequency (dark line) as a function of RE minimum/RE maximum ratio in *Asterina burtoni* Pyloric caeca index (PCI) (thin dotted line) as a function of RE minimum/RE maximum ratio in *Coscinasterias acutispina* (compiled and redrawn from Achituv and Sher, 1991, Haramoto et al., 2007).

nutritive resources by the 4-arm regeneration (RE) in a sea star with one non-regenerating arms (RE) shall be four-times higher than in the four-armed non-regenerating (RE) sea star with a single regenerating arm (RE). Hence, fission in asteroids and ophiuroids is more complicated than holothuroids. This account recognizes at least three internal factors that determine the interval between successive fissions in asteroids and ophiuroids. Fission ratio (re = R minimum divided by R maximum) is calculated as an indicator of the interval between successive fissions. This ratio is grouped into 10 grades with a minimum of 0.0–0.1 and maximum of 0.9–1.0. The grade 0.1 indicates the recent fission in an animal and 0.9–1.0, the recent completion of regeneration or the non-occurrence of fission (cf Seto et al., 2013). Accordingly, the animals with 0.4 to 0.6 grades are fast approaching fission, but those at < 0.4 and > 0.7 grades shall split after a long time. For example, the interval in *C. acutispina* is extended from 100 days in those at 0.6 grades to as long as 400 days in those at 0.8 grades (Seto et al., 2013). In *A. burtoni*, the highest fission frequency occurs at 0.5 grades (Fig. 4.5). In the field too, fission frequency in *Asterina burtoni* increases to a peak at grades between 0.4 and 0.6. Interestingly, Haramoto et al. (2007) have shown that at these grades (0.4–0.6), the sea star splits, even when pyloric caeca index (PCI) is < 10%. However, the fission in *C. acutispina* requires > 12% PCI, when it is at the grades < 0.4 or > 0.6 (Fig. 4.5). The second factor is the health status of the asteroid. Infection by an endogenous parasite *Dendrogaster okadai* almost totally suppresses fission in *C. acutispina* by decreasing pyloric caeca reserves (Haramoto et al., 2007). The third internal factor is the presence of large number of non-regenerating

(RE) arms suppressing fission in asteroids and ophiuroids, and is brought to light for the first time in the following analysis.

Being multivariates, the number of arms vary in the asteroids from 3 to 9 peaking (66%) with 6 RE arms in *Allostichaster capensis* (Rubilar et al., 2005) and 1 to 8 peaking with 55% of 3 RE arms in *S. albula* (Mladenov et al., 1986) as well as 1 to 6 peaking with (55%) 3 RE or 1 to 9 peaking with (45%) 7 RE in the asterinids *N. belcheri* (Kenny, 1969, Ottesen and Lucas, 1982). In the ophiuroid *Ophiocomella ophiactoides*, it ranges from 2 to 5 arms (Mladenov et al., 1983). There are also reports indicating 4 (RE) non-regenerating arms and regenerating (RE) 3 arms, 5 + 5, 6 + 5 and 7 + 5 in *C. calamaria* (Crump and Barker, 1985), 2 + 1 and 3 + 4-5 in *C. acutispina* (Shibata et al., 2011) and usually 3 + 3 but rarely 3 + 4 in *Ophiactis* sp (Boffi, 1972). Nevertheless, the required information could be extracted only for the asteroids *A. capensis* (Fig. 4.6C) and *S. albula* (Fig. 4.6A), the asterinid *N. belcheri* (Fig. 4.6B) and an ophiuroid *O. ophiactis* (Fig. 4.6D). In Fig. 4.6, the thick line represents the number of RE arms as a function of the number of RE arms and the thin

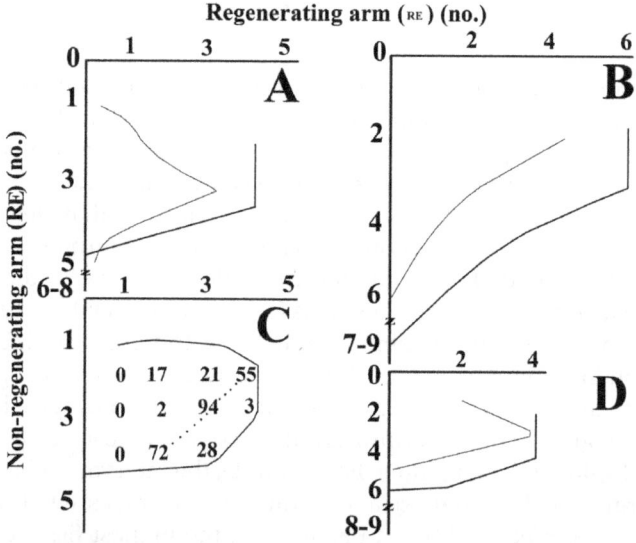

Figure 4.6

Number of regenerating arms (RE) as a function of non-renegerating arm (RE) in asteroids A. *Stephanasterias albula* (drawn from data reported by Mladenov et al., 1986), B. *Nepanthia belcheri* (drawn from data reported by Kenny, 1969, Ottesen and Lucas, 1982) and C. *Allostichaster capensis* (drawn from data reported by Rubilar et al., 2005) as well as D. ophiuroid *Ophiocomella ophiactoides* (drawn from data reported by Mladenov et al., 1983).

line relative levels in the number of RE arms at each RE arm. To make the trends more understandable, the percentage values of RE arm at each RE are indicated in Fig. 4.6C. From the figure, the following inferences can be drawn: 1. In star fish/brittle star with higher number of RE, for example, 6 to 8 in *S. albula*, 7 to 9 in *N. belcheri*, 8 to 9 in *A. capensis* and 5 in *O. ophiactoides* do not undergo fission at all. Some internal factor like the presence of large gonad and pyloric caeca (see Lawrence, 2010) seems to resist further fission in them and/or the stem cells responsible for fission do not express any longer in them. 2. In 0.1, 0.7–0.8 and 3.6% of *A. capensis*, *N. belcheri* and *S. albula*, respectively, the single armed star fish regenerates 1 to 3 buds but fails to regenerate 'comets', as in autotomized asteroids. Hence, the potential to split by the sea star with a single and higher number of arms does not exist. 3. The highest potential for fission occurs in 2 to 4 RE armed asteroids and ophiuroids especially in 3 RE armed ones.

Gonad and sex ratio: A couple of asteroids are reported to reproduce obligately by cloning. However, no invertebrate species is thus far known to exclusively reproduce asexually (Pandian, 2016). For example, *S. albula* and *C. acutispina* reproduce solely by clonal fission in the Mediterranean Sea and Uozu (Japan), respectively but their populations in the Red Sea and Kurosaki are heterogonic and reproduce clonally and sexually. Hence, the distinction between obligate and facultative fission at species level may not be valid.

Due to clear separation between clonal and sexual reproduction by body size and/or season, sex ratio is disturbed in heterogonics especially ophiuroids (Table 4.7). The presence of gonad has been reported from the ramets of *H. atra* and *S. chloronotus* (Doty, 1977, Uthicke, 1997, 1998). In the latter, gonad index is 0.08 and 0.19% in the ramets, in comparison to 0.34% in the intact greenfish (Conand et al., 1998). From a sample of 20 specimens, Lee et al. (2009) have recorded female ratio of 0.3 in *H. difficilis* but they have suggested that a larger sample is required for confirmation. Female ratio switches from 0.33 in < 100 g size to 0.62 in > 1,000 g size in the 'protandric' *H. atra* (Harriott, 1982).

Contrastingly, both gonad and sex ratio of asteroids are reduced (Table 4.6). In fact, no trace of gonad, gonoduct, genital branch of aboral hemal ring and aboral coelomic ring is found in *S. albula* (Mladenov et al., 1986). Apparently, the stem cells responsible for the development of reproductive system also regulate the existence of hemal and coelomic rings associated with the reproductive system. In *N. belcheri*, the gonads proximate to the regenerating arms are regressed or retarded (Ottesen and Lucas, 1982). This is true also in *Ophiactis savignyi* (Fig. 5.3D). Hence, fission occurs at the cost of gonad. As a consequence, female ratio in asteroids and in an ophiuroid is reduced with increasing fissionability (Tables 4.6, 4.7). The ratio is reduced from 0.51 in *C. calamaria* to zero in *Stephanasterias albula*, *Asterina burtoni*, *Allostichaster capensis* and *O. savignyi*, in which fissionability remains at 70–100%.

4.4 Induced Fission

In holothuroids, fission can be induced by ligating the body at one or more levels with band(s) made up of rubber, bicycle tube and plastic, or by cutting the body at desired level by a blade (e.g. Laxminarayana, 1996, Dolmatov, 2014, Dolmatov et al., 2012). Incidentally, the anterior and posterior halves of *Holothuria pervica* and *Massinium magnum* do not survive following a transverse cut (Dolmatov, 2014, Dolmatov et al., 2012). After a surgical cut, only the posteriors regenerate and develop into ramet in *Apostichopus japonicus, H. scabra, Ohshimella ehrenbergi* and *Colochirus quadrangularis* (Dolmatov, 2014, Dolmatov and Mashanov, 2007, Dolmatov et al., 2012). Rarely, but ingeniously, pyloric caecotomy has been surgically made to investigate the role played by this storage organ in reproduction. The surgery involves cutting off the ray tip in an anesthetized sea star and subsequent removal of the pyloric caeca by introducing a hooked probe through the ray tip hole deep into the coelom (Harrold and Pearse, 1980). In *Ohiocomella ophiactoides*, Mladenov et al. (1983) have made one, two or three incisions in the mouth region through the interradi or in the opposite interradii in the brittle star with different arm configurations and potencies. The brittle star passing through 0.50 to 0.75 RE/RE grades split into the number of incision-dependent ramets after different durations of regeneration. However, others, who are at < 0.5 RE/RE grades fail to split, even after two or three incisions (Table 4.10). Notably, with increasing number of incisions, a single genet can yield four ramets.

In holothuroids, two different approaches have been made to know (i) the minimum genet size required to ensure the completion of regeneration, survival and growth of ramets and (ii) the effect of genet's body size on the potential to regenerate and survive (Table 4.11). Though many have reported natural (e.g. Asha and Diwakar, 2015) and induced (e.g. Purwati and Diviono, 2005, Razek et al., 2007) fission, they have not provided details on the size of

TABLE 4.10

Effect of incision number and the number of ramet generated in *Ophiocomella ophiactoides* (condensed from Mladenov et al., 1983), m = optimal grade for fission

Incision (no.) and Position	Arm Configuration (no.)	Products
1 incision in mouth via interradial	2 RE + 4 RE m	19 hours, 2 ramets each with 2 or 4 arms
1 incision in mouth via interradial	6 RE	2 days, 2 each with three arms
2 incisions in opposite interradii	3 RE + 3 RE m	4 or 21 hours, 2 with 3 arms or two with 2 or 4 arms
3 incisions in opposite interradii	6 RE	3 hours, 3 each with two arms

TABLE 4.11

Minimum body size required to induce clonal reproduction in holothuroids

Species, Body size, Reference	Observations and Inferences
Single plane fission	
Holothuria parvula, H. surinamensis, 50% of anterior, Crozier (1917), Kille (1942)	Both the anteriors and posteriors underwent successful cloning
H. atra, 44% of anterior, Chao et al. (1993)	Both the anteriors and posteriors underwent successful cloning
H. leucospilota, 33% of anterior, Purwati (2004)	Both the anteriors and posteriors underwent successful cloning
H. atra, 12–48 g, Purwati et al. (2009)	Both the anteriors and posteriors underwent successful cloning
H. atra, 40–87 g, Laxminarayana (2006)	96–100% survival and cloning, ~ 0.63 g growth/day
Bohadschia marmorata, 43–50 g, Laxminarayana (2006)	93–100% survival and cloning, ~ 0.53 g growth/day
H. fuscogilva, 5.5% of adult weight, 16% of *Stichopus variegatus*, 17% of *H. nobilis*, Reichenbach et al. (1996)	Both the anteriors and posteriors survived and underwent successful cloning
Double plane fission	
Holothuria atra, 12–48 g, Purwati et al. (2009)	The anteriors (85%) and posteriors (95%) underwent and completed cloning on 70th day. But the middle developed tentacles (85%) and anus (55%) on 77th day
Stichopus herrmanni, 117 g, Hartati et al. (2016)	The anteriors (50%), posteriors (48%) and middles (64%) successfully cloned on the 60th, 65th and 73rd day, respectively
Treble plane fission	
Stichopus herrmanni, 88 g, Hartati et al. (2016)	The anteriors (50%), middle 1 (30%), middle 2 (40%) and posteriors (60%) successfully cloned and grew to 11.5, 2.6, 2.0 and 2.4 g on the 16th week

the holothuroid that underwent binary fission. However, admirable efforts have been made to induce single, double and treble fissions by a few authors. From Table 4.11, the following may be inferred: 1. *Bohadschia marmorata*, *H. atra* and *S. herrmanni* are amenable to single, double or treble fissions and yield two, three or four ramets from a single genet, i.e. the minimum genet's size required for fission is 25% of body length. Clearly, all the four ramets arising from *H. herrmanni* genet have inherited equal proportion of dermis, coelom and radial nerves among other organs and systems. Hence, the coelom and radial nerves seem to harbor adequate number of stem cells to ensure successful cloning (see Candia-Carnevali et al., 2009). That the differences in the time required for completion of cloning, however, indicate other factors like the presence of mouth and dorsal mesentery (Kille, 1939, 1942, Mary

Bai, 1994) may be responsible for the observed differences. 2. Ramet survival remains high (85–100%) in *B. marmorata* and *H. atra* following the induction of a single or double fission. But, it decreases to 30–60% in *S. herrmanni*, which has suffered treble fissions resulting in the yield of four ramets. 3. Following double fissions, more number of posterior ramets of *H. atra* (100%) regenerates oral tentacles than the anteriors. It is also true of *S. herrmanni*, which has undergone double fissions. Irrespective of the number of fissions, the middles regenerate slower than either the anteriors or posteriors. Within a period of 77 days following fission, the recovered anteriors grow faster, as they already have a mouth.

With regard to the effect of body size on clonicity, the induction of a single (217 g, Laxminarayana, 2006) or double fission(s) (143 g, Purwati et al., 2009) and the presence of gonads in the recently fissioned ramets indicate that *H. atra* retains its cloning potential even after attaining 70 g size (Fig. 4.2B) and sexual maturity (Table 4.5). In about 300 g size, the body profile of *H. atra* is transformed from a sauage to stout shape (Chao et al., 1993), which progressively prevents its ability to twist and revolve, a behavior that obligatorily precedes the formation of a constriction for the ensuing fission. Incidentally, *B. marmorata*, *Actinopyga mauritiana* and *H. nobilis* (Table 4.12) have retained the fission potential until attaining body size of 96, 170 and 300 g, respectively. Hence, these three holothuroids retain their cloning potential, a little more upto 500 g. Reichenbach et al. (1996) have undertaken simple but a more meaningful experiment to examine the cloning ability of holothuroids weighing upto 4,550 g. Their findings are summarized in Table 4.12 from which the following may be inferred: 1. Amazingly, the posteriors of *H. fuscogilva* and *S. variegatus* retain their fission potential upto the maximum sizes of 4,450 and 4,550 g, respectively. Survival of the posteriors is up to 80% in 3,650 g weighing *S. variegatus*, in comparison to 20% for 3,165 g size in *H. fuscogilva*. But the anteriors retain the potential only upto 350 and 1,300 g in *H. fuscogilva* and *S. variegatus*, respectively. 2. Duration required for regeneration is extended with increasing body size and in all the anteriors. For example, it is only 41 days for the posterior small (600 g) *S. variegatus* but 104 days for the larger (3,650 g) ones. The following may also be noted: (i) The posteriors possessing a greater portion of respiratory tree may more readily meet the oxygen demand by the regenerating halves. (ii) The holothuridids go through twisting and revolving to initiate a constriction but the stichopodids initiate a constriction without involving this behavior. This may be responsible for the higher fission potential of *S. variegatus* than the other holothuridids. (iii) The dermis constitutes 8.9% of body weight of *H. fuscogilva* but that of *S. chloronotus* is 2.7% only (Reichenbach et al., 1996). Unfortunately, no data on dermis weight is available for other holothuroids. Conand (1993) has listed body and drained weights for a number of holothuroids; the drained weight may provide an idea on the dermis weight in relation to body weight. Interestingly, these values are 68%

TABLE 4.12

Effect of body size on survival and regeneration time in holothuroids, which were induced to split by rubber banding the body to initiate constriction (compiled from Reichenbach et al., 1996)

Weight Class (g)	Survival (%)		Regeneration (d)	
	Anterior	Posterior	Anterior	Posterior
Holothuria nobilis				
300	43	86	83	93
1796	0	0	–	–
H. fuscogilva				
175	67	92	83	83
350	30	80	83	104
3165	0	20	–	–
Stichopus variegatus				
600	100	100	55	41
1300	100	100	83	55
3650	0	80	–	104
Actinopyga mauritiana				
170	0	38	–	65
1125	0	0	–	–

The largest fissioned *S. variegatus* weighed 4,550 g and *H. fuscogilva* 4,450 g

for *H. atra* and *A. mauritiana*, 73% for *H. nobilis*, 77% for *H. fuscogilva* and 80% for *S. variegatus*. However, these values do not correlate with the maximum fissionable body size, as they are 300 g in *H. atra*, *A. mauritiana* and *H. nobilis*, 3,165 g in *H. fuscogilva* and 3,650 g in *S. variegatus*. Hence it is the mechanical property rather than dermis thickness that determines the fission potential in holothuroids.

Pyloric caeca index (PCI): In carnivorous asteroids, Pyloric Caeca (PC) serves as a storage organ to provide the required nutrients for sexual and/or clonal reproduction. They also supply digestive enzymes. A definite reciprocal correlation observed between Gonad Index (GI) and PCI in *C. calamaria* (Fig. 3.4) and *A. capensis* (Rubilar et al., 2005) demonstrates that PC supply nutrients in support of sexual reproduction in these sea stars, in which clonal and asexual reproduction co-exist throughout the size range. However, there are others like *A. insignis*, in which clonal asexual and sexual reproduction are clearly separated from each other by body size; the small sea stars of < 20 mm are clonal but in the larger ones of > 20 mm, sexual reproduction dominates. The 15 and 20% PCI in the small and larger sea stars clearly indicate that PC supply the required nutrients to support clonal reproduction in small ones and sexual reproduction in larger ones (Barker and Scheibling, 2008). There are also other reports to confirm this. For example, sandwiched between two peaks of PCI, oogenesis occurs in *N. belcheri*. Of the two peaks, the major (30%) one occurs in July and the minor (20%) in December. A clear reciprocal cycle between oogenesis and PCI indicates that the former is supported by PC possibly between June and July. But clonal reproduction is supported by PC after spawning in August, if it occurs (Ottesen and Lucas, 1982). In *C. acutispina*, fission is initiated in the presence of > 12% PCI in those at < 0.3 and > 0.8 RE/RE grades but fission is initiated even with < 10% PCI in those at grades between 0.4 and 0.7 RE/RE (Haramoto et al., 2007). Hence, PC plays a definite role in initiation of fission. Alves et al. (2002) have found that though PCI is higher in sexuals than clonals of *C. tenuispina* and the reciprocal cycles between GI and PCI are not pronounced. Another factor that complicates the relation between GI and PCI is the ration level. Year long experiments on feeding and starving *A. insignis* have shown that (i) fission frequency increases in fed animals but decreases with increasing body size and (ii) unfed sea stars neither increase GI nor PCI but prolong the interval between successive fissions (Table 4.13, Fig. 4.7A). To assess the role of feeding and starvation on the reciprocal cycles of GI and PCI, Harrold and Pearse (1980) have surgically pyloric caecotomized *Pisaster giganteus*. Feeding enables the PC to supply nutrients required for growth of body and GI as well as gametogenesis in this non-fissiparous sea star (Table 4.14). Notably, food intake decreases from 15 snails/d in the intact sea star to < 0.6 snail/d in pyloric caecotomized sea star. Clearly, the sham and surgical sea star, despite the presence of a mouth and other structures, are unable to eat in the absence of PC that secrete digestive enzymes. Hence, PC is responsible for food intake as well as supply of nutrients to clonal and sexual reproduction.

Experimental evidence for the role of density comes from a single publication. In a four-month long experiment, Sterling and Shuster (2011) investigated the effect of selected densities on fission in *Aquilonastra corallicola*. Fission increased from ~ 0.05% at the high (30 no./19 l) to ~ 0.2% in low (5 no./19 l) density.

TABLE 4.13

Effect of feeding for a period one year on fission in *Allostichaster insignis* males of small, medium and large body size (compiled from Barker and Scheibling, 2008)

	Small		Medium		Large	
	Fed	Unfed	Fed	Unfed	Fed	Unfed
Undivided (no.)	1	2	1	6	5	5
Fissioned (no.)	3	1	5	0	1	0
Days required for						
I fission	150	–	60	–	–	–
II fission	200	–	220	–	–	–
Gonad/GI (%)	0	0	4.5	0	~ 5	0
Pyloric caeca (%)	15	–	19	–	21	–

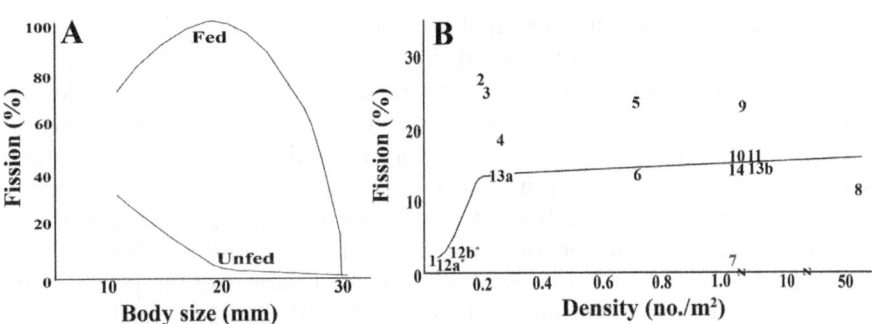

FIGURE 4.7

A. Effect of feeding and starvation on fission frequency in *Allostichaster insignis* (drawn from data reported by Barker and Scheibling, 2008), B. Effect of density on fission frequency in holothuroids and asteroids (drawn using data reported for *Holothuria atra*: 1. La Reunion, reef front, Conand, 1996, 2. Heron Island, Lagoon, Harriot, 1982, 3. Guam, Pago Bay, Doty, 1977, 4. Taiwan, Nanwan, Chao et al., 1993, 5. New Caledonia, Conand and De Ridder, 1990, 6. Taiwan, Wanlitung, Chao et al., 1993, 7. La Reunion, back Reef, Conand, 1996, 8. Eniwick, reef-flat, Lawrence, 1979, 9. Papua New Guinea, Conand and De Ridder, 1990, 10. Saline, back Reef, Conand, 2004, 11. *Stichopus chloronotus*, Conand et al., 1998, 12. *Aquilonastra corallicola*, Sterling and Shuster, 2011, 13. *H. atra*, Lee et al., 2008a, b, 14. *Allostichaster polyplax*, Overdyke et al., 2015).

TABLE 4.14

Effect of feeding and starving for ~ 18 weeks on intact, sham (ray tip removed) and pyloric caecotomized (ray tip + pyloric caeca, surgical) starfish *Pisaster giganteus* (compiled from Harrold and Pearse, 1980, approximate values)

Parameter	Intact + Fed	Intact Starved	Sham + Fed	Surgical + Fed	Surgical + Starved
Movement (no./week)	1	4	1	1	1.5
Bodyweight (%)	+ 20	− 10	+ 5	− 1	− 25
Snails eaten (no./day)	15	–	0.8	0.6	–
PCI	7.0	3.0	6.5	3.0	1.5
GI	3.8	0.0	1.8	1.0	0.4
Gamtegonesis (%)	100	32	80	65	0

4.5 The Trigger

Proposing a model for clonal and sexual reproduction, Mladenov (1996) identified (i) food availability, (ii) population density and (iii) predation pressure as major environmental factors that individually or in combination may trigger clonal reproduction. He has also indicated that other minor factors like desiccation, hypoxia, mechanical disturbance, temperature and photoperiod may also trigger fission. However, the following must be noted prior to identifying the trigger capable of inducing clonal reproduction: 1. In none of the clonal species of echinoderms, cloning is a synchronized event at population and species level, and 2. In almost all these echinoderms, cloning occurs almost throughout the year. Although a level of synchronization of cloning of *Dendraster excentricus* larvae occurs with abrupt changes in food supply (McDonald and Vaughn, 2010), fission is not a synchronized event in the adults. Sexual reproduction occurs during warmer spring and summer months but clonal reproduction peaks during cooler months (Tables 4.5, 4.6). The reported 'r' values for correlation between fission and temperature are 0.603 for *Coscinasterias tenuispina* (Alves et al., 2002) but 0.42 for *Stephanasterias albula* (Mladenov et al., 1986). Those for fission and photoperiod are 0.89 for *S. albula* but 0.510 for *C. tenuispina*. Apparently, the trigger seems to be temperature for *C. tenuispina* but photoperiod for *S. albula*. However, it is difficult to comprehend why one species is triggered to undergo fission, while the other is not, although both of them flourish in the same habitat, sharing the same level of temperature and photoperiod. For example, *Holothuria atra* is triggered to fission in Rangelap Atoll but *H. leucospilota* inhabiting the same atoll is not. Similarly, *Stichopus chloronotus* undergoes fission on Herron Island but *H. leucospilota* does not. This phenomenon occurs in nature with *H. atra*, *H. leucospilota* and *S. chloronotus*, which share the same habitat on Fanning Island as well as *H. atra*, *H. edulis* and *S. chloronotus* maintained under the same laboratory conditions (see Uthicke, 1997). Obviously, the major factors namely food availability, population density and predation pressure may serve as a trigger at an individual level but not at population or species level. Not much information is yet available on predatory pressure triggering fission in adults of clonal echinoderms, although predation is experimentally shown to induce larval cloning in echinoid larvae (Vaughn, 2009, 2010). In the following paragraph, the triggering role played by population density and food availability on fission is briefly analyzed.

Experimental investigations have shown that with increasing density, fission frequency is decreased in a holothuroid and an asteroid. Fission is also increased from 10% at 1.0 no./m^2 to 13% at 0.25 no./m^2 in *H. atra* (Lee et al., 2008b) and 0.05% at 30 no./19 l to 0.2% at 5 no./19 l in *Aquilonastra corallicola* (Sterling and Shuster, 2011). Figure 4.7B shows the effects of fission frequency as a function of population density in a few clonal holothuroids and asteroids. Firstly, the limited available information relates the frequency to density in units of number, although the one with biomass could have provided more precise information. Secondly, the values given in the figure are not means for an equal size range of species or complete size range of a species. The third factor is the patchy distribution. For example, the distribution of one of the smallest holothuroids *H. difficilis* (4.5 g) is reported as patchy. Where *H. difficilis* occurs, its density ranges from 200–600 no./ m^2 (see Lee et al., 2009). With these limitations, the following generalization can still be made: fission frequency rapidly increases up to 0.3 no./m^2 but beyond which it levels off. Notably, with increasing density, food availability may decrease (see Fig. 1.5). Consequently, density may impose its effects through food availability. Hence, more than density, food availability may serve as prime triggering factor.

As fission is triggered by a factor that lies at an individual level rather than at population and species level, internal factors like the proximate composition and its possible effects on fission and regeneration have to be searched. 1. The ash (mineral) content of arms constitutes 60–70% and 75–95% dry weight of asteroids and ophiuroids, respectively. Of the organic matter, protein constitutes 21–37% in asteroids and 11–23% in ophiuroids (see Lawrence, 2010). Notably, protein is an important constituent, and constitutes as much as 30% in arms of the carnivorous asteroids, but 17% only in the almost herbivorous/omnivorous ophiuroids. 2. The resource allocation for regeneration occurs at the cost of somatic growth in *Acrocnida brachiata* (Pomory and Lares, 2000) and with significant reduction of gonad in *Ophiothrix fragilis* (Morgan and Jangoux, 2004). 3. Consequent to the loss of smaller or larger fraction of the oral disk, food consumption in an arm-regenerating asteroid like *Luidia clathrata* is decreased to ~ 2% body weight, in comparison to 4% body weight in an intact sea star (Lawrence and Ellwood, 1991). These findings from autotomic but non-clonal asteroids may explain the results reported by authors, who have estimated the effect of feeding on fission and regeneration in asteroids.

Limited available information on the effects of feeding and starvation, low, medium and *ad libitum* rations on small immature, medium maturing and large mature *A. insignis* is summarized in Table 4.13. The information is also classified to show the effects of feeding on fission, arm regeneration and PC-mediation in arm regeneration (Table 4.15). 1. Fission is suppressed by starvation for 6 to 12 months in *C. acutispina* and *Allostichaster insignis*. If it rarely occurs, it does in 1% of *C. muricata* alone. Clearly, food availability is the most important factor that allows the asteroid to split. 2. More number of medium sized maturing sea stars receiving medium ration underwent fission, as their body and PC (20% in *A. insignis*) hold adequate energy to trigger fission. With less resource in the body and PC (15% in fed *A. insignis*, 0% in starved *C. muricata*), the star fish can ill-afford to split. In the larger sea star, the presence of larger fractions of PC and gonad inhibits parting off of the valuable resource and reserves them for sexual reproduction (see also p 112). 3. With feeding in *A. insignis*, arm regeneration is more than doubled in small immature and medium maturing sea stars and the large (15 to 20%) PC nourishes *C. muricata*. On starvation, the arm length itself is decreased to ensure survival in *A. insignis* and reduces the PC to 0% in *C. muricata*. 4. With the loss of small or larger fractions of mouth and oral disk, food consumption is reduced to < 50% for a long duration (e.g. *Stichopus striatus*, Diaz-Guisado et al., 2006). Briefly, when internal body (including PC) resource has reached a critical level, fission is triggered followed by successful clonal reproduction. In turn, food availability facilitates the accumulation of resources to the critical level that triggers fission in carnivorous asteroids.

Contrastingly, food availability may not be the prime factor triggering fission in the almost 'vegetarian' holothuroids and ophiuroids with no storage organ like the PC in asteroids. Holothuroids are sediment-feeders and ophiuroids feed on small animals and microbes. In these slow motile holothuroids and ophiuroids, predatory pressure must act as a trigger. This aspect is elaborated in Chapter 5. In holothuroids, predatory pressure including harvest by humans may trigger fission. Although many publications describe the fast declining holothuroids stocks, no one has yet reported whether 'overfishing' has induced fission in holothuroids.

TABLE 4.15

Effect of experimental feeding and starvation as well as ration levels on fission-arm growth and pyloric caeca index (PCI) in asteroids

Species, Ration, Duration, Reference	Reported and Inferred Observations
Effects of starvation on fission	
Coscinasterias acutispina, 16 mo, Seto et al. (2013)	Survived starvation for 6 months. But none of the starved ones ever split, while the fed ones divided at least once
Allostichaster insignis, 12 months in small (11 mm) immature, medium (20 mm) maturing and large (30 mm) matured. Barker and Scheibling (2008)	Fission frequency decreased in fed ones in the following order: medium < small < large. None of the starved medium and large ones ever split
C. muricata, 125 days, low, medium *ad libitum*, Skold et al. (2002)	Fission was induced in 2 on 8 days, 3 on 32 days and 3 on 25 days out of 48 sea star receiving low, medium and *ad libitum* ration, respectively
Effects of ration on growth of regenerating arm	
A. insignis, as above	Arm length increased from 50 to 125, 100 to 200 and 175 to 200 mm in fed small, medium, large ones but decreased from 60 to 25, 100 to 80 and 175 to 150 mm in starved ones for 400 days
C. muricata, as above	Arm length increased by 25 and 50 mm in those fed medium and *ad libitum* ration but decreased by 40 mm in those fed low ration
Effects of PCI-mediation on arm growth	
A. insignis, as above	PCI level was 15, 20 and 22 in small, medium and large sized sea star that were fed. But it was 0 and 1% in small ones and others that were starved
C. muricata, as above	PCI increased from 0 to 1% in those fed low and medium ration
Effects on food intake	
Luidia clathrata, Lawrence (2010)	Food consumption decreased from 4% body weight in intact sea star to 2% body weight in those with regenerating arm(s)

4.6 Clonal Autotomy

The echinoderms are unique to possess collagenous connective tissues that are capable of rapid, neurally-mediated changes in their tensile strength (Wilkie et al., 1984, Motokawa and Tsuchi, 2003, Motokawa et al., 2012). When induced by predators (e.g. *Heliaster helianthus* autotomizes provoked by the predatory attack of *Myenaster gelatinosus* and *Luidia magallanica*, Lawrence and Gaymer, 2012) and others, the tensile strength of the key tissues rapidly decreases and results in autotomy in a few species of sea stars, in which regeneration may go up to the formation of new individuals. This sort of clonal reproduction, for some unknown reason, occurs only in six asteroid species belonging to two genera *Linckia* and *Ophidiaster* of the family Ophiasteridae (Table 4.2). Not surprisingly, detailed descriptions of these autotomic asteroids are also limited to Monks (1904), Edmondson (1935) and Rideout (1978). Incidences reported for clonal autotomy are 17% for *L. columbiae* (Monks, 1904), 26% for *O. robillardi* (Marsh, 1977) and 32% in *L. guildingi*. In the latter, 9.2% are comets. Notably, the autotomy occurs in all the size ranges from 4 to 17 mm in *L. columbiae* (Monks, 1904) and from 1 to 28 mm arm length in *L. multifora* (Rideout, 1978).

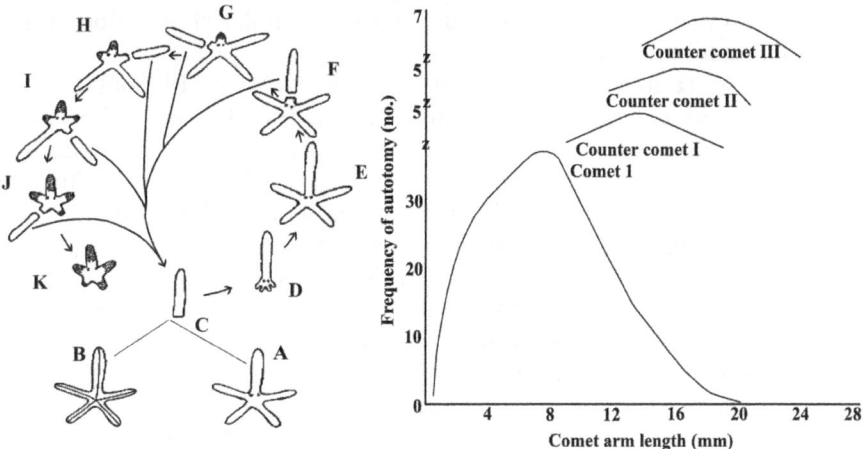

FIGURE 4.8

Left panel: Autotomic clonal reproductive cycle in *Linckia multifora*. A: comet: aboral surface, showing two madreporites (mp) and the numbering system; B: comet, oral surface, with lines indicating ambulacral grooves; C: autotomized arm; D: young comet; E: comet; F: counter-comet (c-c); G: post c-c I; H: post c-c If; I: post c-c III; J: disk -parent; K: disc-parent showing regeneration in all arms. Counter-comet (F) casts off either Arms 2 or 5 with equal occurrence. Right panel: Frequency of autotomy as a function of arm length in comet I and counter comets I, II and III in *L. multifora* (modified and redrawn from Rideout, 1978).

From a well designed and intense field survey as well as laboratory experiments, Rideout (1978) discovered that the autotomy in *L. multifora* can be divided into categories, each of which forms a part of clonal reproductive cycle. Accordingly, a parent disk is the one, which has retained the disk but has lost two to all arms. In the presence of the mouth, it may be able to feed and grow. The comet has inherited a major fraction of a single arm, in which buds have arisen and regeneration is in progress. It invariably develops two madreporites, one on each side of the principal arm. Because of this consistency of madreporite position, the principal arm is numbered as 1 and the other arms in a counter-clockwise order (Fig. 4.8 left panel). Regeneration of the lost arm is a function of time and is independent of the number of arms autotomized. At the advanced stage of regeneration, about 95% of the comets tend to autotomize again (cf Fig. 4.2, p 105). The remaining counter-clock individuals display a decreasing order of autotomy with efflux of time. Hence, only < 80% of the counter-clock individuals II and III are autotomized. Consequently, the mean length of splitting arm is 7.7 mm for the comet, in comparison to 14.8, 19.4 and 19.9 mm for the counter-clock comets I, II and III, respectively. As a result, the cloning or autotomic potency is not only progressively decreased from comets to counter-clock comets I, II and III but also the size range, in which the autotomy occurs is progressively increased (Fig. 4.8, right panel). For example, the principal comet can repeatedly autotomize to generate ~ 20 ramets, in comparison to 5 by counter comets. Incidentally, this progressive decrease in clonal autotomy with increasing number of counter arms recalls a similar observation for the fissiparous asteroids (Fig. 4.6).

The process of autotomy is described by Monks (1904) and Rideout (1978). It commences with narrowing of the arm at ~ 2.5 cm from the disk, followed by separation of the ambulacral plate and is completed by splitting of epidermal and dermal tissues, due to their stretching up to 25 mm. Experimental autotomy has been induced by ligating the arm with a silk thread or copper wire. These studies indicate that the presence of Multipotent Stem Cells (MSCs) responsible for autotomic cloning decreases from proximate to distal end of the arm. For example, the minimum arm length required is 20 mm from the arm tip for successful induction of autotomic cloning but 10 mm only from the disk side (see Emson and Wilkie, 1980).

From his extensive field and experimental observations on autotomy in healthy and parasitized *L. multifora*, Davis (1967) brought to light the possible chemical intervention by a parasite. When infected by the endoparasitic gastropod *Stylifer linckiae*, only one out of ~ 2,500 sea stars was autotomized. None of the other parasitized sea stars was ever autotomized. Autotomy is induced in all the 57 ligated healthy sea stars. However, even with ligation only 30% parasitized sea stars underwent autotomy. Hence, *S. linckiae* inhibits autotomy in *L. multifora* by releasing a chemical, which interferes against a mechanical property involving reduction in the tensile strength of collagenous tissues.

4.7 Larval Cloning

According to Eaves and Palmer (2003), larval cloning is an intriguing new dimension to aquatic invertebrate life histories and confers one or more of the following potential advantages: (i) increased offspring, (ii) prolonged larval duration in the pelagic realm facilitating the bed-hedged distribution in timing of attaining competence to settle, (iii) enhanced scope for dispersal and settlement, and (iv) recycling of otherwise discarded or resorbed larval tissue (e.g. *Luidia sarsi*, Wilson, 1978). Clones can arise at different larval body profiles like posterior end (e.g. Fig. 4.9C *Dendrasterias exentricus*, Balser, 1998) and lateral arms (e.g. *Luidia* sp, Bosch et al., 1989).

Bosch et al. (1989) startled the world of reproductive biologists by the discovery of larval cloning by budding in the asteroid larvae of *Luidia* sp collected from plankton in the Sargassa Sea during July 1987 (Fig. 4.9B). Close to heel, Rao et al. (1993) reported the incidence of budding in bipinnaria larvae from the plankton collection in 1987 from the Vizag coast of Bay of Bengal. Confirming the incidence of larval cloning, Jackle (1994) indicated that the buds arose from any arm of bipinnaria of *Luidia* and brachiolaria of presumably *Oreaster*, which underwent transverse fission. Brachiolaria of *Distolasteriasis brucei* also underwent cloning in laboratory cultures (Kitazawa and Kamatsu, 2000). In 2003, Eaves and Palmer reported budding in auricularia of *Parastichopus californicus* (Fig. 4.9A) and plutei of *Strongylocentrotus purpuratus* and presumably *D. excentricus* (Fig. 4.9C). The incidence of budding in these laboratory cultured echinoderms was 3.5% for *D. excentricus*, 5% for *S. purpuratus* and 12% for *P. californicus*. Confirming the earlier observations of Mortensen (1921) in *Ophiopholis aculeata*, Balser (1998) reported that clones arising from the posterolateral arms are released after the juvenile settlement in *O. aculeata* (but not by the settled juveniles of *Ophiothrix oerstedi*). The arms consist of an intact ciliary epidermis, mesenchymal cells, skeletal rods and blastocoels of the primary larval arms. The inner 'V' undergoes to establish an archenteron, which gives rise to other coeloms. The released arms undergo gastrulation and development similar to that of primary embryo. Hence, the development of secondary pluteus involves "the reorganization of larval tissues instead of cloning from embryonic cells" (Balser, 1998). As field collected larvae can rarely be identified to family but not to genus, Knot et al. (2003) used DNA sequence similarity to explore the possibility of identifying the larvae collected from the subtropical western Atlantic Ocean. But they could assign the recognized four groups of larvae into two genera *Luidia* and *Oreaster* and the other two into the family Oreasteridae. Briefly, larval cloning occurs in all the echinoderm classes except crinoids. It is likely that the non-feeding doliolaria larva also undergoes cloning, when it is bisected, as in larvae of asteroids and echinoids

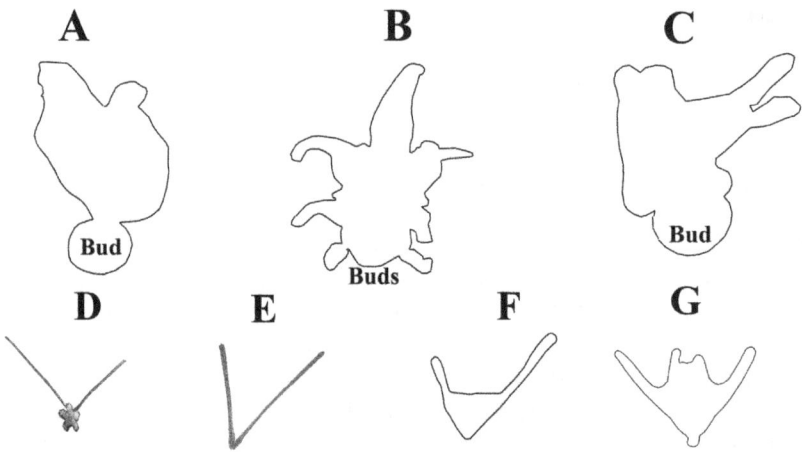

FIGURE 4.9

Cloning in echinoderm larva: Budding in A. auricularia larva of *Parastichopus californicus* (simple sketch drawn from Eaves and Palmer, 2003), B. bipinnaria larva of *Luidia* sp (simple sketch drawn from Jaeckle, 1994) and C. echinopluteus larva of *Dendraster excentricus* (simple sketch drawn from Vaughn, 2010). Cloning in ophiopluteus larva of *Ophiopholis aculeata* D. newly settled juvenile showing posterolateral arms and E. freshly released posterolateral arms, F–G. selected stages of cloning (from Balser, 1998) (all are free hand drawings).

(Barbaglio et al. cited in Candia-Carnevali et al., 2009). Larval cloning is an ancient mechanism of harboring stem cells for cloning in echinoderms and other deuterosomes namely the colonial hemichordates and urochordates (Eaves and Palmer, 2003).

Descriptions on the course of events in development of buds and posterolateral arms are available. In *P. californicus*, a single bud arises from the posterior end of auricularia and is continued to be attached by a thin tether into the benthic pentactula. Eventually, it develops into normal auricularia (Eaves and Palmer, 2003). In *Luidia* sp, buds arise from one or two of the posterolateral or preoral arms either singly or doubly but from different arms at the same time. Being an outgrowth of the soma, the diameter is first expanded, accompanied by proliferation and stratification of the epithelium into multiple cell layers. From laterals of the arm, ciliation arises in the bud. An invagination of epithelium establishes a typical gastrular archenteron. On one side of the gastrula, the presumptive oral surface is flattened and subsequently bilobed. The epithelium joining the primary and secondary larvae separates releasing the motile gastrula, which subsequently develops into bipinnaria (Bosch et al., 1989). However, the bipinnaria of *Oreaster* goes through brachiolaria prior to settling (Vickery and McClintock, 2000, Vickery et al., 2002). In *D. excentricus*, clones are generated from bud or paratomic fission. A preoral abnormality, short postoral arms and arm symmetry in clones arising from paratomic fission indicates that "cloning occurs through more than one mechanism" (Vaughn and Strathmann, 2008, McDonald and Vaughn, 2010).

Food and predators: Food supply is an important driving force to support larval growth, induce fission/budding and subsequent regeneration. For example, the surgically bisected anterior of *Luidia foliolata* late larva possessing the mouth and juvenile rudiment metamorphosed into juvenile on the 3rd day, while the posterior with no mouth required 7 days to metamorphose (Vickery et al., 2002). Hence, algal species and their combinations, and supply level play an important role in the larval growth. Given the high ration, 5 out of 10 bipinnaria of *P. ochraceus* grew, became brachiolaria, developed juvenile rudiment and attained metamorphic competence on the 50th day, while those receiving medium and low rations failed to survive beyond 60–65 days and none could attain metamorphic competence (Fig. 4.10A). Briefly, larvae fed on high rations grew larger and produced more clones. Likewise, those fed on mixed algae or *Isochrysis galbana* supported larval growth, competence and transformation into brachiolaria at the age of 25 days (Fig. 4.10B). However, the larvae fed on single alga *Chaetoceros calcitrans* succumbed to death on the 25th day. Up to 24% of larvae consuming mixed algae produced significantly more clones than those receiving single algal species. The experiments of Vickery and McClintock (2000) indicate that cloning invariably occurs after the larvae have attained asymptotic body growth and only, when food is abundant and of high quality.

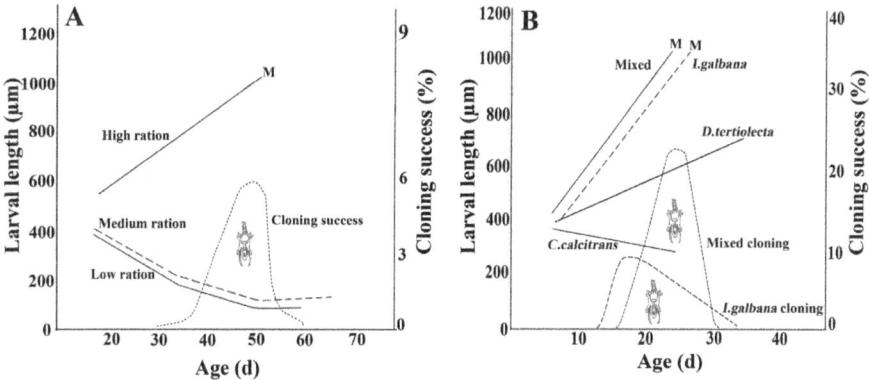

FIGURE 4.10

Larval growth as a function of age of bipinnaria larva *Pisaster ochraceus* fed A. low, medium or high ration and its effects on cloning success. B. effect of mono (*Iscochrysis galbana, Dunaliella tertiolecta, Chaetocerous calcitrans*) and mixed-algae on larval growth and cloning success (compiled and redrawn from Vickery and McClintock, 2000).

Whereas starvation postpones/suppresses fission in adults (Tables 4.13, 4.15), even an abrupt change in food supply induces cloning. Rearing the plutei of *D. excentricus,* McDonald and Vaughn (2010) introduced abrupt changes in algal density from 5,000 cells/ml of *Dunaliella tertiolecta* to 250 cells per ml and from 10,000 cells/ml of *I. galbana* to less density. Of different plutei stages, the 4 to 6-arm transformation stage is sensitive to abrupt changes in algal density. Firstly, high ration induced a level of synchronization

TABLE 4.16

Morphogenesis and organogenesis in surgically bisected clonal larvae of asteroids and echinoids at different larval stages (condensed and compiled from Vickery et al., 2002)

Anterior	Posterior
Asteroid larvae bisected at early larval stage	
within 48 hours mesenchyme cells with pseudopodia aggregate at the site of the blastema	
Retains complete preoral and incomplete postoral ciliary bands and feeding continues. But it is left with no coleom. Thickened undermouth/upper esophagus invaginates into gut. Body wall invaginates to form the anus. Having mesenchyme cells (MCs), it also invaginates to form the anterior portion of the gut, new pair of coelomic pouches. On the 11th day, growth: 233–307 µm	Except for the mouth, gut is retained. From aggregated MCs, new pairs of colemic pouches arise. On the 7th day, the mouth is formed from the esophagus; the gut is also differentiated into the esophagus, stomach and intestine. On 11th day, growth: 376–495 µm
On the 14th day, regeneration is completed both in anteriors and posteriors	
Mid larval stage	
Already has an axocoelom. Appearance of MCs is accompanied by invagination of body wall and elongation of coelom. A new gut is formed on 7th day	As in early larval stage but with accelerated regenerative process
Late larval stage	
As in mid larval stage but an accelerated rate	Presence of juvenile rudiment accelerates metamorphosis even on the 3rd day 500 µm dia. In *P. ochraceus*, regeneration produces first bipinnaria. Juvenile 500 µm
Removal of brachiolar arm tips in *P. ochraceus*	
Unable to regenerate the arm tip even after the 14th day	Regenerates the last tips already on the 7th day
Echinoid larvae	
Aggregation of MCs at the site of bisection. On the 5th day, anterior and posterior plutei regenerate almost all the respective missing body components, including the digestive system and larval mouth, respectively	

(50–100%) cloning event. Secondly, the satiated plutei accelerated juvenile rudiment growth and developed shorter and stubby arms. Finding the presence of short and stubby 4–6 armed plutei and doubling density in the culture, McDonald and Vaughn concluded that an abrupt reduction in algal density induced an *en mass* cloning of plutei at 4 to 6-arm transformation stage.

Not only the stress due to abrupt change in food supply but also due to predation induced larval cloning. Exposure to stimuli from the predatory fish mucus induced cloning by both budding and fission in 40% of early plutei of *D. excentricus* (Vaughn and Strathmann, 2008). Buds released as gastrulae developed into smaller (~ 300 µm height) plutei after a delay, in comparison to plutei (~ 750 µm height) reared in the absence of mucus stimulus (Vaughn, 2009, 2010). The smaller larvae escaped from the visual spectrum of predators. However, the timing and incidence of cloning and size reduction differed among the genets. On exposure to fish mucus for 9-days during early development of plutei from half sibling families (same father but different mothers), rate and success of cloning differed significantly among the larval families indicating the maternal influence (Vaughn, 2009). Interestingly, Carrier et al. (2015) showed that this sort of developmental plascticity is regulated by genes and their expressions in *S. droebachiensis*.

To gain more details on morphogenesis and organogenesis in presumptive ramets, Vickery et al. (2002) transversely bisected clonal larvae at equidistance point between anterior and posterior poles. They selected clonal larvae of two asteroids *Luidia foliolata* (bipinnaria only) and *P. ochraceus* (bipinnaria and brachiolaria) as well as two echinoids *Lytechinus variegatus* and *D. excentricus*. The bisection was also made in the early, middle and late larval stages of asteroids. Their findings are briefly summarized in Table 4.16. They reveal that the clonal larvae of the echinoderms have an extensive capacity to regenerate all the missing tissues and organs, regardless of their development stage. Notably, proliferation of cells (epimorphic) is initiated by the larval epithelial (ectodermal) cells (see also p 126) adjacent to the bisection site. During the process of regeneration, these cells seem to redifferentiate into Mesenchyme Cells (MCs) that later give rise to many regenerated organs, especially coelom. Missing parts of the gut are regenerated by the invagination of the epithelium (see p 126). Briefly, these epithelial cells along with MCs differentiated into all missing tissue types and organs in the anterior and posterior presumptive ramets, although the posteriors showed an edge over the anteriors. Hence, the fate of these larval cells is not limited and their proliferation capacity is also not limited. Implicatively, it suggests that these cells are versatile pluripotents. That would take us to the experimental studies on the potency and regulative process in eggs and blastomeres of echinoderms.

4.8 Eggs and Embryos

In echinoids, rapid embryogenesis, simplicity in shape and structure with just five major tissue types and 14 known cell types (Angerer and Angerer, 2001) and transparency of larva through the entire body profile have facilitated experimental manipulations and observations (Zito and Matranga, 2009). The experimental embryological studies in echinoid have revealed the progressive reduction of totipotency of the egg to different cell and tissue types with the formation of organs and systems during embryogenesis. 1. Individual blastomeres are totipotent up to 4 cell stage. 2. Subsequent divisions lead to the split into lineage specific cell programs (Cameron et al., 1987). 3. Unequal cleavages in the fourth and fifth divisions result in the formation of (i) mesomeres (animal cap), (ii) macromeres and (iiia) small and (iiib) large micromeres (Fig. 4.11). 4. The 60-cell embryo includes five polyclonal lineage elements with a different fate, potency and differentiation. 5. Cleavage includes seven rounds of cell divisions and generates 200–250 blastomeres. Towards the end of the cleavage, the cell cycle slows, becomes asynchronized and visibly polarized. For example, following the ingression of PMCs (primary mesenchyme cells), endoderm begins to invaginate from the vegetal plate and thereby initiates gastrulation (Angerer and Angerer, 2001). 6. The highly dynamic periods of gastrulation, during which the basic three germ layers are formed, is marked by mass migration of cells in different directions. 7. All further developmental transitions are marked by changes in migration routes, cell cycle and transcriptional regulation (see Geneviera et al., 2009).

Tracking of the lineages of mesomeres, macromeres and micromeres has revealed the fate and potency of these blastomeres (Cameron et al., 1987). Accordingly, veg 1 develops into ectoderm and archenteron, veg 2 into the gut, Secondary Mesenchyme Cells (SMCs) and coelomic cells, small micromeres into PMCs and skeletogenic cells (Fig. 4.11). However, it must be indicated that the fate of these blastomeres is not irrevocably fixed. Herbst (1982) first reported the vegetalizing effect of lithium chloride (LiCl) on sea urchins. He has shown that an animal cap cultured in isolation with LiCl develops into pluteus-like larva. In *Peronella japonica*, isolated animal cap from 16-cell stage develops into blastula with blastocoels. Treatment of the cap with LiCl vegetalizes the endoderm and subsequently develops into pluteus like larva (Kitazawa and Ameniya, 2001). Clearly, LiCl can be used to reverse and reprogram the direction of early development. Similarly, almost every lineage of the sea urchin can be reprogrammed to develop. For example, microsurgical removal of PMCs induces some SMCs to become skeletogenic and completion of development of a typical pluteus (see Zito and Matranga, 2009). The small micromeres divide only once during cleavage and the eight cells passively remain until gastrulation, when they migrate to

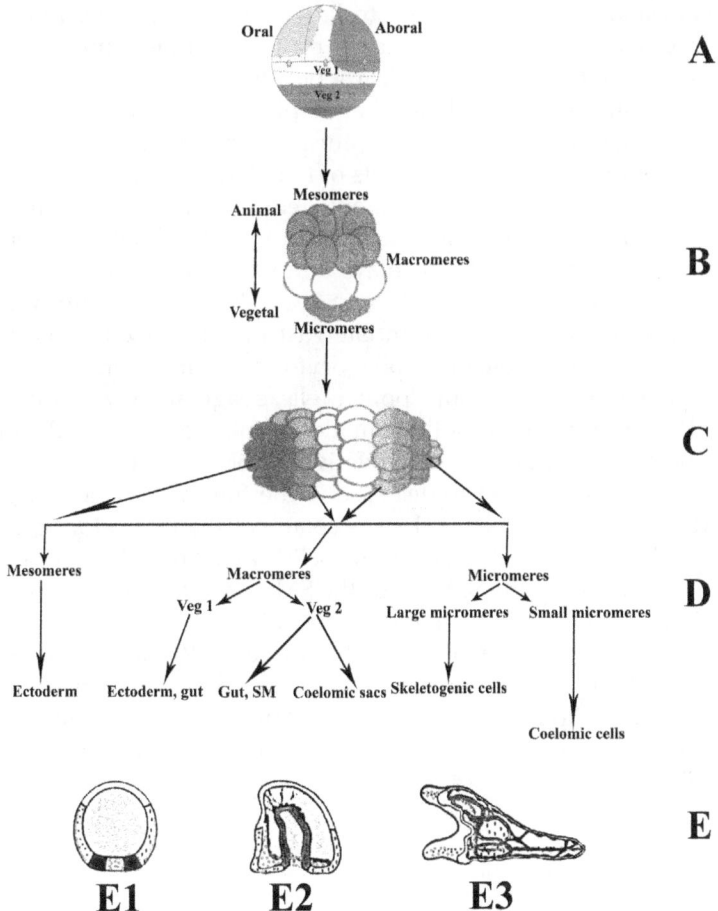

FIGURE 4.11

A. Fate map of a typical echinoid egg, B and C. 16- and 32-cell stage showing antero-posterior axis, mesomeres, macromeres and micromeres, D. Major 5 cell lineages arising from the blastula, showing the tentaive fate of cell lineages, ▨ SM = Secondary Mesenchyme, SMCs = Secondary Mesenchyme Cells. Sections of E1 blastula, E2 gastrula and E3 pluteus, showing the locations of SMCs, ☐ PMCs = Primary Mesechyme Cells, ■ endoderm, ▨ aboral and ▦ oral ectoderm (sketches redrawn from Davidson et al. (1998), Angerer and Angerer (2001), Ransick et al. (1996), Zito and Matranga (2009).

the tip of archenteron, from which the coelomic pouches bilaterally arise at the base of esophagus. Suspecting the micromeres are the germ cell lineage, Ransick et al. (1996) microsurgically deleted them. However, the adults arising from these micromere-deleted blastulae have generated a fertilizable egg and fertile sperm. Incidentally, Voronina et al. (2008) have found that at 16-cell stage, vasa protein accumulates selectively in micromeres and is restricted to small micromeres at gastrulation. However, its accumulation in other lineages sand cells can be induced by, for example, lithium exposure. Briefly, the sea urchin larvae at the early development do not have obligate PGCs. Hence, almost every cell lineage, despite differentiation and split into different lineages, retains its versatile pluripotency for long.

The mesoderm of sea urchin consists of two distinct populations of PMCs and SMCs. Apart from their different ontogenic origin (Fig. 4.11), they differ with respect to developmental fate and turning of epithelial-mesenchyme transition. Interestingly, SMCs displays a number of characteristics typical of Embryonic Stem Cells (ESCs). From SMCs, the following lineages arise: (i) pigment cells, recognizable from late gastrula, dispersed into ectoderm, responsible for photo-protection, phagocytosis (against harmful microbes) and wound healing, (ii) coelomic pouch cells, evagination from the anterior tip of archenteron during mouth formation, proliferation after the onset of feeding and (iii) others. Of these derivatives of SMCs, self renewability and multipotency are suggested to decrease in the following order: pigmented cells and skeletogenic cells < blastocoelar cells and coelomic pouches < muscle cells. Incidentally, the blastocoelic cavity may be considered as the niche for harboring the SMCs. For the SMCs reside in this cavity during gastrulation and the cavity is filled with a fluid containing a large number of undetermined growth factors and other proteins. Regarding the coelomic pouch cells and micromeres, the following has to be mentioned: 1. Larval halves, lacking coelom and hence coelomocytes can still regenerate (see Table 4.16). Therefore, the coelomocytes may have no role to play in regeneration of ramets.

4.9 Searching Stem Cells

Known for their simplicity, the echinoderm larvae have only five major tissue types and 14 known cell types (Angerer and Angerer, 2001). As has been shown, each cell in the five polyclonal lineages arising from a fully developed blastula is versatile and pluripotent, capable of rapid amplification and differentiation into any cell and tissue types that are missing in clonal larvae of echinoderms. However, the number of cell and tissue types is considerably increased that there is a need to search for the niche(s) of the Embryonic Stem Cells (ESCs), which are responsible for clonal reproduction.

Stem cell niches: In animal tissues, stem cells are often located and controlled by special tissue micro-environments called niches (Ohlstein et al., 2004). In clonal echinoderms, body profile and feeding habit seem to play a key role in clonal reproduction. The carnivorous asteroids alone possess the pyloric caeca to store excess resources. All other classes of echinoderms are almost herbivores and continuous feeders. Fission in the antero-posterior body profiled holothuroids usually results in unequal inheritance of organs and systems. Even after an unequal fission, the asteroids and ophiuroids, with an oral-aborally flattened armed disk, inherit a small or large fraction of the disk in the horizontal plane to enable the inheritance of fractions of the oral disk, pyloric caeca and gonads. The gonads are accommodated within the disk in ophiuroids but are usually extended and accommodated within the spacious arms. The globose- and dollar-shaped body profile with no arm in regular and irregular echinoids perhaps do not facilitate fission.

In holothuroids, many authors have induced fission on a single plane; of them, some have increasingly reduced the length from 50% in *Holothuria parvula* to 33% in *H. leucospilota* (Table 4.11). Still others have increasingly reduced the body weight of the anterior half from 50 g in *Bohadschia marmorata* to 12 g in *H. atra*. Strikingly, Reichenbach et al. (1996) indicated that both the anterior and posterior halves are successfully cloned from fragments weighing as small as 5.5% body weight of *H. atra* genet. A couple of publications (Emson and Mladenov, 1987, Purwati, 2004) have listed the organs and systems inherited by posterior and anterior halves. For more details on the ontogeny of development of posteriors and anteriors, Kamenov and Dolmatov (2015, 2016) may be consulted. A few others have also ventured to induce fissions at double or treble planes (Purwati et al., 2009, Hartati et al., 2016). They too have not described the organs and systems inherited by the anteriors, posteriors and middles. Briefly, irrespective of the smallest length (25% in *S. herrmanni*, which underwent fissions at treble planes) or 5.5% of the body weight of the genet, or the anteriors, posteriors or middles, the fragments successfully underwent clonal reproduction in good numbers. The common organs that the smallest fragments inherited are the corresponding fractions of dermis, coelom and radial nerves. Resident Stem Cells (RSCs) stored in the brachial (= radial) nerves and Circulating (Coelomocytes) Stem Cells (CSCs) have been identified as blastema and pluripotent cells in *Asterias rubens*, respectively (Candia-Carnevali et al., 2009). Hence, it is likely that the radial nerves and coelom serve as niches for ESCs (RSCs + CSCs) in clonal holothurians.

Understandably, the MSCs are dispersed in adequate numbers in the oral disk and arms of asteroids and ophiuroids. Even the two armed fragments (e.g. ophiuroids: Fig. 4.6; asteroids: Fig. 4.6) have adequate number of MSCs to enable them to successfully undergo clonal reproduction. Incidentally, single armed fragments are not usually generated. When generated, at the lowest frequencies (0.1 to 3.56%, see p 113) and when regenerated, despite the protruding buddings, they are unable to successfully undergo clonal reproduction, as they have not inherited adequate numbers of ESCs. Yet, some six autotomic asteroids species, which do not inherit any fraction of the oral disk but only a major fraction of one long arm, are able to survive and successfully undergo clonal reproduction (Table 4.2). Unaware of the importance of the ESCs, Edmondson (1935) undertook an interesting experiment to induce autotomy by ligating the arm of *Linckia multifora*. His finding that the minimum ligated arm length required to induce autotomy is 10 mm from the proximal end but 20 mm from the distal end, is indeed a remarkable discovery. This discovery recognizes that MSCs are dispersed in the decreasing order from the oral disk to the arm tip. Hence, the second niche, where the MSCs are concentrated, is at 2.5 cm of the arm from the oral disk in *L. multifora*.

Regarding Primordial Germ Cells (PGCs), the publications by Kille (1939, 1942) are not adequately recognized. An anteriorly eviscerated sea cucumber *Thyone briareus* contains dorso-muscular sac containing cloaca, respiratory trees, gonads and mesentery alone. Further, 95% of the eviscerated cucumber, after one-fifth of longitudinal cut at the anterior end and turning inside out and thereby exposing the gonads, can still be alive. Using this unique ability of the cucumber, Kille (1939) has surgically extirpated partial or complete gonodectomy. His findings may be summarized: (i) the gonad is a lateral projection of a chamber located within the dorsal mesentery, i.e. the gonad-basis and (ii) a mass of tubules originate from the gonad-basis (Fig. 4.12). Following complete gonodectomy, no gonad is regenerated up to 4 months, as the entire basket of PGCs is removed. However, with partial or complete removal of the tubules, new tubules arise from the gonad-

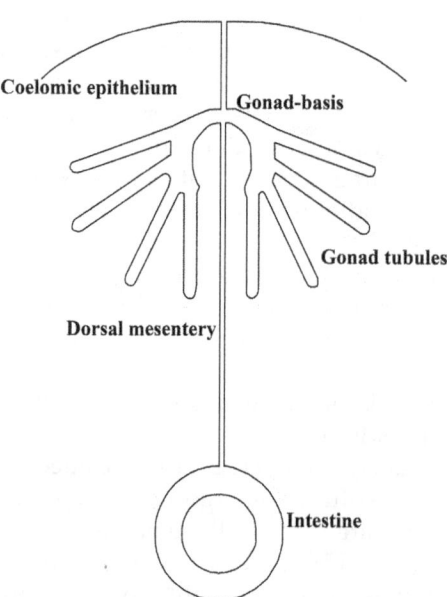

FIGURE 4.12

Free hand sketch to show the locations of gonad-basis attached to the dorsal mesentery arising from coelomic epithelium in *Thyone briareus* (redrawn from Kille, 1939).

basis, indicating that the nest of PGCs is embedded at the gonad-basis in the dorsal mesentery. From further investigations, Kille (1942) found that the gonad arise in the cucumber of < 60 mm body length. Following fission, the gonads are developed in the anterior half but the posterior may develop into a sterile. However, Conand et al. (1998) reported the reduced presence of gonads in 7 and 9% of the field collected posterior and anterior regenerating ramets of *S. chloronotus*. Presumably, the dorsal mesentery is also inherited by both anterior and posterior halves, although the posteriors may develop the gonads a little slower. The earliest appearance of PGCs has been located in the left somatocoel of bipinnaria in *Asterias* (asteroid) and late plutei of *Echinus* (echinoid) and *Amphipholis* (ophiuroid) (see Ransick et al., 1996).

Vasa gene, a molecular marker to identify PGCs, is inherited from maternally supplied mRNA and located at the animal pole in fishes (see Pandian, 2011) and crustaceans (see Pandian, 2016) but the vegetative pole in molluscs (see Pandian, 2017). Subsequently, *vasa* are limited only to germ cell lineage in metazoans except in coelenderates (see Pandian, 2016). Sui et al. (2008) have cloned a full-length *vasa* cDNA sequence from the testes of sea cucumber *Apostichopus japonicus*. *Aj-vasa* is expressed only in the gonads of the male and female. The full-length *Aj-vasa* is 2,167 bp and contains a 3' untranslated region (UTR) of 449 bp, a 5' UTR of 125 bp and an Open Reading Frame (ORF) of 1,593 bp encoding 530 amino acids. From Blast analysis, Sui et al. have also found that > 58% homology in the amino acid sequence between *A. japonicus* and *Strongylocentrotus purpuratus*.

Duration of expression: Using the trends reported in Fig. 4.2, two models are proposed for the expression durations of ESCs for clonal reproduction and PGCs for sexual reproduction. Model 1 includes the holothuroids (see Table 4.1) like *Stichopus variegatus* and asteroids like *Nepanthia belcheri*, *Coscinasterias tenuispina* and *Allostichaster capensis*, in which clonal and sexual reproduction continue to co-exist throughout the entire respective range of body size. In model 2, PGCs may not be present or if present, may not express, as clonal reproduction ceases around 60% body size in *A. insignis* and *A. polyplax*. Similarly, clonal reproduction also ceases within 5 (*Holothuria atra*) –20% (*S. chloronotus*) body lengths and is followed by sexual reproduction alone. Expectedly, PGCs in these size ranges of holothuroids may no longer be present or if present, may not express any longer. However, Spermatogonial Stem Cells (SSCs) and Oogonial Stem Cells (OSCs) may continue to be present and express in protandric *H. atra* up to 1,400 g size and others.

5

Regeneration

Introduction

Regeneration is the potential to repair and replace cells, tissues, organs and systems of an organism (see Candia-Carnevali et al., 2009). In 1898, Morgan classified regeneration into two types: 1. Morphallaxis involving remodeling the existing tissues into missing ones without extensive cell proliferation. 2. Epimorphosis involving massive proliferation of undifferentiated cells and stem cells (Rycel and Swalla, 2009). However, convincing evidence goes to show that both mechanisms are functioning and are interchangeable during regenerative process of echinoderms (Candelaria et al., 2006, Candia-Carnevali, 2006). Yet, the mechanism of regeneration is predominantly epimorphic with blastema formation in crinoids, holothuroids and possibly ophiuroids, but morphallaxic in asteroids. In echinoderms, regeneration is regarded fundamentally as a distinct process and includes (i) intervention of stem cells and/or redifferentiated cells, (ii) extensive or limited cell migration and proliferation and (iii) contribution by growth factors (e.g. Candia-Carnevali et al., 1996) like neuropeptides and neurotransmitters (Candia-Carnevali, 2006).

Echinoderms represent a phylum with exceptional autotomic and regenerative potentials. 1. Autotomy occurs as a voluntary sacrificial mutilation of small or larger body biomass in response to sublethal predation or intolerable environmental stress and is usually followed by regeneration (Lawrence and Vasquez, 1996). 2. Besides serving as an effective escape mechanism from sublethal predation, it is also a means to get rid of infected body tissues. 3. In many echinoderms, it naturally occurs annually to rejuvenate specific body parts, for example, evisceration by temperate holothuroids (see Emson and Wilkie, 1980, see p also 147–148). 4. It occurs in all five classes of echinoderms and the loss ranges from external tissues (e.g. spines, pedicillariae) to organs (e.g. arm) and system (e.g. evisceration involving digestive system and associated organs). Incidentally, the loss of biomass can be as high as 75%, as in *Asterias rubens* (Ramsay et al., 2001). 5. Following direct loss of a small or larger body biomass, the autotomic echinoderms also suffer from indirect loss of motility, feeding ability, and reproductive and mating potential, social status and/or defensive structures (see Barrios et al., 2008). 6. With reduced food intake, regeneration occurs at the expense of body growth and/or sexual reproduction.

5.1 Incidence and Prevalence

Among aquatic invertebrates, echinoderms display an amazing ability to undergo autotomy and regeneration. The credit for assembling the widely scattered information on autotomy and regeneration goes to Emson and Wilkie (1980) and Lawrence and Vasquez (1996). In the present chapter, an effort has been made to consolidate relevant information published between 1996 and 2016. In general, stellate classes with projected arms as in crinoids (Fig. 5.1M, N, O), asteroids (Fig. 5.1G, I, J, K) and ophiuroids (Fig. 5.1L) are often susceptible and suffer the loss of body mass. The large coelomic cavities and rigid test (Table 1.1, Fig. 5.1A, B) in echinoids eliminate the scope for sublethal predation in regular echinoids and irregular spatangoids (Lawrence and Vasquez, 1996).

Echinoids: Surprisingly, the regeneration potential of adult echinoids is limited to replacement of tissues like spines, pedicillariae and podia (Emson and Wilkie, 1980). Hyman (1955) indicated that damages to the test are repaired by aggregation of coelomocytes accompanied by calcareous deposition. Lawrence and Vasquez (1996) reported the loss of tube feet and pedicillariae by *Echinometra lucunter* and edges of test by *Dendraster excentricus* due to predation. The globose body profile of regular (Fig. 5.1A) and dollar-shaped (Fig. 5.1B) irregular echinoids with no arms render them not to lose a large fraction of their body. Briefly, the echinoids do not suffer sub-lethal predation and display the lowest regeneration potential among echinoderms.

Crinoids: The globose body profile with circumferential oral cirri and arms of comatulids (Fig. 5.1N, O), and the stalked isocrinids with a series of arms and their branches bearing pinnules (Fig. 5.1M) facilitate the loss of body parts due to predation or imposed stress. The loss is limited to arms alone in 5 isocrinid species and 20 comatulid species but both arms and visceral fractions including the gut in 11 comatulid species (Table 5.1). In all, 36 out of 700 crinoid species are thus far reported to undergo sublethal predatory loss of pinnules, arms and/or viscera.

Asteroids: On predatory attack, the stellate asteroids escape predation readily by losing one or more arms. However, with stout and stumpy arms,

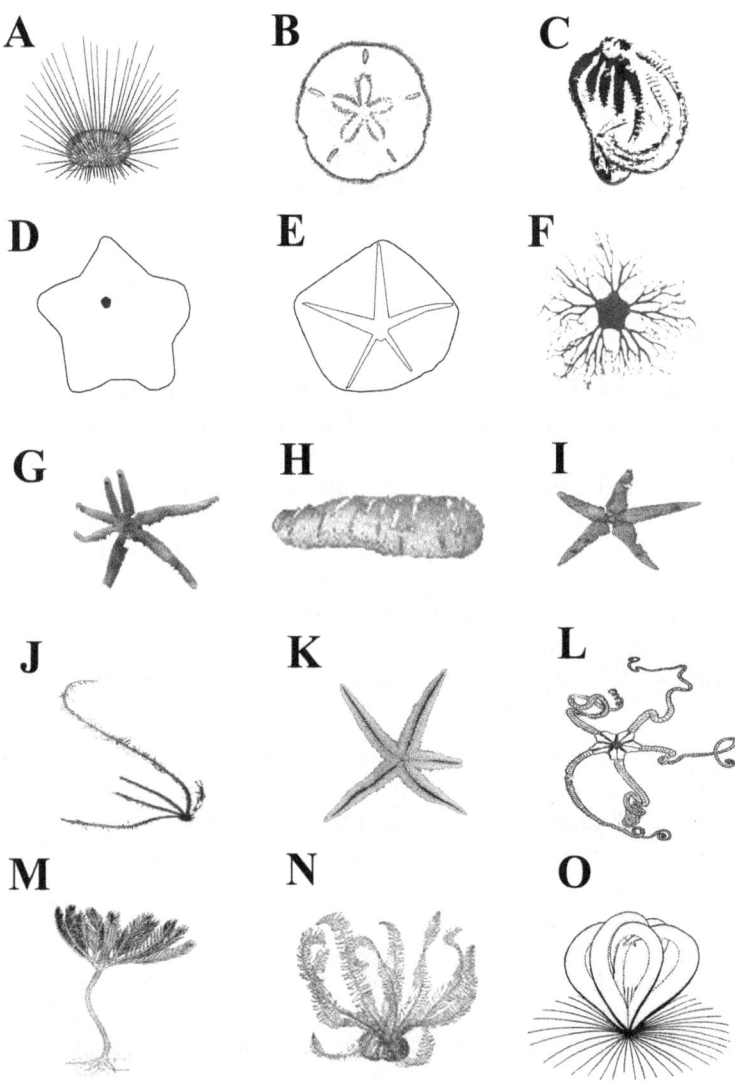

FIGURE 5.1

A. *Diadema antillarum*, B. Sapatangoid (aboral view), C. *Odina antillensis*, D. *Pteraster acicula* (aboral view), E. *Peltaster placenta* (oral view), F. The basket star *Gorgonocephalus* (free hand drawing), G. *Coscinasterias tenuispina*, H. *Holothuria scabra* (from Asha et al., 2015), I. *Henricia sanguinolenta*, J. *Luidia senegalensis*, K. *Luidia clathrata*, L. *Asternyx excavata*, M. Stalked crinoid (from an unknown Google source), N. *Antedon bifida*, O. *Pentametrocrinus varians*. Most pictures are from Downey (1973) and the remaining are free hand drawings from Hyman (1955).

Pteraster acicula (Fig. 5.1D) and *Peltaster placenta* (Asterinidae) (Fig. 5.1E) may not be susceptible to sublethal predation. So are the bizarre-shaped *Odina antillensis* (Fig. 5.1C). On the other hand, those with projected, fragile arms invariably lose the arm, either from the base (B), as in *Ophidiaster ophidianus* and *Asterias rubens* (Table 5.2) or as small fraction of the arm as possible (A) (e.g. *Luidia ciliaris, Astropecten irregularis* (Table 5.2). Others like *Brisighella coronata* readily lose a fraction or the entire arm. Arm autotomy is reported from species belonging to seven asteroid orders. In their review, Emson and Wilkie (1980) summarized the autotomic incidences from 26 species. Lawrence and Vasquez (1996) added another 11 to it. Between 1996 and 2015, reports on three more species were added. Most species belonging to the family Brisingidae are known to undergo arm autotomy. A search indicated that at best 10 more species can be added. In his Smithsonian monograph, Downey (1973) provided figures for the series of asteroids. Some 24 asteroid species were shown to have broken arm(s). But it is not clear whether they are autotomic species or their arm(s) were broken, while they were collected and brought to the institute. Of 24 species, not a single species shows an arm broken from the base. Hence they are not considered as autotomics. In all, 50+ out of 1,800 asteroid species are reported to undergo arm autotomy.

Ophiuroids: As the name suggests, the stellate long fragile arms (e.g. 24 cm in *Acrocnida brachiata*, Bourgoin and Guillou, 1994) are more prone to brittle losing not only arms but also the disk. Yet, Emson and Wilkie (1980) could list reports on the incidence from 32 autotomic species, belonging to eight families, of which ~ 10 species (31%) lose both arms and disk (Table 5.3).

TABLE 5.1

Reported autotomy in crinoids (compiled from Emson and Wilkie, 1980, Lawrence and Vasquez, 1996, Mozzi et al., 2006, Candia-Carnevali et al., 2009)

Order/Species	Autotomy	Incidence (species no.)	Prevalence (%, species no.)
	Isocrinida	5	
Metacrinus rotundus	Arms		
Heterometra savignyi	Arms		89
	Comatulida	20	
Antedon bifida	Pinnules		
Himerometra magnipinna	Arms		100, 4 species
Dichrometra flagellata	Arms		67–80, 7 species
Stephanometra oxyacantha	Arms		< 50, ~ 9 species
Colobometra perspinosa	Arms + viscera	11	9% viscera lost
Comaster bennetti	Arms + viscera		1.8% viscera lost

TABLE 5.2

Reported autotomy in asteroids (compiled from Emson and Wilkie, 1980). A = autotomy at any level in the arm, B = autotomy only at the base of the arm

Order/Species	Autotomy	Incidence (species no.)	Prevalence (%)
Phanerozonidae/Paxillosida/ Platyasterida		5 + 5 ?,	
Luidia ciliaris	Arms A		
L. clathrata	Arms		12–59
L. magallanica			51–69
Astropecten irregularis	Arms A		0–40
A. atriculatus	Arms A		
Ophidiaster ophidianus	Arms B		
	Spinulosida	1	
Henricia sanguinolenta	Arms B		
	Euclasterida	1 +	
Brisighella coronata	Arms A, B		
most Brisingidae	Arms B		
	Forcipulatida	12	
Pisaster ochraceus	Pedicillariae		
Asterias rubens	Arms B		
Stichaster striatus	Arms		> 38
Heliaster helianthoides	Arms		> 62
	Valvatida	1	
Acanthaster planci	Arms		
	Velatida	1	
Solaster simpsoni	Arms		
Newly added by Lawrence and Vasquez (1996), Pomory and Lares (2000), Lawrence (2010), Diaz-Guisado et al. (2006), Barrios et al. (2008), Rubilar et al. (2015)		26 + 11 + 3 + 10? = 50	

The most speciose families Amphiuridae (467 species), Ophiuridae (344 species) and Ophiacanthidae (319 species) undergo autotomic loss of arms + disk in 3.4, 1.2 and 0.3%, respectively. Notably, report on the incidence of arm loss in euryalous basket stars like *Gorgonocephalus* is not yet available. In these basket stars, "the arms branch repeatedly, producing a bewildering maze of intermingled branches" (Hyman, 1955, see also Fig. 5.1F). It is not clear whether the maze of branching arms pose a threat to the predators. Into the list of 32 species of Emson and Wilkie (1980); Lawrence and Vasquez (1996) added another 18 species. Subsequently, there are reports only for five species. Briefly, 55 out of 2,604 ophiuroid species alone are thus far known to autotomize.

TABLE 5.3

Reported autotomy in ophiuroids (compiled from Emson and Wilkie, 1980)

Family/Species	Autotomy	Incidence (species no.)
Ophiomyxidae		1
Ophimyxa pentagona	Arm	
Ophiacanthidae		1
Ophiocantha eurythra	Disk	
Amphiuridae		16
Acrocnida brachiata	Arm	
Amphiura grisea	Disk	
A. filiformis	Arm + disk	
Ophiactidae		3
Ophiactis virens	Arm	
Ophiotrichidae		2
Ophiothrix fragilis	Arm	
Ophiodermatidae		3
Ophioderma cinereum	Arm	
Ophiuridae		4
Ophiolepis elegans	Arm	
Ophiocomidae		2
Ophiocoma nigra	Arm	
Ophiopsila aranea	Arm + disk	
Newly added by Bourgoin and Guillou (1994), Lawrence and Vasquez (1996), Chinn (2006), Yokoyama et al. (2010)		+ 23 = 55

Holothuroids: The number of holothuroids reported to undergo evisceration was 48 in 1980 (Table 5.4). Lawrence and Vasquez (1996) added two more to the list of Emson and Wilkie (1980). Between 1996 and 2016, the reported incidence increased by another 11 species, i.e. hardly 0.4 species is added annually to the list. Briefly, 61 out of 1,000 holothuroid species are reported to undergo evisceration/autotomy. In the antero-posteriorily cylindrical body profile of holothuroids (Fig. 5.1H), only the oral tentacles (e.g. *Athyonidium chilensis*) are lost and regenerated. However, an unbearable predatory attack or other shock evokes an almost violent response of evisceration from the seemingly not susceptible holothuroids—a phenomenon not known from any other autotomic aquatic invertebrates. Evisceration usually involves the expulsion of the digestive system and one or more other organs/systems, depending on the location, through which the expulsion is enacted. In many aspidochirotids, the evisceration through the posterior (PE) or rupture through body wall (ME) (Table 5.4) involves the entire digestive tract from the esophagus to cloaca (leaving the water-vascular system, mysentery and

TABLE 5.4

Reported evisceration in holothuroids (compiled from Emson and Wilkie, 1980). AE = anterior evisceration, PE = posterior evisceration, PES = posterior seasonal evisceration, ME = body wall ruptured evisceration, D = autotomy of non-regenerative posterior end during fission, † = indicates the incidence, ‡ = indicates the incidence at two body profile

Order/Family/Species	AE	PE	PES	ME	D	Total
Aspidochirotida						
Holothuriidae						23
Holothuria atra		†				
H. impatiens				†		
H. floridana			†			
Stichopodidae						11
Stichopus naso				†		
S. mollis		†				
S. regalis			†			
S. chloronotus		‡		‡		
Dendrochirotida						
Cucumariidae						6
Thyone fusus	†					
Cucumaria frondosa		‡		‡		
Phyllophoriidae						3
Mensemaria intercedens		†				
Neothyonidium magnum	†					
Apotida						
Synaptidae						5
Synapta maculata					†	
Newly added by Byrne (1986), Lawrence and Vasquez (1996), Dolmatov et al. (2011, 2016), Kamenev and Dolmatov (2016)						+ 13 = 61

anus unhurt), one or two respiratory trees and sometimes the gonads (e.g. Mary Bai, 1971). On the other hand, the anterior evisceration, which occurs in most dendrochirotids, includes tentacle crown, aquapharyngeal bulb, nerve ring, hemal vessels, water-vascular system, digestive tract, right respiratory tree and variable quantum of gonads. It hardly leaves the sea cucumber with a body wall and its muscles, intestinal mesentery and left respiratory tree (see Emson and Wilkie, 1980). Hence, the anterior evisceration may demand more resources and time than the posterior one. In non-fissiparous synaptids, which cannot regenerate the oral disk, autotomy (D) is limited to the posterior end.

Table 5.5 represents a comparative summary of clonal reproduction and autotomy in the five classes of echinoderms. More reports on autotomy are awaited but not at an explosive rate. Moreover, there are many species

TABLE 5.5

Reported incidences of clonal reproduction and autotomy in echinoderms

Class	Species (no.)	Clonal Species		Autotomic Species	
		(no.)	(%)	(no.)	(%)
Crinoidea	~ 700	–	–	36	5.1
Holothuroidea	> 1,000	37 +	3.7	61	6.1
Asteroidea	~ 1,800	47 +	2.6	50 +	2.8
Echinoidea	> 800	3 +	0.3	~ 2	0.3
Ophiuroidea	2,604	50 +	1.9	55	2.1

like *H. atra*, which are clonal as well as autotomic. Surprisingly, autotomic incidence thus far reported is limited to 204 species. With this, 1.98 and 2.95% of echinoderms are clonal and autotomic, respectively. Hence, not more than 5% of echinoderm species are reported to have the potential to clone and/or autotomize. Yet, these puzzling echinoderms are so curious and fascinating that they have attracted the attention of hundreds of scientists during the last few centuries and shall continue to do so in the years to come.

Prevalence: Whereas the autotomic incidence is limited to 204 species, their prevalence within populations of each autotomic species is high and varies widely. Though much information is available on the prevalence of clonal fission, less information is available on the prevalence of evisceration in holothuroids. Only limited information is available for crinoids and asteroids. For example, of 31 comatulid species, prevalence of arm loss ranges from < 50 to 67–80 and 100% for 9, 7 and 4 species, respectively (Table 5.1). The values for the arm loss in asteroids also range from > 38% in *Stichaster striatus* to 51–69% in *Luidia magallinica* (Table 5.2). Relatively more information is available for the autotomic ophiuroids. Of 23 species, for which details are available (Table 5.6), the prevalence ranges from 13 to 94%. Firstly, the values range from > 50 to 98% for 15 species in their populations in the USA and Europe. Specifically, comparable data are desired for the tropical Asian species. Secondly, the standing biomass of regenerating *Amphiura filiformis* is 29.75 g (live weight)/m^2, which is equivalent to 22% of the total population in Sweden (Skold et al., 1994). For *Acrocnida brachiata*, this value is as high as 33 g dry weight/m^2, which is equivalent to 11% of the total population in France (Bourgoin and Guillou, 1994). The estimate by Skold and Rosenberg (1996) suggested injection of 300 mt of ophiuroid arm biomass into the trophodynamics of Swedish coastal aquatic system. Not surprisingly, 10–60% of the stomach contents of the dab *Limanda limanda* in the Netherland coast, Norwegian lobster *Nephrops norwegicus* and American plaice *Hippoglossoides* were constituted by the predated arms of *A. filiformis* and *A. chiajei* (see Skold and Rosenberg, 1996).

TABLE 5.6

Prevalence of arm regeneration in ophiuroids

Species	Location	Prevalence (%) Arms + Disk	Reference
Ophiura albida	Sweden	13	Skold and Rosenberg (1996)
	Scotland	14	Emson and Wilkie (1980)
Ophiothrix quinquemaculata	Adriatic	22–47	Skold and Rosenberg (1996)
O. ophiura	Sweden	24–62	Skold and Rosenberg (1996)
Ophioderma longicauda	Adriatic	40	Skold and Rosenberg (1996)
Hemipholis gracilis	–	25	see Lawrence and Vasquez (1996)
Ophiolepis impressa	–	28	see Lawrence and Vasquez (1996)
O. ailsae	–	43	see Lawrence and Vasquez (1996)
Ophiocoma nigra	Scotland	44–93	Emson and Wilkie (1980)
Ophioscolex nutrix	Beagle	54	Skold and Rosenberg (1996)
Ophioderma cinereum	–	34	see Lawrence and Vasquez (1996)
O. appressum	–	44 + 50–85	see Lawrence and Vasquez (1996)
Ophiura sarsi	–	39 > 98	see Lawrence and Vasquez (1996)
Ophiacantha bidentata	S.E., USA	60	Brooks et al. (2007)
Microphiopholis gracillima	S. Carolina	20–70	Stancyk et al. (1994)
	Florida	77	Singletary (1980)
Ophiocoma scolopendrina	California	73	Chinn (2006)
Acrocnida brachiata	France	75	Bourgoin and Guillou (1994)
Amphiura filiformis	Ireland	78 + 3	Bowmer and Keegan (1983)
A. chiajei	–	93 + 0.5	see Lawrence and Vasquez (1996)
Ophiactis aspersula	Beagle	80	Skold and Rosenberg (1996)
Ophiothrix fragilis	Scotland	91	Emson and Wilkie (1980)
Amphipholis squamata	–	80 + 3	see Lawrence and Vasquez (1996)
Amphioplus coniortodes	Florida	87 + 3	Singletary (1980)
Ophiophragmus filograneus	Florida	52–94	Clements et al. (1994)

5.2 Induction of Autotomy

Autotomic echinoderms are readily amenable to chemical/surgical manipulations and electrical stimulations. In the crinoid *Antedon mediterranea*, Mozzi et al. (2006) removed the visceral mass from the calyx by gently pulling it after a superficial incision of the tegmen at the point of arm branching (Fig. 5.2A). This procedure is simple and almost harmless due to the loose connection of mesenterial layers anchoring the visceral mass to its calyx. In *A. bifida*, electrical stimulation (8 v) at any site of the arm induces breakage usually at a point slightly distal to the base. Even squeezing of an arm with

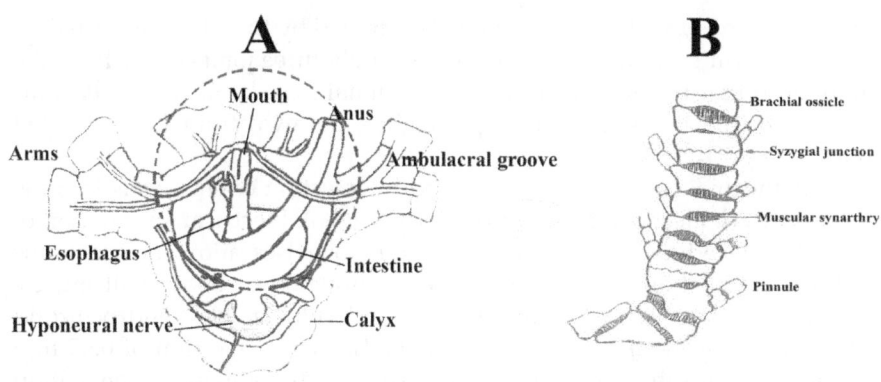

FIGURE 5.2

A. External and internal organs of *Antedon mediterranea*. The viscera is shown by a circle by broken lines (free hand drawing from Mozzi et al., 2006), B. proximal arm region of *Antedon bifida* (free hand drawing from Hyman, 1955).

forceps led to breakage at the nearest syzygy (Fig. 5.2B). However, not all the syzygies have the potential to undergo autotomy (Emson and Wilkie, 1980).

In asteroids, autotomy of an arm can be induced by electrical stimulation (12 v) by placing one electrode on the arm tip and the other at the base of the same arm. The method is 100% effective and is neither lethal nor induces any other side effect (Diaz-Guisado et al., 2006). The duration of stimulation can be short (5 minutes) for juveniles but longer (10 minutes) for adults (Barrios et al., 2008). In *Asterias rubens*, autotomy is induced by applying pressure half way down the arm with a pair of pliers (Ramsay et al., 2001). A cut on the arm accompanied by gentle pulling which causes autotomy at the arm base of *Leptasterias hexactis* (Bingham et al., 2000). However, attempts to induce autotomy in *Asterina gibbosa* and *Pentaceros hawaiiensis* have failed (see Emson and Wilkie, 1980).

There are other factors that are known to induce autotomy. For example, heavily infected asteroids *Sclerasterias neglecta* and *S. richardi* naturally undergo autotomy to reduce the load of parasitic annelid *Myzostomum asteriae* (Riggenbach, 1903). Chat (1962) has observed that an injection of body fluid drawn from an injured conspecific induces autotomy in *Asterias forbesi*. Extending the observation of Chat, Mladenov et al. (1989) discovered an Autotomy Promoting Factor (APF) in *Pycnophodia helianthoides*. Injection of APF drawn from autotomic sea star induces autotomy but the same from an intact sea star fails to induce autotomy. The APF is water soluble, thermolabile peptide derived from the body wall.

In ophiuroids, autotomy can be evoked even by squeezing an arm with forceps and the break occurs within a second or less at a joint proximal to the forceps. But the exact level of break is determined by the site of stimulation. Mostly the arm is ruptured one or two or rarely three joints away from the stimulus so that the animal suffers the minimal loss. In *Amphiura filiformis* and *A. chiajei*, the breakage occurs up to eight joints away from the stimulated point.

In holothuroids, evisceration is elicited by injection of chemicals into the coelom and/or electrical stimulation (Byrne, 1986). Electrical stimulation of the epidermis induces muscle contraction but not evisceration. Intracoelomic electrical stimulation (80–100 v) ensures contraction of body wall muscle accompanied by evisceration in 88% individuals of dendrochirotid *Eupentacta quinquesemita* within 30 seconds. Likewise, injection of 0.45 m/1 KCl or CsCl into the intact specimen results in 100% evisceration within 2.4 minutes. Intracoelomic injection of acetylcholine antagonists tubocurarine chloride (10^{-4} m/1) or atropine (10^{-2} m/1) followed by electrical stimulation (60 v) also successfully induces 100% evisceration. Injection of 2% KCl into the body also induces posterior evisceration in the aspidochirotids *Holothuria atra, H. edulis, H. hilla, H. impatiens, H. pervix, H. scabra, Pearsonothuria graeffei, Stichopus chloronotus* (+ ME also) and *S. herrons* (ME only) and in dendrochirotids *Pseudocolochirus violaceus, Colochirus robustus* and *Ohsimella ehrenbergi*. Notably, the injection fails to induce evisceration in *Colochirus quadrangularis* (Dolamatov, 2014, Dolmatov et al., 2012) and in molpadid *Paracaudina australis* (Silver cited in Emson and Wilkie, 1980).

Extending her investigation, Byrne (1986) discovered an Evisceration Factor (EF) in *Eupentacta quinquesemita*. Injection of extracts made from the body wall, peritoneum and hemal system of eviscerating *E. quinquesemita* induces evisceration but that from coelomic fluid fails to do it. These findings hold true also for *Parastichopus californicus* and *Cucumaria miniata*. The former eviscerates, even when roughly handled but the latter rarely eviscerates. Hence the thermostable small peptide molecule in the EF may be a widespread substance and responsible for initiation of evisceration in holothuroids.

5.3 Causes and Consequences

Natural autotomy occurs due to stress evoked by the wave action in a rocky coast and predatory shock. From field investigation on *Leptasterias hexactis* from five locations in Washington state, USA, Bingham et al. (2000), however, could not correlate the arm damage index to the amplitude of waves from −1.0 to + 0.5 m. Hypoxia is also reported to evoke arm autotomy in *Ophioderma longicauda* (see Emson and Wilkie, 1980). Nilsson (1999) found that *Amphiura filiformis* is more sensitive to hypoxia. Another important factor is trawling, especially beam trawling, because of its deeper penetration into the sediments that kills, autotomizes, transplants and/or discards the sediment-inhabiting holothuroids and ophiuroids (cf Pandian, 2016). Of six potential crab species tested in the laboratory, neither the presence nor the predator's body size has any relevance to the damage. Given 27 sea stars, *Hemigrapsus oregonensis* (1.6 cm carapace width, CW) and *Telmessus cheiragonus* (3.7 cm CW) fail to attack even once. *Cancer productus* (5.4 cm CW) and *C. oregonensis* (2.3 cm CW) induce tip loss on two arms in a single attack, and tip loss in four and entire arm loss in three in 21 out of 27 attacks, respectively. Sea stars deliberately autotomize the arm tip(s) and arm(s) in an escape reaction rather than arms being forcibly detached by the crabs (Bingham et al., 2000). Clearly, the autotomic echinoderms autotomize more due to intolerable shock elicited by predatory attack and other environmental hazards.

Crinoids autotomize their arms more readily than their viscera. Of 31 incidences summarized by Lawrence and Vasquez (1996), only in nine incidences are both arms and viscera lost (Table 5.1). In them too, fractions of visceral mass ranging from 1.8 in *Comaster bennetti* to 9% in *Colobometra perspinosa* alone is lost. Of 17% loss of pinnule incurred by *Antedon bifida* due to predation, the loss is confined to the genital pinnules alone (Nichols, 1994). The quantum of gamete loss from these pinnules remains to be estimated. Lawrence and Vasquez (1996) summarized the then reported incidences of arm autotomy in 29 comatulid species. Of them, they have provided information on prevalence and multiple (all) arm regeneration for 17 species. Leaving a couple of them (e.g. *Comantheiria briareus*: prevalence 75%, multiple regeneration 83%), the remaining values for the multiple regeneration are plotted against prevalence (Fig. 5.3A). The emerging trend reveals an interesting relation between these parameters. With increasing prevalence, incidences of multiple regeneration also increase but only up to 15% at 100% prevalence. It remains to be known whether the increasing prevalence (and corresponding competition for resources) in the comatulid population 'inhibits' multiple arm regeneration beyond the limit of 15% at 100% prevalence. Interestingly, Skold and Rosenberg (1996) reported that 13.3% of total production in *Amphiura filiformis* population is allocated for arm regeneration.

Holothuroids: In high latitudes, holothuroids almost regularly undergo natural evisceration once a year during the autumn-winter season. Of 119 specimens of *Stichopus regalis* collected from the French coast during September–October, 117 (> 98%) were either eviscerated or regenerating. Year-round collection of *Parastichopus californicus* confirmed the regular evisceration in October. In California, 55% *S. parvimensis* were reported to have undergone evisceration during October–November. Understandably, *S. tremulus* (40%) eviscerated during November in Norway. However, natural evisceration became infrequent and occasional in *Holothuria floridana, S. mollis* (New Zealand) and *Actinopyga agassizi*. The evisceration frequency ranged from 0 to 18% in *H. scabra*, perhaps collected from the subtropical coast (see Hamel

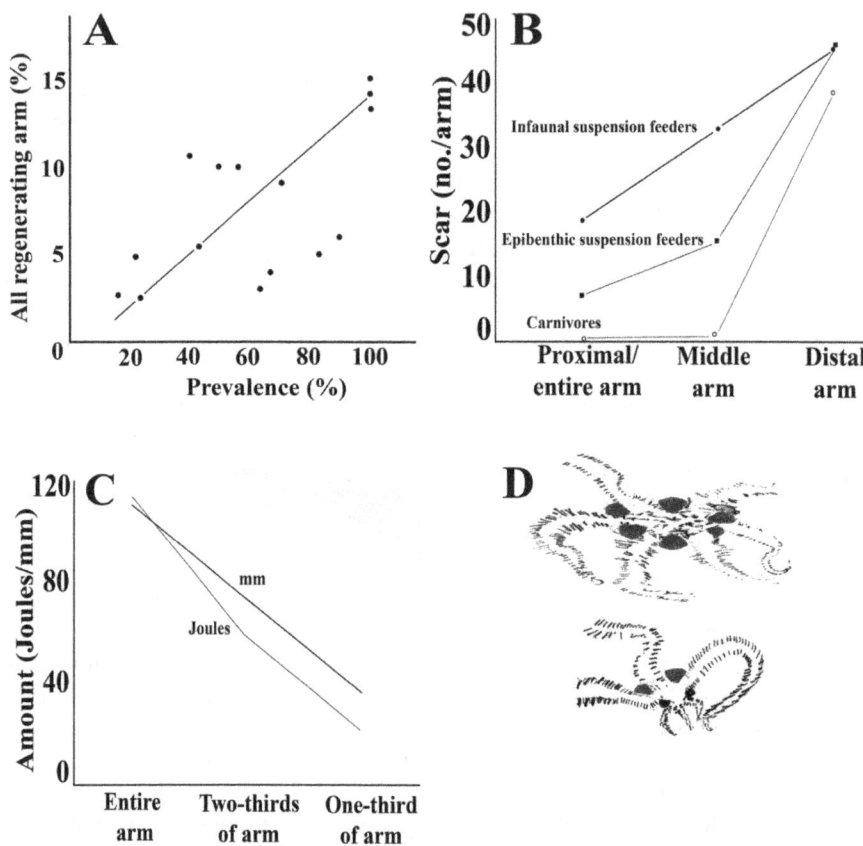

FIGURE 5.3

A. Multiple arm regeneration as a function of prevalence of autotomy in comatulids (drawn from the data [shown] reported by Lawrence and Vasquez, 1996), B. Scar frequency in an arm as a function of arm fraction in suspension feeding and carnivorous ophiuroids (drawn calculating the mean values for data reported by Skold and Rosenberg, 1996), C. Relation between arm loss length as well as energy lost as a function of the length of arm lost by an ophiuroid *Ophiophragmus filograneus* (from Lawrence, 2010), D. Sketches of ovigerous *Ophiactis savigyni*. Upper: Hexamerous intact female with matured gonads. Lower: 3-armed regenerating female. Note the absence and partially developed gonads between regenerating arms and non-regenerating arms, respectively (redrawn from Boffi, 1972).

et al., 2001). However, Mary Bai (1971), who examined 4,000 specimens of *H. scabra* from the Gulf of Mannar (8° N) encountered not a single eviscerated holothuroid. Byrne (1986) discovered the Evisceration Factor (EF) in the Canadian *Eupentacta quinquesemita*. Interestingly, Silver (cited in Emson and Wilkie, 1980) found no evidence for the presence of EF in the coelomic fluid of eviscerating *H. scabra*.

Ophiuroids: The modes of food acquisition and habitat play a key role in prevalence of autotomy at the distal, middle or proximal (basis of the arm), leaving a scar. Counting these scars in carnivores (e.g. *Ophiura ophiura*), epibenthic suspension/deposit-feeders (e.g. *Ophiothrix fragilis*) and infaunal suspension/deposit-feeders (e.g. *Ophiura filiformis*), Skold and Rosenberg (1996) brought to light interesting observations. The arms of suspension feeding infaunal ophiuroids are more prone to attack and suffer autotomy than the epibenthics, as they can wholly or partly withdraw into holes and crevices. The arms of carnivorous ophiuroids do not undergo autotomy at proximal or middle of the arm. At the best, 36% of them lose the distal fraction of the arm, while attacking the prey or being attacked by predators (Fig. 5.3B). Consequently, the prevalence of autotomy in the carnivorous ophiuroids is 22% only (see Table 5.6). Contrastingly, the infaunal suspension feeders display scars for having lost 17 (range = 12–36), 31 (16–53) and 45% (32–56) of their entire arm, middle + distal and distal fractions of the arm, respectively (Fig. 5.3B). These values for the epibenthics are 7, 14 and 45%. Correspondingly, the prevalence of arm loss is also 45 and 85% in these infaunal and epibenthic suspension/deposit feeders, respectively. Incidentally, these findings also explain the range of prevalence values reported for ophiuroids (Table 5.6). Ophiuroids suffer progressively more loss with increasing loss of arm length and require more energy to regenerate the missing fraction of arm (Fig. 5.3C). Incidentally, corresponding data for asteroids remain to be collected. Along with the arm loss, autotomized ophiuroids incur the loss of gonads. For example, all the six bursal sacs are filled with mature gonads in *Ophiactis savignyi* but only the two sacs located between unbroken arms are filled with mature gonads but those two sacs between the regenerating arms do not hold any trace of gonad. However, the partial gonads fill the sacs located between unbroken and regenerating arms (Fig. 5.3D, Boffi, 1972). Incidentally, this observation of Boffi (1972) amply demonstrates that regeneration occurs at the cost of the gonad.

Asteroids: Being carnivorous with pyloric caeca as storage for reserve nutrients, asteroids have been subjected to many investigations. For the almost 'herbivorous' crinoids, holothuroids and ophiuroids share the limited incoming nutrients between regeneration, growth and/or reproduction. With the presence of pyloric caeca and gonad extension into arms, autotomy of either a fraction of an arm or one or more arms imposes variable consequences on juvenile and adult male and female asteroids. Tracking sex-dependent survival of intact and 5-, 4- and 3-armed *L. hexactis* for a period of 28 months, Bingham et al. (2000) have brought to light one of the rarest information. In intact hexamerous sea star, 50% females survive at the end of 28 months, in comparison to ~ 20% males (Fig. 5.4). With reduction in the number of autotomized arms from 5 to 3, a definite shift in survival occurs from equal one in the 5-arm sea star to 0% in 3-armed female sea star in the 28th month.

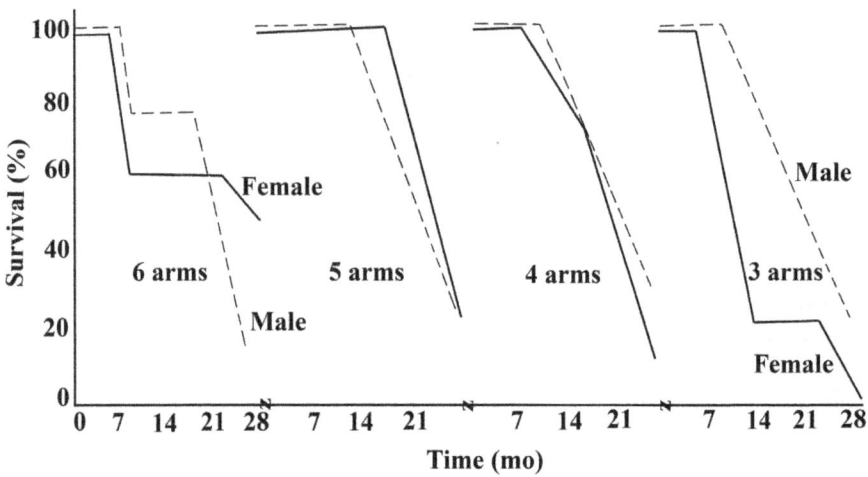

FIGURE 5.4

Effect of the number of autotomized arms on survival of females and males of *Leptasterias hexactis* during 28 month period following autotomy (compiled and redrawn from Bingham et al., 2000).

Unable to share the very limited incoming food energy between regeneration, growth and perhaps vitellogenesis, the females encounter a drastic reduction in survival from the 12th month onwards. Apart from its contribution, the publication of Bingham et al. (2000) has implications for those, who intend to survey the frequency of autotomic arms in sea star population(s).

The autotomy restricted to a single arm in pentamerous *Luidia clathrata* (Pomory and Lares, 2000), hexamerous *L. hexactis* (Bingham et al., 2000) and 40-armed *Heliaster helianthus* (Barrios et al., 2008) accounts for 20.0, 16.7 and 2.7% loss, respectively. Hence, it may impose far different consequences on the functionality and regenerative potential of these sea stars. Information available on this aspect is limited to *L. clathrata* and *L. hexactis*, in which the autotomic prevalence is 60% for their populations. In both of these sea stars, the frequency decreases with increasing arm loss. For example, it decreases from 41% in an intact *L. hexactis* to 4.1% in that with just two arms. Clearly, with increasing arm loss, the star fishes are more prone to death Fig. 5.5B represents simplified trends obtained using recalculated data for the frequency of intact and 1-, 2-, 3-, 4-, 5- and 6-arm regenerating sea stars. The frequency drops faster with increasing number of autotomized arms in the hexamerous sea star than the pentamerous ones. In other words, the frequency of a 4-arm regenerating pentamerous sea star is higher than that of 4-armed hexamerous ones, despite the fact that the former incurs 20% loss and the latter 17% only.

Despite conferring survival, autotomy also inflicts many negative effects such as reductions in locomotion, food consumption, growth and reproductive

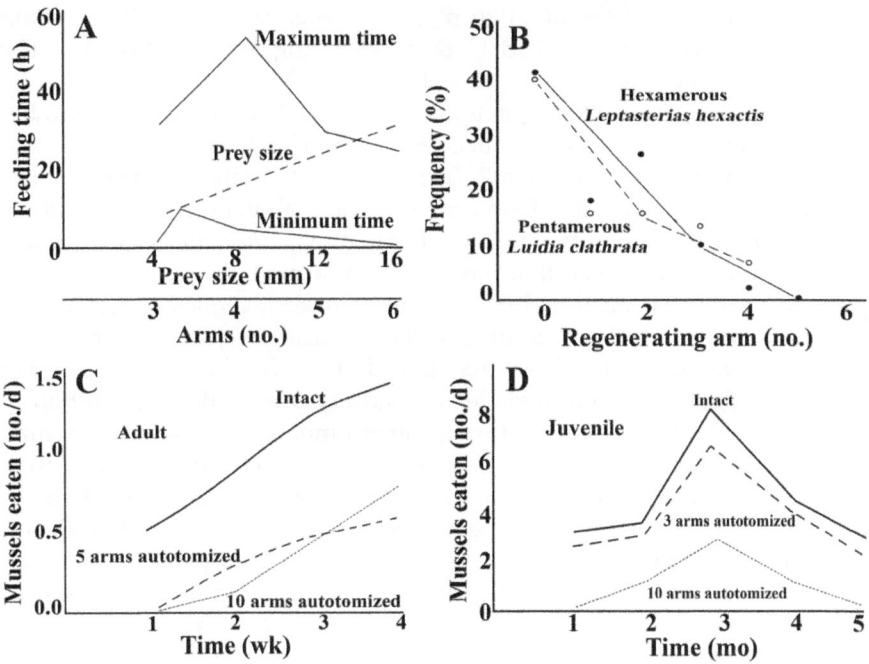

FIGURE 5.5

A. Feeding time as function of prey size and minimum and maximum time required to consume food by intact and autotomized *Leptasterias hexactis* offered 60 *Littorina sitkana* (redrawn from Bingham et al., 2000), B. Simplified trends for autotomic frequency as a function of number of regenerating arm in hexamerous *L. hexactis* and pentamerous *Luidia clathrata* (recalculated and compiled from Bingham et al., 2000, Pomory and Lares, 2000) C and D. Food (*Semimytilus algosus*) consumption by C. juvenile and D. adult *Heliaster helianthus*. Note the difference in the levels of food consumption by sea stars that have lost 3, 5 and 10 arms (redrawn from Barrios et al., 2008).

output. For example, movement of *L. hexactis* was reduced from ~ 50 cm/ hour in intact sea star to < 5 cm/h in the autotomized ones. Irrespective of decreasing arm number from 6 to 3, *L. hexactis* consumed the snail *Littorina sitkana* of 5–15 mm size, although the duration of handling the prey was extended from ~ 10 hours for 5 mm sized snail to ~ 35 hours for the 15 mm *L. sitkana* (Fig. 5.5A). However, when the feeding duration was limited to 24 hours, the 3- and 4-armed *Asterias rubens*, given fairly larger *Mytilus edulis*, only 31 and 47% of autotomized sea stars could eat the mussels. It must be indicated that some parameters like the prey consumed and time taken for feeding also depends on the choice, size and shell thickness of the prey. In general, given free time, and smaller and preferred prey, the handicapped with autotomized sea stars may have the same ability to feed, as that of intact ones. Information on food consumption by intact autotomized juvenile and adult as a function of regeneration period is available for *H. helianthus* and *Stichaster striatus*. From Fig. 5.5B, C, D the following may be inferred: 1. The number of snails (*L. sitkana*) or mussels (*Semimytilus algosus*) consumed by the autotomics is always less than that of the intact. For example, the number of mussels consumed by autotomized *S. straitus* is reduced to ~ 33% of that consumed by the intact sea star. 2. Increase in the number of arm loss also correspondingly reduces the food consumption both in juveniles and adults of *H. helianthus*. The number of mussels eaten by *H. helianthus* that has lost 10 arms is less than those that have lost five arms only. Surprisingly, the number of mussels consumed by the autotomized juveniles is nearly two-three times more than that of adults, i.e. the handicapped juveniles are able to handle the prey more effectively than the adults. 3. The feeding responses of autotomics differ between juveniles and adults. The peak consumption of the 3-armed *S. striatus* reaches during the 3rd month (not shown in figure). But following it, the 3- and 10-arm autotomized *H. helianthus* juveniles, food intake begins to decrease until the 5th month. However, the 5- and 10-arms autotomized adults continue to feed more and more even after 4 weeks, as food has to provide additional resources for growth of pyloric caeca and gonad. 4. The theoretical consideration too suggests continued increase in food consumption until all the arms are regenerated (Lawrence, 2010).

Barring *S. striatus*, food consumption is decreased during the initial period of regeneration in autotomized sea stars and with loss of more and more arms, the consumption becomes less and less. Notably, the 10-arms autotomized juvenile *H. helianthus* consumes only 40 and 20% of that consumed by the 3-arms autotomized and intact juveniles, respectively (Fig. 5.5D), i.e. out of the mean of 18 arms, the 3-, 5- and 10-arms autotomized sea stars account for reductions of 17, 28 and 55% of pyloric caeca, respectively. Moreover, the protein (digestive enzymes) content of the Pyloric Caeca (PC) precipitously decreased from ~ 38 mg/dry arm mass in intact *H. helianthus* to > 5 mg/dry arm mass in the autotomized sea stars that have lost five or 10 arms (Fig. 5.6A). This is also true of *S. striatus*, in which protein content of PC decreases from 1.2 mg in intact to 0.25 mg in autotomized sea stars (Diaz-Guisado et al., 2006). For more information on steroids, Table 5.7 may be consulted. During the 18-week feeding, the pyloric caecotomized *Pisaster giganteus* allocated more resources to the caecal growth than for gonadal growth, as the need for digestive enzymes is greater to sustain consumption of more and more of food (see Table 4.14, Harrold and Pearse, 1980). *Luidia clathrata*, fed a maintenance ration, allocates more resources to the pyloric caecal growth in the intact arms in preference over the regeneration of lost arms (Lawrence et al., 1986). Briefly, this analysis has brought to light for the first time that the reduced food consumption in autotomized sea stars is more due to the inadequate digestive enzyme from the reduced pyloric caeca rather than the ability to handle the prey.

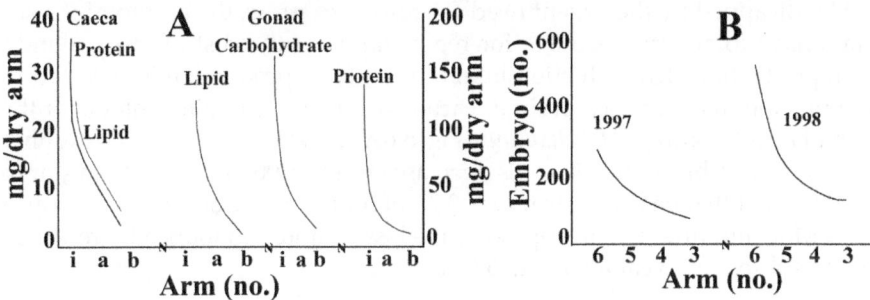

FIGURE 5.6

A. Protein and lipid contents of pyloric caeca in intact (i) and autotomized *Heliaster helianthus*. The figure also shows lipid carbohydrate and protein contents of gonad in intact (i) as well as 5-arm (a) and 10-arm (b) autotomized *H. helianthus* (compiled and redrawn from Barrios et al., 2008). B. Effect of the number of arms lost on the number of brooded embryos during successive years of 1997 and 1998 in *Leptasterias hexactis* (simplified and redrawn from Bingham et al., 2000).

TABLE 5.7

Characteristics of intact and autotomized asteroids

Parameter	Intact	Autotomized
Stichaster striatus **(Lawrence and Larrain, 1994)**		
Radius (cm)	11.2	9.8
Live weight of pyloric caeca (g)		
(i) in an arm	0.8	0.2
(ii) in all arms	3.6	0.9
(iii) lipid (% dry weight)	47	28
(iv) energy (kJ) in an arm	9.8	1.8
in all arms	46.7	6.7
S. striatus **(Diaz-Guisedo et al., 2006)***		
Pyloric caeca (mg dry weight/arm)	0.85	0.20
Protein (mg/arm)	1.2	0.25
Energy (kJ/arm)	0.85	0.15
Heliaster helianthus **(Barrios et al., 2008)***		
Pyloric caeca (mg protein/g arm)	37.5	
5 arm autotomized		~ 5
10 arm autotomized		~ 6
Pyloric caeca energy	2	
5 arm autotomized (kJ)		~ 0.3
10 arm autotomized (kJ)		~ 0.4
Luidia clathrata **(Lawrence et al., 1986)****		
Pyloric caeca (mg dry weight)	80	105

* fed for 5 months, ** 3 months after autotomy

Handicapped by the urgent need for resources *per se*, the autotomized sea stars have to allocate resource for regeneration at the cost of growth and/ or reproduction. The reduction in allocation for reproduction is reflected in the precipitous decreases in lipid, carbohydrate and protein contents of the gonad in autotomized *H. helianthus* (Fig. 5.6A). As a consequence, the number of brooded embryos of *L. hexactis* decreases from ~ 300 in an intact sea star to ~ 100 in that that has lost 3 arms, i.e. 33% of reproductive output for 50% arm loss. More importantly, the reproductive loss is more pronouncedly reflected in the subsequent year also (Fig. 5.6B).

5.4 Growth and Differentiation

Being an epimorphic process, regeneration in autotomic echinoderms involves extensive proliferation of cells resulting in growth and differentiation of internal and external tissues, organs and systems. For example, the simplest arm regeneration requires growth of the arm in length and differentiation of external tissues like podia and spines. Regeneration of the simplest pinnules of crinoids involves not only growth but also differentiation of genital tissues. In asteroids, it requires growth of the arm or fraction of the arm, and differentiation of external tube feet and the ambulacral groove as well as internal pyloric caeca and gonads. The most complicated regeneration following evisceration in holothuroids requires growth of existing tissues and differentiation of an array of organs/systems that are missing in the posteriors or anteriors.

Crinoids: Whereas visceral regeneration is a rapid process and requires, for example, just 3 weeks in *Antedon mediterranea* (Mozzi et al., 2006), the regeneration of stalk and crown in the sea lily takes a longer duration. Amemiya and Oji (1992) have reported that the terminal fraction of the stalk including the crucial basal plates of the sea lily *Metacrinus rotundus* has the potential to regenerate the entire crown. Apparently, the terminal fraction of the stalk seems to harbor the stem cells to accomplish the entire regenerative process. The process requires a long duration of > 350 days, similar to that required for arm regeneration in asteroids. The aboral nerve center, indispensable for the arm regeneration of feather stars, is located at the oral end of the stalk and within the circle of basal ossicles in the lilies. Incidentally, the genital pinnules (37.8%) and pinnules (52.8%) constitute as much as 90.6% of the body of *Promachocrinus kerguelensis*, while that of viscera is 5.8% only (McClintock and Pearse, 1987).

Experimental studies on autotomic crinoids have identified four cell types as responsible for regeneration of the arm and viscera. They are (i) amoebocytes or Resident Stem Cells (RSCs) arising from the brachial nerve, (ii) migrating coelomocytes (CSCs), (iii) phagocytes (PCs) and (iv) granulocytes (GCs). The epimorphic RSCs are responsible for the blastema formation. Subsequently, the RSCs and CSCs, as progenitor pluripotent cells, are jointly responsible for differentiation of all missing tissues and organs (Candia-Carneveli et al., 2009). Mozzi et al. (2006) have shown that visceral regeneration, which includes mostly the digestive system (see Fig. 5.2A), is a rapid and effective process. After a brief healing process resulting in the formation of a blastema, tissues and organs develop as a result of extensive migration and transdifferentiation of coelothelial cells.

Holothuroids: Unlike the other echinoderms, evisceration involves almost all internal organs and systems. Not surprisingly, the process is far more complicated than in the others. Earlier studies are limited to the macro-anatomical level of regeneration (e.g. Emson and Mladenov, 1987). More detailed electron microscopic studies on posterior and anterior regeneration of *Cladolabes schmeltzii* have been undertaken by Kamenev and Dolmatov (2015, 2016). The major events of clonal regeneration are usually described in five stages (see also Emson and Mladenov, 1987). Table 5.8 provides

TABLE 5.8

Comparative account on the major events in clonal regeneration in *Cladolabes schemeltzii* (compiled from Kamenev and Dolmatov, 2015, 2016)

Stage	Anterior Regeneration	Posterior Regeneration
Inherited	Posterior gut, cloaca, respiratory tree	Aquapharyngeal complex, gonad, anterior descending gut
1	Anterior body end is healed and closed. Proliferation of coelomic epithelial cells	Posterior body end is healed and closed
2	Radial water-vascular system is formed. Radial nerve cords appear, being formed by ectoneural part of the cord	Posterior rudiment gut anlage appears and grows to 50% of its maximal length. Luminal epithelium (LE) of the gut anlage and mesenchymal amoebocytes appear
3	LE is formed from the enterocytes. Anterior gut with pharynx and esophagus are formed	Gut is fused to cloaca. Microvilli appear in gut
4	Polian vesicles and stone canals are formed. Mouth orifice appears	Respiratory tree (RT) is formed from endodermal tissues and appears from the dorsal side of cloaca. RT consists of LE, connective tissue and external mesothelium
5	All missing organs are formed and their structure resembles normal ones. But gonad is not yet formed	Posterior body is completely formed

a comparative account on the major events that occur during these five stages. The primordium of the regenerative digestive tract develops from a (proliferation) thickening of the free edge of the mesentery, which plays a key role in regeneration of the digestive tract (Garcia-Arraras, 2001). Essentially, a gut is formed in *Eupentacta fraudatrix* from two rudiments located at the anterior and posterior (cloacal) anlagen. The anterior anlage extends posteriorly and the posterior towards the anterior. With their fusion, a functional gut is formed including the descending and ascending loops on the 27th day after fission. The gut begins to function from the fourth stage onwards. Notably, the Luminal Epithelium (LE) is formed from mesodermally derived mesothelium and endodermally derived epithelium of cloaca (Mashanov et al., 2005). For details on regeneration of muscle, hemal system and Cuverian tubules in the ramets of holothuroids, Candelaria et al. (2006) and Garcia-Arraaras and Dolmatov (2010), Garcia-Arraras and Greenburg (2001) and Vandenspiegel et al. (2000), respectively, may be consulted.

Ophiuroids: A forcible breakage of the arm produces jagged surfaces. But the deliberate 'casting off' of the fraction of an arm at any plane facilitates the escape from the predator at the lowest cost. The deliberate autotomy occurs due to profound (but reversible) relaxation of the 'tonus' of the inter-ossicular muscles. Obviously, the stem cells responsible for growth and differentiation in regenerating arm are harbored in the inter-ossicular muscle (cf Pandian, 2016) and/or in the brachial nerve (Candia-Carneveli et al., 2009) in ophiuroids and the sygial junction in crinoids (Fig. 5.2B). Arm length attained by ophiuroids ranges widely and *Orchasterias columbiana* has one of the longest arms in relation to its disk size (Hyman, 1955). As a consequence, the potential to regenerate the lost arm or arm fraction differs (0.14 to 0.40 mm/d, Dupont and Thorndyke, 2006). The potential, for example, decreases in the following order: *Ophiocoma scolopendrina* > *Ophiocoma* sp (not shown) > *O. elegans* > ophiodermatids (Fig. 5.7A). Unexpectedly, *Macrophiothrix longicauda* with simultaneously regenerating multiple arms displays the fastest arm growth. However, Chinn (2006) has not indicated whether the observed arm length is limited to arm growth or also included differentiation of podia and spines. The trends reported for the arm growth and differentiation are inverse in *Amphiura filiformis* (Fig. 5.7B). In sea lily *Metacrinus rotundus* too, similar inverse trends have been reported for the rapid regenerative growth of the arm length but slower differentiation of stalk diameter (Amemiya and Oji, 1992).

Asteroids: The carnivorous asteroids differ from all other 'herbivorous' echinoderms on two counts. Firstly, their normal number of arms may vary from 5 to 25, which has varying implications on the routine functions of autotomized sea stars. For example, the two arm (40%) loss in pentamerous *Stichaster striatus* may severely reduce its ability to capture and 'handle' the

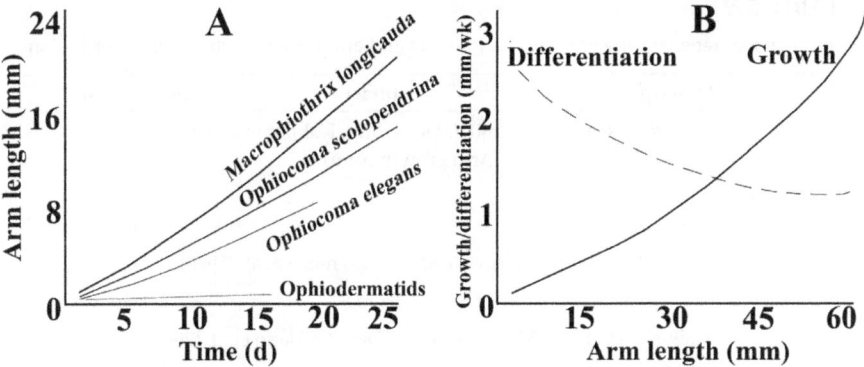

FIGURE 5.7

A. Regenerative growth of arm in ophiuroid species from different habitats (simplified and redrawn from Chinn, 2006), B. Growth and differentiation of a regenerating arm in *Amphiura filiformis* (compiled and redrawn from Dupont and Thorndyke, 2006).

prey (Diaz-Guisado et al., 2006). Contrastingly, following the two (8%) arm loss, the 25-armed *H. helianthus* may function normally (Barrios et al., 2008). Secondly, with reduction in arm number and the corresponding quantitative loss, the pyloric caeca may not be able to supply adequate digestive enzyme and thereby reduce food intake in autotomized sea stars (see p 118). As a storage organ, the caeca can support arm regeneration in fed and to some extent in starved sea stars (e.g. *L. clathrata*, Lawrence et al., 1986, see also Table 5.7). Understandably, autotomy of proximal (containing the caeca) arm is rapidly regenerated at 11.0%/month, in comparison to 7.5 and 3.9% in middle and distal fractions of regenerating arm(s) (Lawrence and Pomory, 2008). Table 5.9 summarizes the available information on growth of intact and regenerating arms of asteroids. The following may be inferred: 1. Following fission, arm regeneration occurs in ramets of *Allostichaster capensis*. In fact, nutrient storage plays a key role in the initial regeneration of arms in fissiparous species (Rubilar et al., 2011) but food consumption in non-fissiparous species (Ramsay et al., 2001). 2. In length, growth of the regenerating arm exceeds that of intact arm in *L. clathrata*, *Asterias vulgaris* and *A. rubens*. For example, it is 3.2 and 5.2 cm in the intact and regenerating arms of *S. striatus*, respectively (Table 5.9). 3. However, increase in arm weight serves as a better and representative index. Arm growth ranges from 2.7 to 3.0 g dry weight in intact but 1.1 to 1.4 g in regenerating *A. rubens*. This observation presents the reverse of that reported considering arm length. 4. Once the arm growth reaches 50% of the total length, pyloric caeca and gonads begin to be accommodated. That demands allocation of more resources for every unit of arm growth.

TABLE 5.9

Summary of regenerating arm growth in intact and autotomized asteroids. * after fission

Description	Intact	Autotomized
*Allostichaster capensis**, 80 days feeding, Rubilar et al. (2015) Arm growth (mm)		
Size 1.5 mm disc	0.0	0.8
Size 2.5 mm disc	0.0	1.1
Luidia clathrata, 91 days feeding, Lawrence et al. (1986)		
	55 mg	27 mg
Stichaster striatus, 150 days feeding, Diaz-Guisado et al. (2006)		
	0.5 g, 3.2 cm	2.5 g, 5.2 cm
Heliaster helianthus, 150 days feeding, Barrios et al. (2008)		
	3.0 g, 3.8 cm	
Regenerating 3 arms		3.0 g, 3.8 cm
Regenerating 6 arms		1.8 g, 3.6 cm
Asterias rubens, 249 days feeding, Ramsay et al. (2006)		
Growth of intact arms with		
1 autotomized arm	3.0 g	
2 autotomized arms	2.7 g	
3 autotomized arms	2.8 g	
Growth of regenerating		
1 arm		1.1 g
2 arms		1.2 g
3 arms		1.4 g
Luidia clathrata, 360 days feeding, Pomory and Lares (2000) 100% regeneration completed		

During the first 130 days, growth of regenerating arms of *A. rubens* reach the length of 52, 46 and 43% of the maximal length in one, two and three arms regenerating sea stars, respectively. But then it slows to additional 14, 18 and 23% on the 249th day in a single arm regenerating sea star; the corresponding values are 18 and 23% in sea stars regenerating two and three arms, respectively (Ramsay et al., 2001). 5. With increase in the number of the regenerating arm, the growth rate increases in the pentamerous *A. rubens* but decreases in the multi(25)-armed *H. helianthus*.

Pyloric caeca: During the period of arm regeneration, quantity of pyloric caeca, and their protein and energy contents are reduced to varying levels in pentamerous *S. striatus* that has lost 40% arms (Diaz-Guisado, 2006) and the multi-armed *H. helianthus* that has incurred 20% (five arms) or 33% (10 arms) loss even after feeding *ad libitum* in the former and maintenance in the later for 5 months. Protein is the major (27%, see p 121) constituent of the arm. The drastic loss in protein and energy contents indicates that pyloric caeca supply nutrients for growth of regenerating arms. In pyloric caecotomized asteroids, the completion of regeneration lasts for 8 months in *A. rubens* (Sharlaimova and Shabelnikov, 2010), but the repair process lasts for only 5–6 and 8 weeks in *Asterias forbesi* and *Henricia leviscula*, respectively.

5.5 Growth Factors and Genes

Recent studies have focused on the presence and potential role of growth factors like Nerve Growth Factor (NGF) and Transforming Growth Factor-β (TGF) and molecules like ubiquitin and heat shock protein (Hsp 70) (Partuno et al., 2001). Partuno et al. (2003) have indicated the need for extensive alteration in protein component during regeneration. Being 'vegetarians', the crinoid, holothuroid and ophiuroid complex have less protein in their body, especially in the arm (e.g. 11–23% protein in ophiuroid arms, see Lawrence, 2010). However, arms of carnivorous asteroid contain 21–37% protein (Lawrence, 2010). Hence, it is the protein content and carnivorous feeding habit that are responsible for the observed slower regeneration potential of asteroids.

Regarding nerve cells, it has become increasingly clear that glial cells play a key role in neurogenesis. Mashanov et al. (2013) have demonstrated that redifferentiated radial glial cells of *Holothuria glaberrina* accounts for vast majority of cell proliferation. It is from these radial cells that new neurons arise, persist longer and express neuronal markers typical of matured echinoderm.

Echinoids regenerate external appendages like spines, pedicellariae and tube feet. Firstly, the spine regeneration commences with the wound-healing process, in which epidermis is reconstituted around the broken spine. Subsequently, mineralization occurs in syncytium formed by scelerocytes. Tube feet are fleshy extensions of the water-vascular system. There are ~ 1,500 tube feet in a sea urchin. Each foot consists of an outer epidermis, a basiepidermal nerve flex, a connective tissue layer and a horizontal muscle layer (Reinardy et al., 2015). Injection of gamma-secretase inhibitor DAPT (3.5 Diflourophenactyl L) results in a dose-dependent inhibition of regeneration of spines and tube feet, indicating that their regeneration requires Notch signaling, which is known to play a key role in stem cell expression. Across many phyla, the genes *vasa* and *piwi* are known for their role in an array of multipotent stem cell types. Using the Notch signaling inhibitor, Reinardy et al. (2015) have quantified the gene expression of

TABLE 5.10

Expression of stem cell markers in *Lytechinus variegatus*
(from Reinardy et al., 2015)

Gene	q RT-PCR Cycle Threshold (Ct)	
	Tube Feet	Spines
vasa	22.0	25.2
piwi	27.4	24.5
ubiquitin	17.8	17.4
cyclophilin 7	20.3	19.7
rp 18	20.6	19.7
profilin	25.5	22.5

these stem cell marker genes and their control genes in *Lytechinus variegatus* (Table 5.10). Clearly, they have demonstrated the presence of multipotent progenitor stem cells in the spines and tube feet as well as Notch signaling as essential for regeneration. Further, their immunohistochemical studies have revealed the presence of vasa protein throughout the epidermis of arm and disk, esophagus, radial nerve and coelomocytes but its absence in muscle and connective tissues.

Besides these markers, *AnBMP*, a new member of the superfamily TGF-β known from *Antedon mediterranea* shows similarity with other echinoderms and human *BMP*. According to its expression pattern, it is now recognized to play a key role in blastema growth and skeletogenic tissue differentiation (Patruno et al., 2003). *Afuni* has been identified as the first gene responsible for generation in *Amphiura filiformis*. It is expressed in coelomocytes and in differentiation sites of regenerating arms (Bannister et al., 2005a, b). As a centralized structure for recruitment of progenitor cells of any type, the coelomic network is considered to play an important role in regeneration of adult echinoderm (see Candia-Carneveli et al., 2009).

In an important contribution, Candia-Carnevali (2006) summarized the available information on the mechanism, stem cells and recruited cells, growth factors and genes involved in regeneration of lost body parts of echinoderms. Table 5.11 represents an updated compilation of autotomized and successfully regenerated tissues/organs/systems of echinoderms. Whereas epimorphic mechanisms involve extensive cell proliferation and migration as the key feature in regeneration of crinoids, ophiuroids and holothuroids, morphallaxis without blastema formation but with limited cell proliferation is the dominant feature of regeneration in asteroids and to some extent echinoids. This may be the reason for the observed slow regeneration in asteroids. For example, the entire process of regeneration following evisceration requires only 7 days in *Holothuria scabra* (Mary Bai, 1971), 30–40

TABLE 5.11

Autotomized tissues/organs/systems, regenerative process, growth factors and genes involved in restoration of missing body parts in echinoderms (compiled from Candia-Carnevali, 2006 and others)

Tissues/Organs/Systems	Mechanisms	Stem Cells	Regenerative Recruited Cells	Growth Factors	Genes
Crinoids					
Arms, pinnules, cirri, Viscera, Stalk, crown (Ameniya and Oji, 1992)	Epimorphic Blastema, Nerve-dependent regeneration	Amoebocytes, Coelomocytes, Phagocytes, Granulocytes	Transdifferentiation, Extensive migration and proliferation	Neurotransmitters, Neuropeptides, Neural growth factor	*Anbmp*
Ophiuroids					
Arms, Bursal sacs (Boffi, 1972)	Epimorphic Blastema, Nerve-dependent regeneration	Coelomocytes, Dedifferentiated cells?	Dedifferentiation, Redifferentiation, Transdifferentiation, Extensive migration and proliferation	Neural growth factor	*Afuni* (cloned)
Holothuroids					
Digestive tract, Respiratory tree(s), Hemal system, Gonads, Cuverian tubules, Polian vesicles	Epimorphic Blastema, Nerve-dependent regeneration	Myocytes, Coelomocytes	Dedifferentiation, Redifferentiation, Transdifferentiation, Extensive migration and proliferation	Neural growth factor	*EpenHg*
Asteroids					
Arms, Pyloric caeca, Gonads *Sharlaimova and Shabelnikov (2010)	Morphallaxic No blastema, Nerve-dependent regeneration	Coelomocytes Dedifferentiated cells, amoebocytes* Epitheliocytes*	Dedifferentiation, Redifferentiation, Transdifferentiation, Extensive migration, Limited proliferation	Neurotransmitters, Neuropeptides	*ArHox1*
Echinoids					
Spines, Pedicillariae, Tube feet	Morphollaxic Annular blastema	Coelomocytes Dedifferentiated cells	Dedifferentiation, Redifferentiation, Transdifferentiation, Extensive migration, Limited proliferation		*vasa, piwi* (Reinardy et al., 2015)

days in *H. glaberrina* (Garcia-Arraras et al., 1998) and a maximum of 145 days in *Stichopus molli* (Dawbin, 1949). This may be compared with the required duration of 360 days for arm regeneration in *Luidia clathrata* (Lawrence et al., 1986, see p). Incidentally, regeneration of the stalk and crown in *Metacrinus rotundus* seems to be a morphollaxic process, as it requires > 360 days (Amemiya and Oji, 1992). In all the taxons, coelomocytes play a major role in regeneration, in which cells are recruited following dedifferentiation, redifferentiation and transdifferentiation. For each class, one or two genes responsible for regeneration have been identified. A number of neuropeptides and neurotransmitters (e.g. dopamine, serotonine) are also identified as growth factors in the regenerative process.

6

Sex Determination

Introduction

In echinoderms, sexuality includes gonochorism, parthenogenesis, protandric and protogynic sequential hermaphroditism, which implies the presence and operation of diverse mechanisms of sex determination. Notably, gonochorism is limited to 75% in molluscs (Pandian, 2017) and 92% in crustaceans (Pandian, 2016), compared to > 99% in echinoderms (p 76). Hence a single sex determination mechanism may be in operation in a vast majority of echinoderms. Sex ratios represent the cumulative end product of sex determination and differentiation processes. Relatively more information available on sex ratios clearly indicates that sex is almost irrevocably determined at fertilization in most echinoderms. Scanning publications on hybridization, karyotypes and ploidy induction, this account, however, could not find reliable evidence for the presence of sex chromosomes in any echinoderm, although heteromorphic chromosomes have been identified in a few of them.

6.1 Species and Fidelity

Hybridization and resulting sex ratio may provide an idea about the sex chromosomes. For example, hybridization between the Nile tilapia *Oreochromis nilotica* ♀ (XX) and *O. macrochir* ♂ produces 100% male offspring, clearly indicating that *O. macrochir* harbors YY chromosomes (see Pandian, 2011). In echinoderms, the presence of bindin in the sperm acrosome serves as an adhesive responsible for sperm attachment to the vitelline layer of conspecific eggs (see p 81–82). Hence bindin plays an important role in determining whether gametes of two species within a genus are compatible and fertilize each other. However, the role of bindin is not a simple lock-and -key mechanism, in which a molecular change in one of the gametes, a change in sperm need not necessarily result in analogous change in the egg also and vice versa. As a consequence, heterospecific gametic incompatibility may be complete or incomplete. Gametic compatibility is defined as the ratio of mean percentage of eggs fertilized in heterospecific crosses divided by the mean percentage of eggs fertilized in homospecific crosses at a sperm concentration required to fertilize > 90% of eggs in homospecific cross. Selected values summarized by Lessios (2007) are listed in Table 6.2. In such species, analogous molecular change(s) have either occurred or not occurred in eggs and sperm, the ratio of reciprocal heterospecific gametic compatibility is 1.00, as it is between *Lytechinus variegatus* and *L. williamsoni* as well as between *Arbacia punctulata* and *A. incisa*. However, the ratio begins to decrease, when there are slight changes in eggs but not in the sperm. As a result, for example, *Echinometra vanbrunti* sperm are unable to fertilize 100% eggs of *E. viridis* but *E. viridis* sperm are able to fertilize 100% eggs of *E. vanbrunti*. With more molecular changes in eggs alone, *Strongylocentrotus droebachiensis* sperms are not compatible to *S. palidus* but *S. purpuratus* sperms remain compatible to *S. droebachiensis*.

Due to difficulty in maintaining parents, larval stages and hybrids until backcrossing, available information on hybrid fitness is limited to a few echinoids alone. Interestingly, echinoid development is known for their ability for auto-regulation. Besides, time since divergence, other factors like development mode seem to have complicated the level of hybrid fitness. Limitation of sperm density to 10^{-5}/ml facilitates backcross fertilizations in all types of F_1 hybrids of *E. mathaei* and *Echinometra* sp C (Table 6.1). That the differences in survival of > 60% and < 60% in conspecific and heterospecific crossings, respectively indicate that the divergence since 1.2 million years (My) alone cannot account for the reproductive isolation (Rahman and Uehara, 2004). Despite divergence since 4.5 My, hybrids between *Heliocidaris erythrogramma* and *H. tuberculata* are viable perhaps more due to the direct mode of development (Raff et al., 1999). In *Strongylocentrotus*, the interspecific hybrids that survive to sexual maturity are all females and their eggs can be fertilized by *S. pallidus* but not by *S. droebachiensis* (Strathmann, 1981). The limited data suggest that the divergence since 2.3 My may not be adequate to genetically isolate sympatric species (Lessios, 2007). Briefly, fidelity of these echinoid species remains vague, even after the divergence since ages. However, no information is provided on the sex ratio of the progenies arising from hybridization of these echinoids.

Hotchkiss (2000) reviewed inheritance of arm number in a few asteroids that were hybridized. Of them, one may be mentioned, although it may not immediately be concerned with sex determination. The number of arms is not a stable trait in the predominantly hexamerous *Patiriella gunni* and

TABLE 6.1

Survival of hybrids generated from reciprocal heterospecific crosses in selected sea urchins in relation to time since divergence (from Lessios, 2007, modified)

Time (My)	Genus	Parental Species		Reported Observation
		Egg	Sperm	
~ 1.2	*Echinometra*	sp *A*	*oblongata*	Normal
		oblongata	sp *A*	Normal
~ 1.3	*Echinometra*	sp *C*	*mathaei*	Normal
		mathaei	sp *C*	Depressed survival
~ 1.3	*Strongylocentrotus*	*droebachiensis*	*pallidus*	Normal
		pallidus	*droebachiensis*	Normal
~ 4.4	*Strongylocentrotus*	*purpuratus*	*franciscanus*	Normal
		franciscanus	*purpuratus*	Dies at gastrulation
~ 1.6	*Pseudochinus*	*albocinctus*	*huttoni*	Normal
		huttoni	*albocinctus*	Normal
7.0	*Pseudochinus*	*huttoni*	*novaezealandiae*	Dies at 8 arm pluteus
		novaezealandiae	*huttoni*	Low settlement
~ 4.5	*Heliocidaris*	*erythrogramma*	*tuberculata*	Viable
		tuberculata	*erythrogramma*	Dies at gastrulation
7–14	*Diadema*	*savignyi*	*setosum*	Viable

TABLE 6.2

Ratios of heterospecific gametic compatibility in selected echinoids (from Lessios, 2007, modified)

Genus	A = egg	B = sperm	A × B	B × A
Lytechinus	*variegatus*	*williamsoni*	1.00	1.00
	pictus	*variegatus*	0.72	1.00
Arbacia	*punctulata*	*incisa*	1.00	1.00
Echinometra	*viridis*	*vanbrunti*	0.92	1.00
	sp *C*	*oblongata*	0.96	0.88
	lucunter	*viridis*	0.19	0.97
Strongylocentrotus	*droebachiensis*	*pallidus*	0.82	0.07
	droebachiensis	*purpuratus*	0.75	0.00
Echinometra	sp *A*	*oblongata*	0.00	0.52
	mathaei	*oblongata*	0.00	0.22

octamerous *P. calcar*. The arm number varies widely in *P. calcar* from 7 (3%) to 8 (91%) and 9 (6%) in the field but in the laboratory cultures, it varies from 6 (6.5%) to 7 (35.5%), 8 (55%) and to 9 (3%). The cross between *P. gunni* ♀ × *P. calcar* ♂ yields juveniles with 5 (8%), 6 (67%) and 7 (25%) arms, i.e. majority of juveniles have inherited the maternal hexamerous arm trait. However, the reciprocal cross between *P. calcar* ♀ × *P. gunni* ♂ produces 5- (2%), 6- (33%), 7- (57%), 8- (6%) and 9- (2%) armed juveniles, i.e. neither majority of the juveniles carry the hexamerous paternal nor the octamerous maternal trait. Briefly, hybridization experiments in echinoids and asteroids do not provide any clue on the sex chromosomes.

6.2 Karyotypes and Heteromorphism

Despite the availability of actively regenerating tissues/organs, echinoderm cytologists have usually selected the early embryos for karyotyping (e.g. Eno et al., 2009). Chromosomes of echinoderms are relatively short (1–6 µm, Eno et al., 2009) and tightly clustered (Kondo and Akasaka, 2012). The diploid number ranges from 36 to 46 in echinoderms (Colombera and Venier, 1981, Lipani et al., 1996, Saotome and Komatsu, 2002, Duffy et al., 2007, Eno et al., 2009, Okumura et al., 2009). Only 14 out of 40 species subjected to various cytogenetic analyses until 1991 revealed the presence of recognizable mitotic chromosomes. However, none of these analyses could reveal the presence of sex chromosomes. Yet, the presence of heteromorphic chromosomes were reported for the sea urchins *Paracentrotus lividus* (Lipani et al., 1996), *Strongylocentrotus droebachiensis* and *S. purpuratus* (Eno et al., 2009) and an asteroid *Asterina pectinifera* (Saotome and Komatsu, 2002). In *P. lividus*, the length measured 4.50, 4.03 and 3.42 µm for the autosomes, and the designated 'Y' and 'X' chromosomes. The relatively short submetacentric/metacentric and the shortest subtelocentric chromosomes were claimed as Y and X chromosomes. Their heteromorphic trait was ascertained by the (i) consistent presence of 50% X and 50% Y chromosomes in all the identical blastomeres and (ii) invariable presence of Y chromosome in spermatogonial metaphase of five males subjected to cytological analysis. Still, no experiment was made to confirm the heterogamety from the inheritance of sex by F_1 offspring derived from the cross between the claimed Y males and X females. Until such experiments are undertaken, the identified differences in chromosome size may have to be designated as heteromorphism rather than heterogamety. Searches for heterogametic chromosomes in ophiuroids *Ophiodaphne formatum*, *O. scripta*, *Ophiosphaera insignis* and *Astroclamys bruneus* may prove rewarding, as the 'mini-males' are reported to cling to the female.

6.3 Ploidy Induction

Microsurgery, a procedure required for gaining more information on the development potential of echinoderms (e.g. Ransick et al., 1996), may be hampered by the small egg size. To increase egg size without altering any of its property, techniques for oocyte fusion have been developed. The exposure of eggs to one or other fusogens like polyarginine or polyethylene glycol (PEG) or application of electrical pulse ensures the fusion of two eggs in sea star *Aphelasterias japonica* and sea urchin *Strongylocentrotus nudus*. To achieve the said fusion, removal of the follicular envelop and vitellin membrane by agitation (see also p 45) or centrifugation is required. The denuded oocytes, following a brief exposure to 1 M urea with 1 Mm $CaCl_2$ for 30 seconds, are exposed to a suitable dose of PEG for an appropriate duration. Vassetzky et al. (1986) have indicated that it is possible to fuse two oocytes into a single egg in sea urchins and asteroids. Following fusion, cytoplasm of the two oocytes, however, do not mix. It is also possible to obtain cytoblasts (eggs with no nucleus) but with one or more intact germinal vesicles. The agitation or centrifugation can produce various sizes of enucleated egg fragment and the bisection of eggs may not supply adequate enucleated unfertilized eggs. Hence a modified procedure to harvest enucleated eggs (4–10×10^4 eggs per ml) is described by Saotome (1999). Adequate number of small ($\sim 1/4$th) and larger ($\sim 1/2$th) eggs can be obtained from two different layers at the bottom in a 50 ml centrifuge tube after centrifugation at different rpm for different durations in saccharose density gradient produced by mixing 1 M saccharose to sea water at ratios of $1:2, 1:1, 2:1$ and $1:0$ from the top to bottom. Both the small and larger enucleated eggs of *Hemicentrotus pulcherrimus* are fertilizable. After fertilization, androgenics (andromerogones) are produced (Fig. 6.1C). The presence of 21 chromosomes in these androgenics has been confirmed from the majority of 2- (73%), 8- cell (84%), morula (75%), hatchling (76%) and swimming blastulae (87%) stages (Saotome, 1999) indicating the successful induction of androgenesis, as the diploid number is 42. The presence of (4–16%) aneuploids may be due to polyspermy and/or irregularity in the distribution of chromosomes. Unfortunately, these androgenics have not been reared to sexual maturity to know their sex ratio and to know whether the androgenics carry X or Y, or Z or W chromosomes.

In most animals, meiotic division is completed after the eggs are released from the ovary and following fertilization. Consequently, the polar body is released after the completion of fertilization process. Hence, the eggs are readily amenable to ploidy induction using a shock to retain the body to produce triploid in fishes (Pandian, 2011), crustaceans (Pandian, 2016) and molluscs (Pandian, 2017). Echinoids are unique in that both meiotic divisions occur within the ovary and prior to fertilization. As a result, the scope for ploidy induction in echinoids is very limited (Walker et al., 2005). However, the first triploid echinoderm was produced in *S. droebachiensis* by Bottger et al. (2011). The two oocytes, each containing 21 chromosomes were first fused and subsequently the fused eggs were fertilized (Fig. 6.1B). In their blastulae and gastrulae triploidy was confirmed by the presence of 63 chromosomes. Unfortunately, these triploids could not be grown to sexual maturity. Hence

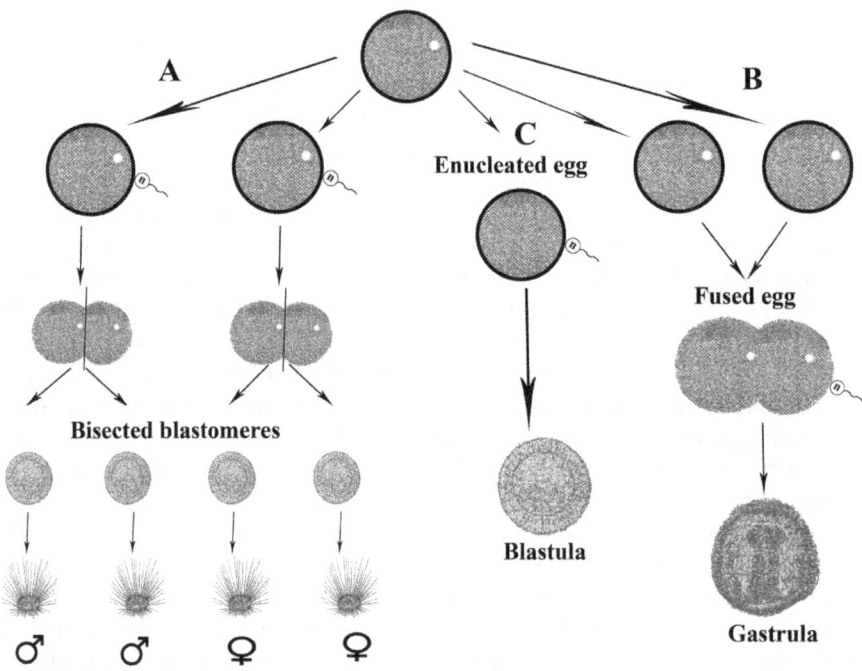

FIGURE 6.1

Schematic representation of ploidy induction and other experiments performed to identify sex chromosomes. A. Bisection of sea urchin blastula resulting in twin progenies of the same sex (see Cameron et al., 1996). B. Induction of androgenesis by fertilizing enucleated eggs of *Hemicentrotus pulcherrimus* (see Saotome, 1999) and C. Induction of triploidy by fertilizing the fused eggs of *Strongylocentrotus droebachiensis* (see Bottger et al., 2011).

sex ratios in these triploids could not be known. Briefly from ploidy induction resulting in the production of androgenics and triploids, no information on sex ratio or sex chromosome could be gained.

At this juncture, it may be relevant to describe a different experiment undertaken in sea urchin by Cameron et al. (1996). The first two blastomeres of the urchins were bisected and the twins were reared to sexual maturity (Fig. 6.1A). Between the ages 1.5–2.0 years, the urchins were induced by electrical stimulation to emit the gametes. Of nine pairs of twins that were successfully induced, five pairs of twins released sperm alone and the remaining four pairs spawned eggs alone. Hence the sex of pairs in each of these twins was the same. Clearly, sex determination in the urchin is chromosomal, which serves as carrier of sex determining genes.

6.4 Genome and Sequencing

Except for crinoids, cellular DNA contents, estimated from all echinoderm classes indicate that the content of haploids ranges from 0.54 pg in *Dernasterias imbricata* to 4.4 pg in the sea cucumber *Thyonella gemmata*. Hence, the genome size differs by about eight times from 500 mb to 4 GB (see Kondo and Akasaka, 2012). The variation in the content range only ~ 5% as in *Strongylocentrotus purpuratus*. Over 1,000 gene models with relevance to immunity and other blood functions and biomineralization, *SPARC* (osteonectin)-related genes namely *Sp-osteonectins* and *SPARC* calcium-sensing receptor gene (*casr*), which are homologous to vertebrates, have been identified. However, genes related to reproduction and developments remain to be discovered.

7

Sex Differentiation

Introduction

In animals, sex determination and differentiation are successive but diverse processes that have evolved independently a number of times (Hodgin, 1990) and echinoderms are not exceptional to this dictum. Expected of primitive dueterostomes, sex steroids are involved in sex differentiation and play regulative functions in echinoderms, as is known for vertebrates like fishes (e.g. Pandian, 2013). However, (i) echinoderms do not possess a well developed glandular endocrine system but have a complex of chemically mediated interactions between cells (see Pinder et al., 1999) and (ii) the differentiation process in echinoderms is not as labile as in fishes (e.g. Pandian, 2000). In echinoderms, the regulative functions of sex steroids are often investigated from the following: 1. Incubation or culture of tissues to trace the biosynthetic pathways and metabolism of sex steroids, 2. Seasonal changes in gonad structure and levels of sex steroids, 3. Steroid injection into asteroids (e.g. *Asterina pectinifera*, Takahashi, 1982) or supplemented diets in echinoids (e.g. *Lytechinus variegatus*, Wasson et al., 2000a). The precision and reliability of results arising from these studies have been improved from employing microscopy (e.g. *A. pectinifera*, Takahashi and Kanatani, 1981) to electron microscopy (e.g. *Asterias rubens*, Shoemarkers et al., 1977), from a thin layer chromatography (e.g. *L. variegatus*, Wasson et al., 1998) to radio immunoassay (RIA) (e.g. *L. variegatus*, Wasson et al., 2000b) and to cDNA cloning (e.g. *L. variegatus*, Brooks and Wessel, 2002). Due to their small size, ophiuroids seem not to have attracted much attention. But it is difficult to comprehend why the larger holothuroids have not received much attention (e.g. *Stichopus mobii*, Donahue, 1940), despite the scope for gonodectomy and their enormous capacity to tolerate induction of evisceration and turning the skin from inside to outside (cf Kille, 1939, 1942). Relatively more information is available for asteroids and echinoids, and it has been reviewed from time to time (e.g. echinoids, Wasson and Watts, 2001). Interestingly, the asteroids have a nutrient storage cum steroidogenic organ, the pyloric caeca. In echinoids, Nutritive Phagocytes (NPs) within the gonads assume the role of nutrient supply to sustain gametogenesis. Incidentally, not much information is yet available for holothuroids and ophiuroids but they are herbivorous and possess no storage organs, as echinoids do. Major Yolk Protein (MYP), which is ubiquitously present in echinoids, is also reported from some holothuroids (Reimer and Crawford, 1995, Reunov et al., 2010). Hence, the role of sex steroids is presented in two models: (i) the herbivorous echinoids, holothuroids, ophiuroids with no storage organ, as represented by echinoids and (ii) the carnivorous asteroids with pyloric caeca as a storage cum steroidogenic organ (Table 7.1).

7.1 Asteroid Model

Ever since Shoemarkers et al. (1977) brought ultrastructural evidence for the presence of steroid synthesizing cells in the ovaries of *Asterias rubens*, evidence for biosynthesis of sex steroids in asteroids began to accumulate. Injection of cholesterol or estrone (E_1) into the coelom increased gonad index of *Asterina pectinifera* (Takahashi, 1982), clearly indicating that the increased availability of cholesterol led to biosynthesis of more sex steroids resulting in increased gonad index. Daily injection of androstenedione (AST) and E_1 over 16 day periods induced gonad growth in female *A. pectinifera*, whereas P_4, T and E_2 did not. Takahashi (1982) presumed that AST, on conversion to E_1, mobilized protein into the ovaries. From their studies on *in vitro* incubation of gonad tissues (both ovaries and testes), Shoemarkers (1979b) and Shoemarkers and Voogt (1980, 1981) demonstrated that *A. rubens* converted cholesterol to C_{21} steroids, progesterone to C_{19} steroids and AST to other androgens (Fig. 7.1). Earlier, evidence was obtained for the presence of the two key enzymes for biosynthesis of steroids namely 3β-hydroxysteroid dehydrogenase (3β-hsd), which converts prognenolone into progesterone (P_4) and 17β-hsd responsible for conversion of AST into testosterone (T) and 17β estradiol (E_2) into E_1 (Shoemarkers, 1979a). E_1 can be converted into E_2 in the ovaries but not in the PC (Shoemarkers et al., 1981). Incidentally, Gaffney and Goad (1974) reported the conversion of P_4 into some intermediates in *Marthasterias gracilis*, which are possibly involved in biosynthesis of astrosaponins, unique steroid-like compounds in asteroids. On incubation, homogenates of Pyloric Caeca (PC) of *A. rubens* also convert progesterone into 17α-hydroxyprogesterone and AST into T (Shoemarkers et al., 1978). From these bits and pieces of information collected from different asteroid species, it has been possible to trace the sequence biosynthetic pathways of sex steroids in the ovaries, testes and PC of asteroids (Fig. 7.1). The pathways are more or less similar to those of vertebrates.

TABLE 7.1

Contrasting reproductive features in asteroid and echinoid models

Asteroids	Echinoids
Gonads consist primarily of gametes	Gonads consist of both NPs and asynchronous population of gametes
Oocytes do not mature within the ovary and remain in the full grown state with germinal vesicle until spawning	Within the ovary, older oocytes complete maturation one by one, while the younger ones await for their respective turns
Nutrients are translocated from PC to gonads via body fluids	NPs directly supply nutrients to gonads via gonadal lumen (Pearse and Cameron, 1991)
Translocation of nutrients from distantly located PC to gonads demands increased levels of hormones for gametogenic regulation	Proximity between NPs and gametes within the gonad reduces the requirement of hormone levels for gametogenic regulation
Consequently, steroids levels are higher, e.g. 140–5,500 pg/g gonad for T and 4–460 pg/g for E_2 in *Asterias vulgaris* (Hines et al., 1992)	Consequently, steroids levels are low, e.g. 60–320 pg/g gonad for T and 5–160 pg/g for E_2 in *Lytechinus variegatus* (Wasson et al., 2000a)
The germinal layer consists of gametogenic cells, which are individually enveloped by follicle cells. Translocation of nutrients from PC to gonad may require an endocrine mechanism (see Barbaglio et al., 2007)	The germinal layer consist of gametogenic cells that are in intimate contact with NPs. Paracrine-like mechanism may sequestered and transfer nutrients from NP to developing oocytes (Pearse and Cameron, 1991)
E_2 stimulates oocytes growth but not ovarian growth	E_2 or its metabolites stimulate ovarian growth
E_2 treatment increases lipid levels of PC (Van der Plas et al., 1982)	Administration of E_2 + P_4 increases lipid levels of gonads (Wasson et al., 2000b)
Vitellogenin is produced and cleaved to form abundant yolk protein in eggs of asteroids with different development modes, indicating the conserved protein function	MYP ubiquitously presents and also in some holothurians. It is synthesized mainly in inner epithelium of digestive tract and in NPs (Shyu et al., 1986) of both the sexes (Unuma et al., 2010)
E_2 treatment affects vitellogenesis and protein incorporation into oocytes (Van der Plas et al., 1982)	MYP expression is not influenced by E_2 exposure
MYP is not reported	No vitellogenin-like molecule is reported, though genome contains *vtg* pseudogene with atypical features (Prowse and Byrne, 2012)

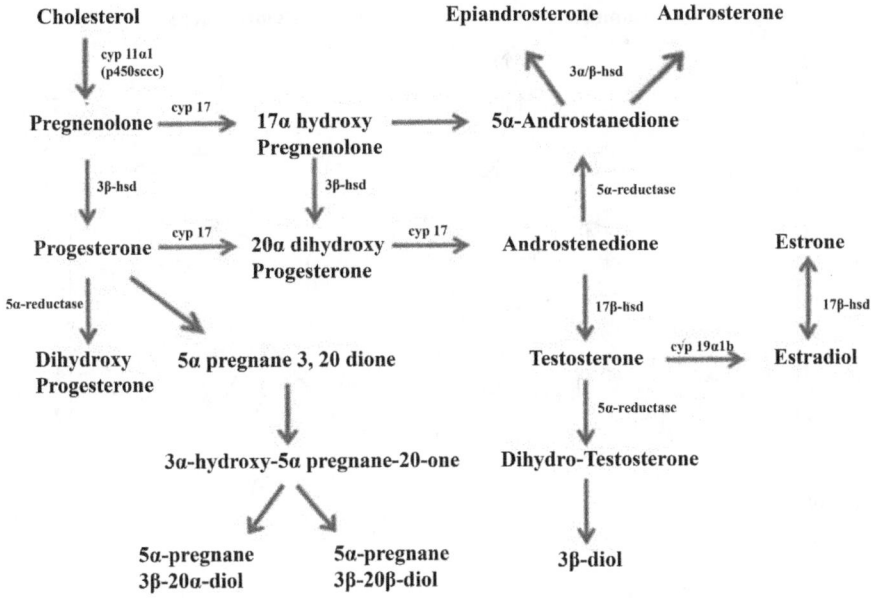

FIGURE 7.1

Steroidogenic pathways in echinoderms (drawn using data reported by many authors).

Investigations of seasonal changes in E_2, T, P_4 levels in the ovaries, testes and PC of *Asterias vulgaris* have brought convincing evidences for (i) the presence of entire sequence of biosynthesis from cholesterol to P_4, T, and E_1–E_2 in the ovaries and testes in an asteroid species and (ii) the transfer of nutrients from PC to ovaries/testes (Hines and Watts, 1992), Fig. 7.2 represents a compiled picture of seasonal changes of sex steroid levels in the gonads and PC of females and males as well as gonad (GI) and Pyloric Caeca (PCI) Indices. It also relates these seasonal changes to mitotic proliferation, maturation and vitellogenesis/spermiogenesis in a representative temperate asteroid as a function of temperature. The peak levels of E_2, T and P_4 are progressively shifted from October to November and December, respectively in testes and from November to December and January in ovaries. However, the peak GI in males and females is observed in February–March prior to spawning. Similar trends are also reported for E_2, T and P_4 in PC of males and females. Remarkably, PCI sharply decreases with meiotic proliferation of spermatogonia in September but only after completion of vitellogenesis in December. Briefly, mitotic proliferation of spermatogonia is regulated by E_2, spermatocyte maturation by T and spermiogenesis by P_4. In ovaries too,

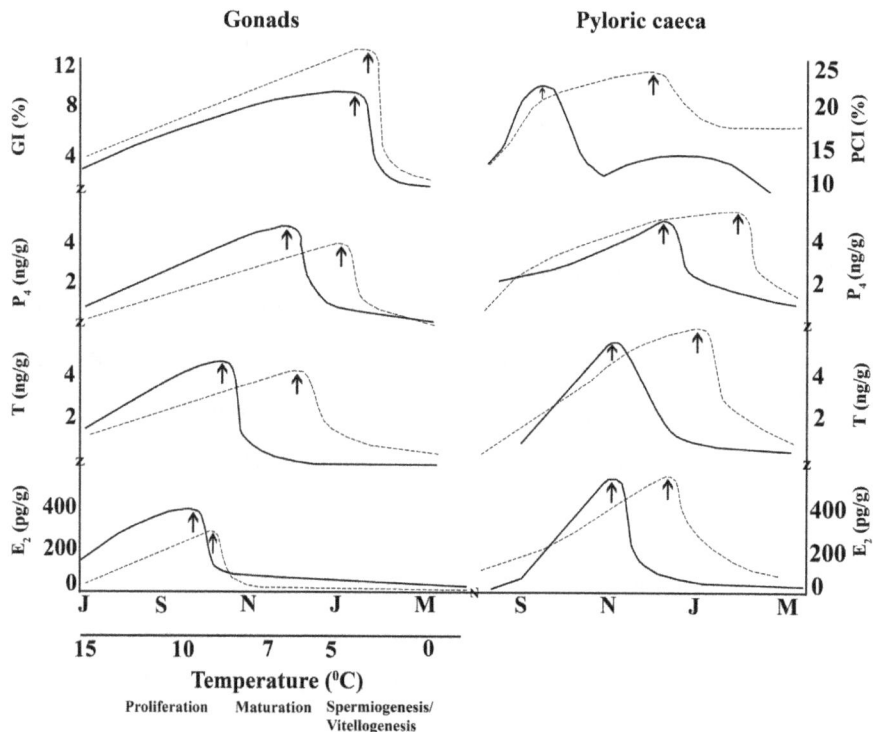

FIGURE 7.2

Seasonal changes in steroid levels in males (thin dotted lines) and females (thick continuous lines) of a representative asteroid *Asterias vulgaris*. Arrows indicate the peaks (simplified sketches compiled from Hines and Watts, 1992).

oogonial proliferation, oocyte maturation and vitellogenesis are regulated by E_2, T and P_4, respectively, although the role of P_4 in vitellogenesis is not clear. E_2 seems to have a greater control over vitellogenesis in *Asterias rubens* (Shoemarkers and Dieleman, 1981) and *Scleroasterias mollis* (Xu and Barker, 1990a, b). Interestingly, the shifts from mitotic proliferation to maturation and spermiogenesis/vitellogenesis in temperate asteroids are regulated by external factors like temperature. For example, the proliferation in *A. rubens* at Maine, USA occurs with declining temperature at 10°C, maturation at 7°C and spermiogenesis/vitellogenesis and subsequent release of gametes at 0°C (Hines and Watts, 1992). Decreasing day length to 12L : 12D increases the production of AST to maximum (Voogt et al., 1991). Investigation on tropical *Echinometra mathaei* in the Red Sea has shown that the urchin spawns in early summer and spawning requires a minimum temperature and minimum nutrient storage in NPs (Pearse, 1969b).

7.2 Echinoid Model

As in asteroids, bits and pieces of scattered information for echinoids indicate almost the same steroidogenic biosynthetic pathways. However, the following may be noted: 1. Unlike asteroids, the echinoid gonads consist of two major cell types: (i) germinal cells that undergo mitotic divisions and complete the entire sequence of meiotic divisions including the release of polar bodies and (ii) NPs that undergo cyclic depletion and synthesis of renewable macromolecules to supply nutrients to the developing gametes (Wasson and Watts, 2001). 2. Barbaglio et al. (2007) brought direct evidence for the occurrence of aromatase activity in *Paracentrotus lividus*. The activity is not sex specific and remains constant (0.24–0.32 pmol/mg protein/ hour) almost throughout the year except during autumn (0.39–0.59 pmol/ mg protein/hour). It is 20-fold higher in the digestive tube than in gonad (Lavado et al., 2006a, b). Incidentally, the presence of *aromatase* gene has also been detected in *Strongylocentrotus purpuratus* (Goldstone et al., 2007). 3. The Major Yolk Protein (MYP) is the most abundant protein in sea urchins. Using antibody specific MYP, Shyu et al. (1986) have shown that a 155-kD polypeptide is synthesized in the intestine of female and male *S. purpuratus*. Subsequently, it is converted into a 175- and 195-kDa vitellogenin and further modified into a 180-kDa MYP in the ovary only. 4. Cloning the cDNA of *MYP*, Brooks and Wessel (2002) have described its structure. MYP is a member of transferring superfamily of iron binding proteins. 5. Wasson et al. (2000b) reported that the metabolization of P_4 yields 5α reduced metabolites (Fig. 7.1). Incidentally, metabolization of P_4 and conversion of ASD into androsterone and epiandrosterone occurs in echinoids only. The remaining steroidogenic pathways are common to both asteroids and echinoids. 6. Investigations on dietary supplementation have shown that feeding T or E_2 induces no effect on growth of juvenile echinoids. But that of ASD or E_1 initiates spermiogenesis in male juveniles. However, oogenesis can be induced by E_1 only, when the juvenile attains sexual maturity (Unuma et al., 1996, 1999).

7.3 Induction of Spawning

Whereas the echinoid oocytes complete the entire process of maturation within the ovary, asteroid oocytes do not mature within the ovary and remain in the full grown state with germinal vesicles until spawning (Table 7.1). Expectedly, the chemicals responsible for maturation of oocytes and induction of spawning may also differ. In 1959, Chaet and McCannaughy reported that an injection of water extract of the radial nerves of star fish induced spawning in ripe star fish. The presence of this active gonad stimulating substance (GSS) has been demonstrated in > 30 asteroid species, in which spawning has been successfully induced at 9.6 pg/ml. It is a thermo-stable polypeptide consisting of 22 amino acids (see Kanatani, 1975) with molecular weight of 5,600 daltons (Cochran and Engelman, 1972). Following its release from the nerves into the coelomic cavity, GSS is transported by coelomic fluid to the gonads. In asteroids, L-methyladenine, a spawning inducer, is synthesized in the follicular cells, only when GSS of neural origin acts on the ovary (Hirai and Kanatani, 1971). Briefly, GSS is responsible for gonadal maturation and L-methyladenine acts as a trigger to induce spawning.

Interestingly, water extract of sea urchin *Anthocidaris crassispina* ovary also induces gonadal maturation in star fish *in vitro*. The presence of the active substance, identified as L-methyladenine (Kanatani, 1974), is also demonstrated from the testes and ovaries of sea urchins (*Pseudocentrotus depressus*, *A. crassispina*, *Hemicentrotus pulcherrimus*) and sand dollars (*Clypeaster japonicus*, *Peronella japonica*). Clearly, L-methyladenine secreted in the ovary of echinoids is responsible for gonadal maturation alone. In sea urchins, the trigger to induce spawning acts on the gonad muscle, as the discharge of mature eggs can easily be achieved by a treatment with potassium chloride, acetylcholine or electrical stimulation. Contrastingly, these agents have little effect in inducing spawning in asteroids. Briefly, GSS, secreted from radial nerves, acts on the ovary to synthesize L-methyladenine. In asteroids, GSS is responsible for gonad maturation and L-methyladenine serves as a trigger to induce spawning. But in echinoids, L-methyladenine secreted in the ovary is responsible for gonad maturation and agents like KCl or electrical stimulation serve as a trigger to induce spawning.

7.4 Endocrine Disruption

More than 60% of the 100,000 man-made chemicals are in routine use worldwide since 1990s. Every year 200–1,000 new synthetic chemicals enter the market (Shane, 1994). Some of these chemicals either mimic or antagonize the action of endogenous hormones. They are known as Endocrine Disruptors (EDs). Mostly, they are insecticides, fungicides, biocides, polyphenyl bisphenyls (PCBs) and organotin compounds. Much is known about their mimicking or antagonistic effects in aquatic animals like fishes (Pandian, 2015), crustaceans (Pandian, 2016) and molluscs (Pandian, 2017). Being primitive deuterostomes, the endocrine sex differentiation process of echinoderms resembles that of vertebrates. However, they provide a unique opportunity to recognize the ED effects on regeneration. There are a few publications on ED effects in arm regeneration of crinoids and ophiuroid and many on spermio- and embryo-toxicity of asteroids and echinoids. Surprisingly, no publication is yet available on ED effects in any holothuroids, despite the fact that they incessantly consume ED-concentrated sediments.

Concentration and metabolization: Lavado et al. (2006b) exposed *Antedon mediterranea* to androgenic (17α-methyltestosterone, MT, phenyltin, TPT) and anti-androgenic (DDE, cyproterone acetate, CPA and fenarinol, FEN) compounds and estimated the effects at the highest effective dose on T and E_2. The effective doses, at which T was increased to the maximum following 2 weeks exposure, were 30 and 1,000 ng/l for CPA and MT, respectively (Table 7.2). At these doses, accumulation of T was 180 and 200%. Maximal accumulation of E_2 occurred at the doses of 10 and 240 ng/l for MT and FEN. Remarkably, the longer duration (4 weeks) of exposure to DDE (from 140 to 360% of T) and MT (280 to 740% E_2) concentrated the levels of T and E_2, respectively. The crinoid metabolized and reduced the residual level of from 100 to 45% of E_2 and from 180% to 110% of T in those exposed to TPT and CPA, respectively. In all, MT is the most potent androgen that induces 200–230% concentration of T and 280–740% concentration of E_2. With extended duration of exposure, DDE and MT concentrated T and E_2 levels but CPA and TPT reduced the residuals of T and E_2.

TABLE 7.2

Changes in levels of testosterone (T) and 17β estradiol (E) in the whole body of *Antedon mediterranea* following 2 or 4 weeks exposure at the highest effective doses of endocrine disruptors. All are approximate percentage values (from Lavado et al., 2006a)

Disruptors	Dose (ng/l)	Levels (%) After Exposure To	
		2 wk	4 wk
Testosterone			
Fenarimol (FEN)	2400	140	180▲
17α methyltestosterone (MT)	1000	200	230▲
p,p' DDE	100	140	360▲
Triphenyltin (TPT)	225	160	160
Cyproterone acetate (CPA)	30	180	110▼
Estradiol			
17α methyltestosterone (MT)	10	280	740▲
Cyproterone acetate (CPA)	30	120	300▲
p,p' DDE	500	110	120▲
Triphenyltin (TPT)	100	100	45▼
Fenarimol (FEN)	240	750	250▼

Arm regeneration: Involving the crucial cell proliferation, migration and differentiation of tissues/organs, regenerating echinoderms serve as a more sensitive bioindicator of ED-induced stress. In European waters, bisphenyl level ranges from 258 to 490 ng per l (see Pandian, 2016). Even the exposure at the low dose of 14 ng PCBs/l, Arochor[R] 1260 induced abnormal arm growth in terms of morphology and anatomy of *A. mediterranea* (Candia-Carnevali et al., 2001). Extending this study, Barbaglio et al. (2006) exposed the crinoid to 50–1,000 ng/l FEN for 3 days to 2 weeks or to ~ 0.025 mg/l TPT and inferred the following: 1. Blastema formation was eliminated by non-blastemal myocytes migrating from distal muscles. 2. The repairing granulocytes were degranulated, especially in the TPT exposure series. 3. Briefly, (a) both the EDs affected growth and development by interfering with the same basic cellular mechanisms of regeneration involving cell proliferation, migration and dedifferentiation. (b) They modified the timing, modalities and pattern of regeneration. Exposing regenerating *Ophioderma brevispina* to 0.1 µg/l tributyltin oxides, Walsh et al. (1986) reported that being neurotoxicants, the oxides inhibited regeneration. Possibly, tributyltins are the most potent inhibitors of regenerative process in echinoderms.

PCBs enter the asteroid both through surrounding water and food. They are not readily metabolizable and hence are bioconcentrated. In an interesting study, den Besten et al. (1990) investigated the effect of the food chain, through which PCBs are bioconcentrated in *Asterias rubens*. The star fish was fed on mussels, which were themselves feeding on algae grown in PCB-contaminated water. Hence, the mussels contained 20 µg PCBs/g lipids. Exposure to PCB-contaminated water and feeding PCB-contaminated mussels to *A. rubens* for 5–7 months resulted in the peak GI of 20% in female, instead of 27% in control. Simultaneously, PCI was reduced from 20 to 10%, in comparison to 20 to 8% in control. Feeding the PCB-contaminated mussel led to the residual PCB level of 43 and 45% in the ovaries and pyloric caeca (PC) of females. These values were 23 and 60% in testes and PC. Being a carnivore the star fish bioconcentrated the PCB in its gonad and PC. Unlike PCBs, cadmium (Cd) enters *A. rubens* through surrounding water but not via food. Hence Voogt et al. (1987) studied the effect of Cd on steroid metabolism in the star fish. Cd stimulated 17β-hsd and thereby increased T biosynthesis. As a result, T levels were elevated in the ovaries and testes as well as PC. Notably, the elevation was seven and nine times higher in the testes and ovaries of star fish exposed to 0.2 µg per l PCB for 5 months (Table 7.3).

Other investigations have highlighted spermiotoxicity and embryotoxicity in asteroids and echinoids. In *Astropecten aranciacus* oocytes, exposure to insecticides delayed germinal vesicle breakdown by 30 minutes. Echinoid development is known for its auto-regulation. Hence, it is interesting to know whether the echinoids can also auto-regulate the development on exposure to one or other ED. Available information listed in Table 7.3 indicates the inability of echinoids to autoregulate the development, when exposed

TABLE 7.3

Spermiotoxicity and embryotoxicity in asteroids and echinoids

Species, ED, Reference	Reported Observations
Asteroids	
Asterias rubens 0.6 µg/l PCBs for 5 months den Besten et al. (1989, 1991)	Elevation of PCB levels to 7 and 9 times higher in testes and ovaries, respectively. Abnormal development of oocytes and embryos
A. rubens, 0.25 µg/l Cd for 5 months den Besten et al. (1989)	Elevation of Cd levels to 17 and 50 times higher in testes and ovaries, respectively. Oocyte maturation was delayed but no abnormal fertilization
Astropecten aranciacus Methychlor (MXC), dieldrin, lindane, Picard et al. (2003)	Germinal vesicle breakdown (GVBD) delayed by 30 minutes. Formation of mitotic spindle and extrusion of polar body altered
Echinoids	
Paracentrotus lividus Triphenyltin (TPT) Moschino and Marin (2002)	At 10 µg/l, sperm fertilizability reduced to 45%. At 10 and 5 µg/l, gastrulation was arrested and no pluteus produced. Sperm are more sensitive than eggs and embryos
P. lividus, Methychlor (MXC), dieldrin, lindane, Pesando et al. (2004)	Fertilization of egg was not affected but polyspermy increased to > 70% by MXC even at sperm dilution of 10^2. 8-minute exposure to lindane reduced fertilization success to 0%. At 100 µM exposure, lindane delayed or arrested first mitotic division. MXC is the most potent of all, as it produced deformed embryos alone
Lytechinus pictus Chlorpropham (herbicide), Holy (1998)	Exposure at 1000 µM resulted in unequal cell division at atypical times. Abnormal formation of skeletal spicules
P. lividus, Tributyltin, triphenyltin, Novelli et al. (2002)	At 1.0–2.6 µg/l tributyltin was more toxic to sperm. At 3–18 µg/l triphenyltin was more toxic to embryos
P. lividus, Mono-(MBT), di-(DBT) and tri-(TPT) butyltin, Marin et al. (2000)	Dose-dependent reductions in development, which was totally (100%) arrested at 2 and 200 µg/l TPT and DBT, respectively. But even at 3 mg/l, MBT arrested development in 18% embryos only
P. lividus, Pentachlorophenol, Oetken et al. (2004)	At 0.1–0.8 mg/l, pentachlorophenol altered embryonic development and differentiation process

to one or other insecticides and one or other form of butyltin. In general, fertilization is not affected by exposure to insecticides, but fertilizablity of sperm is dramatically reduced to < 45% on exposure to 10 µg per l TPT. Among insecticides, methychlor is the most potent, as it exposure results in 100% deformed plutei. The potency of butyltin is decreased in the following order: tributyltin < dibutyltin < monobutyltin.

7.5 Parasitic Disruption

In aquatic invertebrates, parasitic disruption of sex differentiation includes (i) partial or complete sterility in molluscs (e.g. *Bulinus globosa* infected by *Schistosoma hematobium*, see Pandian, 2017) and crustaceans (e.g. *Carcinus maenus* infected by *Sacculina carcini*, see Pandian, 2016), (ii) morphological and anatomical deformation (e.g. lernaepodid and monstrilloid copepods, isopods, see Pandian, 2016) as well as (iii) sex change (e.g. *Ione thoracica*, see Pandian, 2016). In echinoderms, parasitic disruption *per se* includes also (i) sterility either partial (e.g. echinoids infected by gastropod *Robillaria crenica*, see Jangoux, 1990) or complete (e.g. *Entocolax* spp castrating their holothurian hosts, see Jangoux, 1990) and (ii) suppression of asexual reproduction. For more details, Jangoux (1990) may be consulted.

Infection by ascothoracid (cirriped) endoparasite *Dendrogaster okadai* almost totally suppressed asexual reproduction by fission in asteroid *Coscinasterias acutispina* (Haramoto et al., 2007). Similarly, the endoparasitic gastropod *Stylifer linckae* inhibited asexual reproduction by clonal autotomy in *Linckia multifora*. The inhibition was so intense that out of 2,570 infected star fish, only one reproduced asexually (Davis, 1967). As parasitic inhibition of asexual reproduction is already described (see p 111, 125), the ensuing account is limited to parasitic role on partial sterility. However, it must be indicated that parasitic castration is not common among echinoderms (Jangoux, 1990), despite its widespread occurrence in other aquatic invertebrate phyla.

Infected echinoids suffer from partial sterility by the gnathosomatid nematode *Echinocephalus pseudouncinatus* and the mermithid nematode *Echinomermella matsi*. *E. pseudouncinatus* juveniles encyst in the gonad tubules of *Centrostephanus coronatus* and thereby suppress gametogenesis in that part of the host's gonads (Pearse and Timm, 1971). *E. matsi* is a large (> 60 cm long females) nematode and may exceed 10% live body weight of the host with a load of 10/host (Hagen, 1987, 1992). As an example, this account summarizes the incidence and load of *E. matsi* on the reproductive performance of infected and non-infected *Strongylocentrotus droebachiensis* in the Norwegian coast (Hagen, 1996).

In Norway, *S. droebachiensis* grows to a maximum size of 75 mm (test diameter). The urchins were infected at 15 mm size onwards. With increasing body size, abundance of lightly and heavily infected urchins increased up to 35 mm size. Subsequently, it decreased (Fig. 7.3A). Larger urchins of > 65 mm sizes were not available, indicating that they were killed by the parasite. In fact, the declining trend of the non-infected urchins beyond 35 mm size also confirm that more and more larger urchins were dead due to infection. With increasing body size, the gonad weight decreased with increasing parasitic

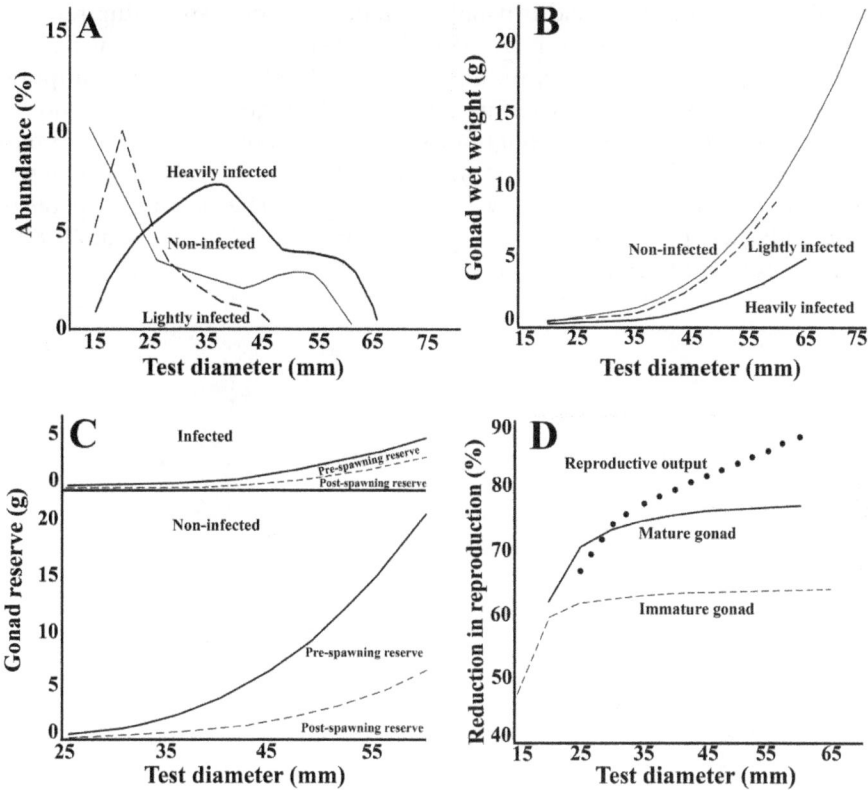

FIGURE 7.3

A. Abundance of nematode parasite *Echinomermella matsi* as a function of body size of green sea urchin *Strongylocentrotus droebachiensis*, B. Relation of gonad weight to body size in infected and non-infected green urchin, C. Gonad reserve in pre-spawning and post-spawning green urchin in relation to body size, D. Reduction in reproduction in infected sea urchin of different body size (simplified, modified from Hagen, 1996, reproductive output data are from Stien, 1993).

load from 1 to 6 g but the information on the relationship was marred by the widely scattered data. Nevertheless, clear decreasing trends became apparent for the gonad weight-body relationships in lightly and heavily infected urchins collected from two sampling sites at Godoystraumen and Vaeroy (Fig. 7.3B). The quantity of gonad consumed by the parasite was so high that in the absence of gonads, sex of 95% of the infected urchins could not be identified. The percentage of such undifferentiated urchins began to decrease with increasing body size but beyond 55 mm size, all of them were undifferentiated.

As echinoids have very little muscle tissues but have more NPs, their gonads serve the dual functions of nutrient storage and reproduction. A certain portion of the gonad is usually retained as a post-spawning reserve. Figure 7.3C shows that the infected urchins have a lower nutrient reserve as well as greatly reduced reproductive output. Briefly, due to consumption of the gonad, reductions in immature and mature gonad mass amount to 60 and 75%, respectively throughout the range of body size up to 65 mm size. Consequently, the reduction in reproductive output remains between 60 to 88% with increasing body size (Fig. 7.3D). The nematode parasite *E. masti* inflicts partial (60%) sterility in urchins of < 30 mm size but > 80% sterility in larger (> 50 mm) urchins.

Section Ib
Hemichordata

8

Reproductive Biology

Introduction

The hemichordates are a relatively small taxon of invertebrate deuterostomes. With ~ 130 species, they comprise the solitary vermiform enteropneustic acorn worms (< 110 species measuring 0.5 mm to > 2 m, e.g. *Balanoglossus gigas*) and sessile colonial pterobranchs (~ 25 species). The former includes the obligate gonochoric harrimaniids and ptychodorids that undergo direct and indirect development, respectively (Fig. 8.1). Their body is divided into protosome, mesosome and metasome that correspond to the proboscis, collar and trunk (Hyman, 1959). Pterobranchs include cephalodiscids and rahpdopleurids; both of them are heterogonic, i.e. reproduce sexually and asexually. They bear one or more conspicuous pairs of tentaculous arms in the mesosome. They inhabit hard benthic substrates, since their tubes build encrusting aggregates on the rocks and shells, whereas enteropneusts are found in burrows of sandy beaches. The body size of an individual pterobranch zooid measures from 1 to 5 mm.

8.1 Life Cycles

With the largest eggs, the harrimaniids display the simplest direct life cycle (Fig. 8.1A). The ptychodorids generate more number of relatively small eggs and pass through tornaria larval stage (Fig. 8.1B). For more details on their indirect life cycle, Urata and Yamaguchi (2004) may be consulted. The elaborate review by Kaul-Strehlow and Rottinger (2015) provides more details including the expression of a number of genes that regulate early development. Being colonials, both cephalodiscids and rhabdopleurids generate eggs and sperm from the constituent female and male zooids. Following fertilization, the yolky egg also passes through an entoproct-

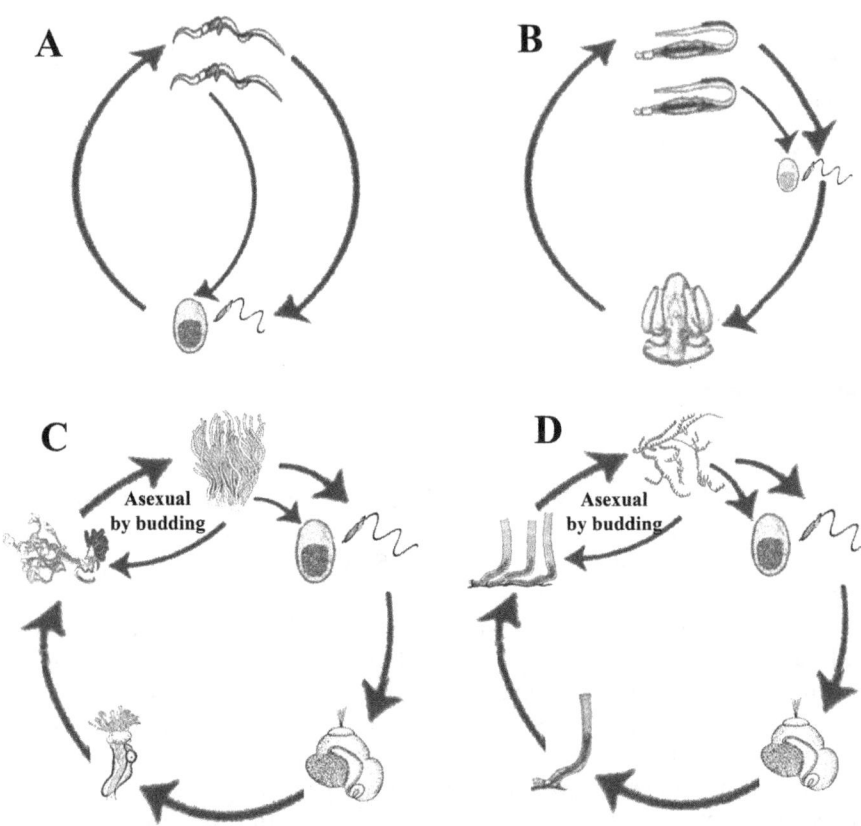

FIGURE 8.1

Life cycles of gonochoric solitary and colonial hemichordates. A and B show the direct and indirect (with tornaria larva) life cycles in harrimaniids (*Balanoglossus*) and ptychoderids (*Ptychodera*), respectively. From pterobranchid (*Cephalodiscus*) and rhabdopleuran (*Rhabdopleura*) colonies, female and male zooids release eggs and sperm (C, D). After passing through an ectoproct type larval stage (see Hyman, 1959), parent individuals are generated. Asexual reproduction by budding results in formation of vertically and horizontally dispersed colonies in pterobranchids and rhabdopleurans, respectively.

like larva (see Hyman, 1959), which metamorphoses to establish the parent individual (Fig. 8.1C, D) and the entire colony arise from this single parent that reproduces asexually (see Kaul-Strehlow and Rottinger, 2015). In the cephalodiscids, 1–14 buds arise from the base to clump the zooids vertically. The asexually produced zooids secrete their own external collagenous tube, the coenecium and remain attached to the parent zooid with or without physical link. Contrastingly, budding in rhabdopleurids occurs along the stolons that link individual zooid to one another, simulating those of colonial tunicates. The differences in the budding pattern in these two sub groups result in zooids being vertically clumped in cephalodiscids but horizontal dispersion along the stolons in rhabdopleurids (Rychel and Swalla, 2009). A stolon consists of an outer epithelium enclosing a pair of cavities lined by peritoneum and separated by a medium mesentery. The young bud is simply a hollow evagination of the stolon consisting of the coelomic cavities and median mesentery (Hyman, 1959).

8.2 Gonads and Consequences

In gonochoric solitary enteropneusts, sex is indistinguishable externally. In *Ptychotera flava*, males may, however, be identified by brown flecks on the genital wings (Rao, 1954). Genital wings of *Balanoglossus misakiensis* are filled with grayish-pink ovaries and whitish–yellow pouches of sperm (Urata and Yamaguchi, 2004). In the colonial *Cephalodiscus hodgsoni*, female zooids are red or reddish with 12 arms and the male zooids pale brown with 10–11 arms. In *C. sibogae*, males have only two arms bearing no tentacles (Hyman, 1959).

A series of paired gonads are located between post-branchial and pre-hepatic trunk of enteropneusts (Fig. 8.2). But the number of gonad is drastically reduced in the colonials. For example, it is reduced to a single symmetric pair located in the anterodorsal part of the trunk in front of the stomach of cephalodiscids (Fig. 8.2). It is further reduced to a single gonad located on the right metacoel in rhabdopleurans. The ovary is round but the testis is elongated.

In enteropneusts, sex ratio seems to be skewed in favor of females. For example, it is 0.98 ♀ : 0.02 ♂ in *P. flava* (Rao, 1954). In the colonials, the female ratio is preciptously reduced on two counts. Firstly, 83–90% of the zooids remain as immature neuters. Secondly, the ratio is reduced to unity and less. For example, sea ratio is 0.09 ♀ : 0.08 ♂ : 0.83 neuter in 20 colonies of *Rhabdopleura compacta* (see Stebbing, 1970). Of 300 rhabdopleuran colonies examined, 25 were males, three females and the remaining were also neuters, i.e. sex ratio is 0.01 ♀ : 0.08 ♂ : 0.91 neuters (Hyman, 1959).

8.3 Regeneration in Enteropneusts

Fortunately, the limited available publications are spread over harrimaniid (e.g. *Saccoglossus kowalevskii*, Tweedell, 1961) and ptychodorid (e.g. *Ptychotera flava*, Rao, 1954) as well as different fractions of the proboscis, collar and trunk. The credit for precise description of the entire sequence of regeneration

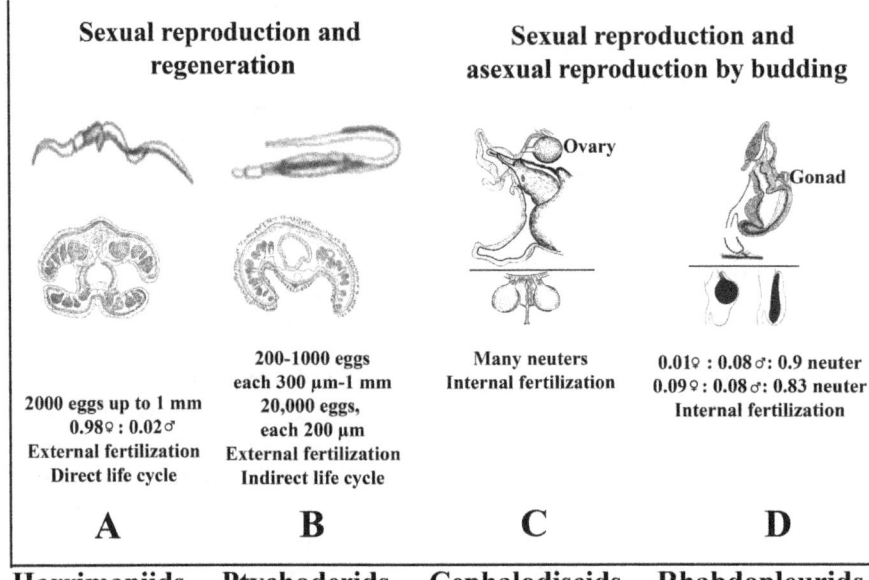

FIGURE 8.2

Some features of reproduction in hemichordates taxons. A and B are cross sections of ovaries of enteropneust and pterobranch. C and D are longitudinal sections of cephalodiscid and rhabdopleurid. In C and D, the lower ones show the appearance of paired ovaries in C and median ovary or testis in D. Note the progressive decrease in the number of gonads from solitary to colonial hemichordates (free hand drawings from Hyman, 1959, Information pooled from different sources).

involving coelomocytes and mesenchymal cells must go to K. Pampapathi Rao of Madras University, India. In anesthetized *S. kowalevskii*, Tweedell (1961) amputated at mid-proboscis, mid-collar and pre- and post-branchial trunk. In the first two, i.e. mid-proboscis and mid-collar, the anteriors survived for sometime but failed to regenerate the missing posterior fractions (Table 8.1). The posterior trunk fraction, however, small or large also failed to regenerate. But the anteriors successfully regenerated the missing pre-branchial and post-branchial trunk, albeit at very low frequency of 5–6%. The experiments of Tweedell have demonstrated that unlike the bidirectional regeneration of echinoderms, regeneration in hemichordates is unidirectional, i.e. only one half regenerates. The regenerative potential of enteropneusts is very low and not comparable to that of echinoderms. Incidentally, there are claims on the possible bidirectional asexual reproduction in a few enteropneust species. However, convincing evidence to substantiate this claim (Packard, 1968, Petersen and Ditadi, 1971) is yet to to be made.

Using molecular techniques, Rycell and Swalla (2009) confirmed the precise observations of Rao (1955). Accordingly, following amputation at the

TABLE 8.1

Survival and regeneration of different fractions of amputated *Saccoglossus kowalevskii* (simplified from Tweedell, 1961)

Amputation at	Amputated (no.)	Anterior (no.)		Posterior (no.)	
		Survival	Regeneration	Survival	Regeneration
	33	24	0	23	21
	58	48	0	9	3
	35	29	0	4	2
	36	32	22	27	0

post-hepatic trunk, the entire posterior was regenerated inclusive of an anus. Following amptutation between the anterior and post-branchial trunk, the newly formed stump first developed the distal proboscis and then the collar. Mesenchymal cells or coelomic cells from the coelomic fluid of the trunk region migrated to the wound site. Within 2 days, the wound was healed. It was followed by blastema formation. On the 6th day, (i) the buccal diverticula originated from the forward evagination of the digestive epithelium and then (ii) the mouth is formed. Between 6th and 8th day, the stomatochord was formed by evagnation of endoderm. The gill apparatus also arose from the digestive tube. The collar-cord was developed from the thickneming (cell proliferation) of the nervous layer. During this period, mesenchymal cells, with participation of ectoderm and endoderm, developed the proboscis and heart-kidney complex. Incidentally, it is not yet known whether the mesenchymal cells are an undifferentiated mass of stem cells or whether they dedifferentiate into stem cells, once the wound was healed (Rycell and Swalla, 2009).

Section II
Chordate Deuterostomia

9

Cephalochordata

Introduction

The phylum Chordata includes three subphyla Cephalochordata, Urochordata and Vertebrata. The first two subphyla are collectively called Prochordata and belong to the primitive deuterostomes. Though this book series is related to reproduction and development in aquatic invertebrates, an account on the subject of prochordates is also provided in the ensuing two chapters to (i) compare their regenerative and clonal potentials with those of echinoderms and (ii) link the subject with another series of five volumes devoted to that of aquatic vertebrate teleostean fishes (Pandian, 2010, 2011, 2012, 2014, 2015).

9.1 Reproductive Biology

The cephalochordates constitute the smallest chordate subphyla and comprise 35 species belonging to three genera: *Branchiostoma*, *Epigonichthys* and *Asymmetron* (Bertrand and Escriva, 2011). Being benthic, suspension-feeders, they burrow into coarse sand particles in shallow coasts from 0.5–40 m depth. However, *Asymmetron inferum* has been found at 229 m depth (Kon et al., 2007). They are spread over tropical and temperate waters around the world (Azariah, 1994).

Most lanceolates are gonochorics; however, anomalous occurrence of simultaneous hermaphroditic amphioxus species is reported (e.g. Chen, 1931). In gonochorics, sex ratio is male-biased, as in *Branchiostomma lanceolatum* (Azariah, 1994). The two series of symmetric gonads are located ventrolaterally in *Branchiostoma* but *Asymmetron* develops gonads on only the right side of the body (Bertrand and Escriva, 2011). In a 25 mm long *B. lanceolatum*, Azariah (1969) has counted as many as 22 ovarian pairs; the number of ova in each ovarian pouch ranges from 10 to 66, the maximum at the center and the minimum at either end with total fecundity of ~ 1,150 eggs. Besides, plankton samples of lanceolate larvae may also serve as an index of fecundity. More than 37,000 larvae per haul from the northwest African coast have been collected (Gosselck and Kuehner, 1973). From the Cape Blanc (northwest Africa), Flood et al. (1976) have reported the presence of 40,000 larvae/m^2; considering the translocation speed of 0.1 m/second over a distance of 110 km during the spawning season lasting for 4 months, Flood et al. have estimated an amazing value of 1.1 million ton larval production by *B. senegalense*. Another value reported for *B. nigeriense* was 27,000 million ton (Gosselck, 1975).

Interestingly, a single spawning season lasting for 2–3 months during spring is reported for *B. lanceolatum* (Fuentes et al., 2007) and *B. belcheri* (Chen et al., 2008) from temperate waters. In tropical waters of the Madras coast, *B. lanceolatum* spawns twice a year, the first one during June–July (Southwest monsoon) and November to January (Northeast monsoon) (Azariah, 1994). Spawning can also be induced in ripe lanceolates by electrical stimulation (e.g. *B. floridae*, Holland and Holland, 1989). The planktotrophic larval duration may last for 2–3 weeks in tropical waters but for 2–3 months in temperate species. Interestingly, the larvae are asymmetrical with the mouth on the left side and gills on the right side. On metamorphosis, bilateral symmetry is restored. It may be recalled that the reverse is true of echinoderms, i.e. the bilaterally symmetrical echinoderm larvae following metamorphosis assume radial symmetry.

Captive culture from egg to egg stage has been achieved only in *B. belcheri* (Yasui et al., 2007). In temperate lanceolates, life span ranges from 2–3 years in *B. floridae* (Futch and Diwineil, 1977) and *B. belcheri* (Chen et al., 2008) to 8 years in *B. lanceolatum* (Desdevises et al., 2011). Reports on regeneration or clonal reproduction are not available. Apparently, the cephalochordates seem to have lost this potential. In fact, tunicates and not cephalochordates are the closest living relatives of vertebrates (Delsuc et al., 2006).

10

Urochordata

Introduction

The subphylum Urochordata (Tunicata) consists of three major classes namely the benthic Ascidiacea and pelagic Larvacea (Appendicularia) and Thaliacea, which includes Salpidae and Doliolidae. Unlike cephalochordates, the tunicates are speciose with ~ 2,300 (Skold et al., 2009) ~ 3,000 (Manni et al., 2007) species. Owing to their spectacular potential for cloning and coloniality, they have attracted much attention and the accumulated publications have formed the base for many reviews (e.g. Cloney, 1990) and a few books (e.g. Bone, 1998). However, it must be indicated that contributions on cloning in pelagic tunicates by planktonologists and those on benthic ascidians by molecular biologists have remained isolated, except for a rare passing remark (e.g. Skold et al., 2009). One objective of this account is to bridge them together and to find whether these spatially separated tunicates share common stem cells to achieve cloning and coloniality.

10.1 Pelagic Tunicates

Their tunics being primarily made up of acidic mucopolysaccharides, pelagic tunicates are highly transparent with a high water content of ~ 95% of live weight (Acuna, 2001). Consequently, their specific gravity is 1.024 g/ml at 22°C and is slightly greater than sea water (Tsukamoto et al., 2009). Hence, they are the most buoyant of all zooplankton (Deibel and Lowen, 2012). They are exclusively marine holoplankton and efficient suspension-feeders; using mucous filters with very fine pores, they are capable of acquiring food particles of wide size range. For example, *Oikopleura dioica* can filter live and dead organic particles spanning several orders of magnitude in size down to 0.2 μm (Flood and Deibel, 1998). With absolute fecundity ranging from

~ 0.1 to 1.9 eggs per day, generation time 2 to 50 days (cf 5.5 days life span of *Miona mucrorura*, see Pandian, 2016) and population doubling time 8 hours to 1 week, the pelagic tunicates build massive population density, especially immediately following phytoplankton bloom. Working with a 23-year dataset, Menard et al. (1998) established that doliolid peaks are coincident with phytoplankton blooms. Maximum pelagic tunicate density ranges from 450/m³ in *Doliolum gegenbauri* (He et al., 1988) to 53,000/m³ in *O. dioica* (Uye and Ichino, 1995). Being the first animal taxon to arrive in massive numbers immediately following bloom, the tunicates have received much attention by planktonologists. Table 10.1 lists some common characteristics shared by all three or any two of the pelagic tunicates. Briefly, the shortest generation time and population doubling time resulting from clonal reproduction has led to rapid build up of population density. However, low fecundity, semelparity and clonal reproduction in pelagic tunicates have limited the scope for genetic diversity and speciation to < 150 species. Semelparity, for example, has limited the species number to 761 in gonochoric cephalopods and 2,000 in hermaphroditic ophisthobranchs (see Pandian, 2017).

Being inhabitants of the oligotrophic oceanic surface waters, pelagic tunicates are exposed to relatively wider variations in algal density and temperature. Hence, it is advantageous for them to reduce generation time, reproduce early and sustain the population level. The best option for them is to maximize reproductive output in a single event (semelparity) at the shortest generation time. Incidentally, generation time of pelagic thaliaceans is the time taken from a birth of a female blastozooids to the birth of the next generation female blastozooids (Heron, 1972), which is different from other animals. In animals, generation time is considered as a time required from

TABLE 10.1

Common and other reproductive features of pelagic tunicates

Parameter	Larvacea	Salpidae	Doliolidae
Habitat	Open ocean	Open ocean	Open ocean
Sexuality	Protandric	Protogynic	Protogynic
	Semelparous	Semelparous	Semelparous
Generation time	Short	Short	Short
Fertilization	External	Internal	Internal
Development	Oviparous	Viviparous	Viviparous
	Direct	Direct	Indirect
Asexual	No	Yes	Yes
Life cycle	Solitary	Sexual solitary alternating asexual colonial	Sexual solitary alternating asexual colonial
Zooids	Monomorphic	Monomorphic	Polymorphic
Fecundity	Sexual > 300 eggs	Sexual 1 egg	Sexual 2–6 eggs
Species (no.)	70		75

hatching to puberty (e.g. crustaceans, Pandian, 2016). In pelagic tunicates, semelparity imposes one of the two alternate reproductive strategies, i.e. either to manipulate generation time or clutch size. According to Deibel and Lowen (2012), the clutch manipulators like the larvacea have a fixed generation time but their fecundity is determined by body size, algal density and temperature. For example, absolute (life time) fecundity of *O. dioica* increases from ~ 100 to 400 with increasing body size from 1,200 to 1,600 µm (Lombard et al., 2009a, b). Likewise, body size attained at the termination of generation time depends on temperature and food density. At algal density of 100 µg C/l, maximum size attained at the termination of generation time can be 1,200 µm body length at 14°C and 600 µm at 28°C. At the same density of 100 µg C/l, fecundity is ~ 200 eggs at 14°C but just 75 at 28°C (see Fig. 3b, c of Deibel and Lowen, 2012). On the other hand, time manipulators like the thaliaceans have a fixed fecundity but the generation time is determined by algal density and temperature. For example, generation time of *Thalia democratica* is 2.5 days in summer but ~ 14 days in winter (Heron and Benham, 1985). However, there are also contradictory reports. It is 6.6 and 7.2 days in winter at Eden and Sydney for *T. democratica* within Australia but is 9.8 and 17.4 days in summer (Henschke et al., 2015).

Larvacea: Despite the small number of 75 species, larvacea are the second most abundant zooplankton taxon with worldover distribution in oceanic waters (Ganot et al., 2007). With a single exception of gonochoric *O. dioica*, larvacea are protandric hermaphrodites. From a single ovotestis, male and female gametes are more or less simultaneously developed from the juxtaposed compartments. In *O. longicauda*, sequential milting and spawning with an interval of 30 minutes (Miller and Cosson, 1997) probably precludes self-fertilization. Their density of > 500 no./m² may ensure high percentage of fertilization success. In them, indirect development involves a less differentiated tadpole larva (Fig. 10.1A). For details on embryonic and larval development, Galt and Fenaux (1990) may be consulted. However, it may be added that many studies cited by Galt and Fenaux are concerned with tracing the cell lineage of ectoderm and nerve cord, endoderm and notochord, muscles and other organs but almost nothing is mentioned about the origin and development of the gonad.

The credit for culturing a larvacea through numerous generations goes to Paffenhofer (1973). In larvacea, the germline of unknown origin becomes apparent at the trunk-tail junction after metamorphosis (Ganot et al., 2007). The 3rd day following metamorphosis in *O. dioica*, oogenesis commences with a founder oogonial cell or cystoblast undergoing repeated mitotic divisions followed by incomplete cytokinesis and forms interconnected germ cells called the germline cyst. The cysts appear as a cluster of germ cells and are also called nests. Thus, the entire female germline is contained in a unique giant multi-nucleate cell, 'coenocyst'. Within this 'one germ cell', two populations of 10^3–10^4 nuclei co-exist. One population provides meiotic

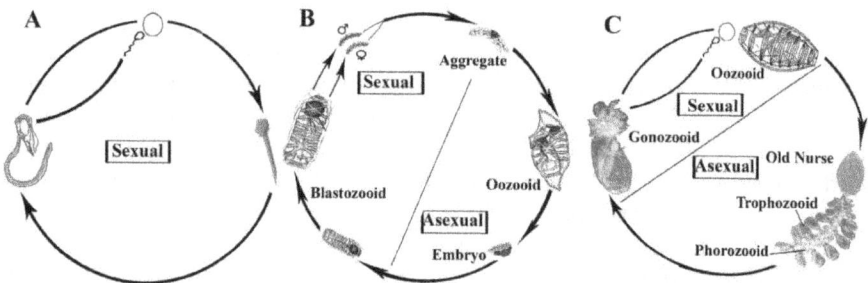

FIGURE 10.1

Life cycle of pelagic tunicates. A. Larvacea: Sexual only; oviparous and indirect life cycle involves less differentiated tadpole (compiled from Google, modified). B, C. Thaliacea: Metagenics, both sexual and asexual are viviparous. B. Salpidae: Sexual solitary blastozooids alternates with asexual oozooid (compiled from Foxon, 1966, Casarets and Nemoto, 1986, Nishikawa and Tsuda, 2001, Loeb and Santara, 2011), C. Doliacea: Sexual solitary gonozooid alternates with asexual old nurse consisting of trophozooids and phorozooids (compiled from Paffenhofer and Koster, 2011 and others).

nuclei for the future oocytes (2 μm diameter each), while the other enters endocycles and becomes increasingly polyploid (100 μm diameter each) to serve as a nutrient source (like the nutritive phagocytes of echinoids) for maturation of oocytes. For the germline marker *vasa* mRNA, the uniform staining indicates the common origin of the two different nuclear populations. To distinguish the homogenous nuclear population in a syncytium from the heterogenous nuclei enclosed in a common cytoplasm, the term coenocyst has been introduced. This novel cellular architecture of multiple nuclei of germline origin that have distinct differentiation fate, is enclosed in the unique cellular compartments of coenocyst (Ganot et al., 2007). Notably, the meiotic nuclei even at this stage are linked to the common cytoplasm via a ring canal.

Between 4th and 6th day following metamorphosis, a sub-set of the precursor oocytes undergo complete cellularization by equally partitioned common cytoplasm being transferred through ring canals. The number of oocytes selected for maturation depends on accumulated resources in the coenocyst (Troedsson et al., 2002). The coenocyst organization is a conserved feature of larvacean oogenesis, allowing efficient numerical adjustment of oocyte production. Interestingly, the ovary namely coenocyst consists of both gametic cells and nutritive polyploid cells in close proximity within a single organ and thereby reduces the cost of translocation of nutrients from a storage organ and hormone levels (see also Table 7.1). This structural organization is similar to that of the echinoid gonad, in which both gametic cells and Nutritive Phagocytes (NPs) are accommodated within the gonad. Notably, both echinoids and larvacea do not clonally reproduce, while their closely related taxa do it.

Larvacea secrete a mucous feeding apparatus called 'the house' with the system of channels and filters, through which the animal pumps water with its tail and thereby filter pico- and nano planktons from the surrounding water (Galt and Fenaux, 1990). The house is discarded and renewed at an average of 16 times per day (see Ganot et al., 2007). The carbon cost of such renewal of the house amounts to two-three times the body carbon (Sato et al., 2001). But it is an adaptive strategy to lower food levels and has enabled them to colonize large areas of oligotrophic oceanic waters (Troedsson et al., 2002). From laboratory reared *O. dioica*, Troedsson et al. (2002) recognized four phases in its life cycle. During the first two, representing periods of somatic growth, *O. dioica* is dependent on temperature but is non-responsive to increasing algal density beyond a threshold level required for survival. Phase 3 is dependent on both temperature and algal density. The differential growth between standard and limited supply of algal food is entirely allocated to reproductive growth. During phase 4, spawning and death occur. At a given temperature, limited food supply reduces fecundity but not the spawning duration and generation time.

Salpidae: Salps are strictly metagenic tunicates, i.e. they obligately alternate between sexually reproducing blastozooids and asexually budding solitary oozoids (Fig. 10.1B) (Henschke et al., 2015). The blastozoid-to-oozoid ratio is used as an index to determine the stage of swarm formation (e.g. Loeb and Santora, 2012). The oozoids are sterile but the clonally produced protogynic hermaphroditic blastozooids have a single ovotestis. Sex ratio of relatively longer/older *Thalia democratica* is 0.93 ♀ : 0.06 ♂ (Henschke et al., 2015). Both blastozooids and oozoids are viviparous. In an oozoid, cells, which become apparent in the mid-ventral line between the posterior part of the endostyle and anterior part of the tightly looped gut, begin to proliferate and give rise to a tube-like stolon. The stolon has a central endodermal canal, a nerve tube, lateral muscle bands and a strand of gonadal tissue (Lacalli, 1999). As the oozoid grows, the stolon begins to segment forming a single row of aggregate/buds. Subsequently, this aggregate is rearranged as a double row that extends from a 'deployment point' producing a single aggregate's block consisting of clonal individuals of the same size, age and genetic make-up. Subsequent development results in the formation of additional blocks. Consequently, the fully developed stolon has three blocks and is extended from the deployment point to the terminals, which hold fully differentiated juvenile blastozooids. On being released, the female blastozooid with a single egg is fertilized by an older male blastozooid (Miller and Cosson, 1997). From the fertilized eggs, the oozoid is parturated.

From descriptions of the stolon structure by Lacalli (1999) and clones arising from the stolon in three successive blocks of *Iasis zonaria* by Daponte et al. (2013), the following may be inferred: 1. In the proximal unsegmented region, the stolon is already in possession of all the important tissue types and the bud/aggregate represents only an evagination of the clonal primordium. During their passage from the proximal unsegmented to the distal segmented zone of the stolon, almost all the organs/systems including an ovotestis with oocytes are all differentiated. Understandably, during this passage, the 'stemness' of the pluripotent/multipotent stem cells is progressively reduced to oligo/unipotency. The stolon deserves more attention of molecular biologists interested in stem cell research. 2. With increasing body size of the oozoid in *I. zonaria*, the first, second and third blocks are formed at the body length of 51, 57 and 61 mm and the number of buds per block also increases with increasing body size in the first, second and third, *albeit* at decreased levels (Fig. 10.2). In all, each oozoid can clonally produce three-four blocks and as many as 300–400 blastozooids (Daponte et al., 2013). The clonal potential ranges from 32–112 buds in a small neretic *Thalia democratica* (Braconnot, 1963) to 800 in the Antarctic *Salpa thompsoni* (Daponte et al., 2001). 3. At the maximum, one or two oocytes mature in *I. zonaria*. Through sexual reproduction, one-two zooids alone are generated but 400 through asexual reproductions. Hence, the prodigeous asexual reproduction adds considerably to the population density. This is also true of *T. democratica* (Heron and Benham, 1985).

Doliolidae: The life cycle of doliolids resembles that of salps. But the asexually produced zooids are polymorphic. According to Braconnot (1970), the

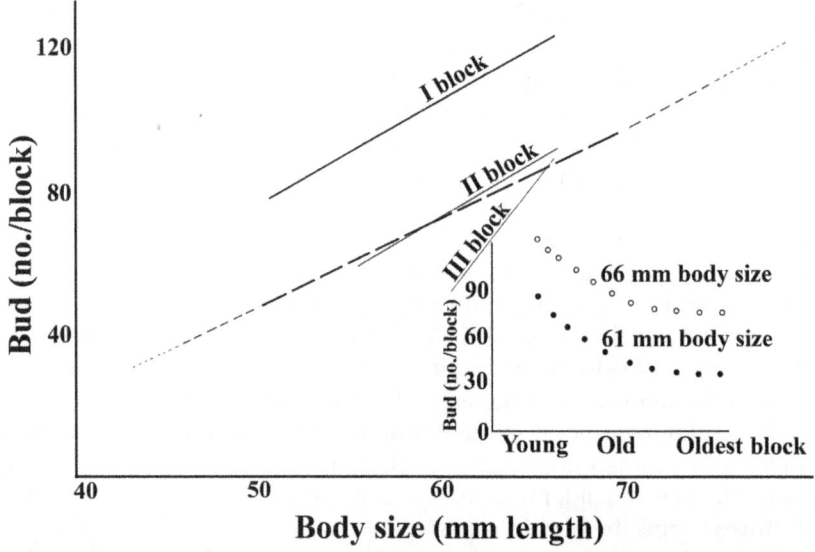

FIGURE 10.2

Number of buds in the blocks as a function of body size of *Iasis zonaria*. In window: Number of buds successfully differentiated in young, old and oldest blocks of *I. zonaria* (drawn from data reported by Daponte et al., 2013).

mature asexual stage called oozoid or Old Nurse generates a large number of buds from a rudimentary ventral stolon, as in the salps (Fig. 10.1C). Leaving the stolon, the buds migrate to a dorsal stolon-like cadophore, where they are aligned in rows of three pairs. It is in these alignments, that the fate of the buds is determined. The two lateral pairs develop into trophozooids, which are considered to nourish the entire colony. The buds in the central pair develop into phorozooids, which remain in contact with cadophore. The phorozooids continue to grow but on reaching a certain size, they are detached from the cadophore and transformed into gonozooids. The gonozooids are protogynic hermaphrodites and release two–six eggs over a period of a few days. Fertilization occurs within the ovary in *Doliolum denticulatum* but after shedding in *Diolioletta gegenbauri*. Self-fertilization is possible in *D. nationalis*. Development is indirect. The doliolid tadpole metamorphoses into an oozoid. The stolon called 'rosette organ' "is very complex and protrudes behind the cardio-pericardium as a ventral outgrowth, covered by an ectoderm fold; it comprises two cellular strands of pharyngeal origin, two of cloacal origin and two more respectively derived from the mesoderm and pericardium". In all, the stolon has tissues types derived from ectoderm, endoderm and mesoderm. After a few days of free-living, the oozoid transforms into a Nurse, a movable blastogenic system. It then loses most of its viscera, namely the endostyle, branchial wall and gut. Only the neural ganglion, peri-cardium and stolon are retained (Godeaux, 1990). The estimated reproductive output of *D. gegenbauri* is in the range of 784,000 progeny over a period of 50 days with a population doubling time of ~ 2.5 days (Paffenhofer and Gibson, 1999).

10.2 Benthic Tunicates

Unlike the pelagics, the benthic, sessile, suspension-feeding ascidians are more speciose (> 2,815 species, see Shenkar and Swalla, 2011). They are divided into three sub orders including 10 families (Abbott and Newberry, 1980). In them, the presence of both solitary and colonial forms indicates that coloniality has emerged several times independently (Brown and Swalla, 2012, Table 10.2). Though colonials are limited to ~ 40 species out of 2,815 species, 22 of 38 genera are colonials. They are metagenics, alternating between sexual and asexual reproduction. Almost all the colonials are ovoviviparous. Among them, *Diazona violacea* is an exceptional oviparous species but *Hypsistozoa fasmeriana* and some species of *Botrylloides* are truly viviparous. The stolidobranchs include solitaries and colonials as well as some intermediate morphologists. Typically, colonials consist of zooids embedded in a common test but in the socials like *Perophora sagamiensis* and *Metrandrocarpa taylori*, the zooids are not completely embedded in the test (Zeng et al., 2006). Clonal reproduction is predominantly by

TABLE 10.2

Distribution of solitary, social and colonical forms in Ascidiacea (compiled from Cloney, 1990, Zeng et al., 2006), ⌒ = solitary, 🐾 = colonial, ⌒≗ = social

---

Suborder Aplousobranchia

Polyclinidae: *Polyclium* 🐾, *Aplidium* 🐾, *Euherdmania* 🐾, *Ritterella* 🐾, *Pseu dodistoma* 🐾

Polycitoridae: *Distaplia* 🐾, *Clavelina* 🐾, *Hypsistozoa* 🐾, *Cystodytes* 🐾, *Eudi stoma* 🐾, *Pycnoclavella* 🐾

Didemnidae: *Didemnum* 🐾, *Diplosoma* 🐾, *Trididemnum* 🐾

Suborder Phlebobranchia

Cionidae: *Ciona* ⌒, *Diazona* 🐾

Corellidae: *Corella* ⌒, *Chelyosoma* ⌒

Perophoridae: *Perophora* 🐾, *P. sagamiensis* ⌒≗, *Ecteinascidia* 🐾

Ascidiidae: *Ascidia* ⌒, *Ascidiella* ⌒, *Phallusia* ⌒

Suborder Stolidobranchia

Styelidae: *Styela* ⌒, *Cnemidocarpa* ⌒, *Botryllus* 🐾, *Botrylloides* 🐾, *Metandrocarpa* 🐾, *M. taylori* ⌒≗, *Okamia* ⌒, *Dendrodoa* ⌒, *Polycarpa* ⌒, *Symplegma* 🐾, *Pelonaia* 🐾

Pyuridae: *Pyura* ⌒, *Boltenia* ⌒, *Halocynthia* ⌒, *Herdmania* ⌒

Molgulidae: *Molgula* ⌒

---

budding; however, fragmentation to generate a new colony is limited to didemnids alone (see Ritzmann et al., 2009). The colonials are prolific budders almost throughout the year except during unfavorable seasons. But sexual reproduction is limited to once or twice a year and results with limited number of 50 eggs/m³ area of the colony (e.g. *Didemnum rodriquesi*, Ritzmann et al., 2009). With a few exceptions, ascidians are simultaneous hermaphrodites (Cloney, 1990). Solitaries are oviparous, reproduce sexually alone and broadcast a large number of smaller (160 μm) eggs that develop into less differentiated tadpole larvae (Fig. 10.3A). A few solitaries, however, brood large highly differentiated larvae. For example, most mogulids and a couple of styelids (e.g. *Pelonaia corrugata, Polycarpa tinctor*) develop directly and are called 'anurans'. *P. tinctor* generates the largest (730 μm) eggs (Cloney, 1990).

Next to echinoids, the solitary ascidians have served as models for the study of autoregulation of early embryonic development. Ortani (1958) has amassed a volume of evidence on their ability to autoregulate the development. In solitaries like *Ciona intestinalis, Phallusia mammillata* and *Ascidia malaca*, both fertilized (diploid) and unfertilized (haploid) eggs are

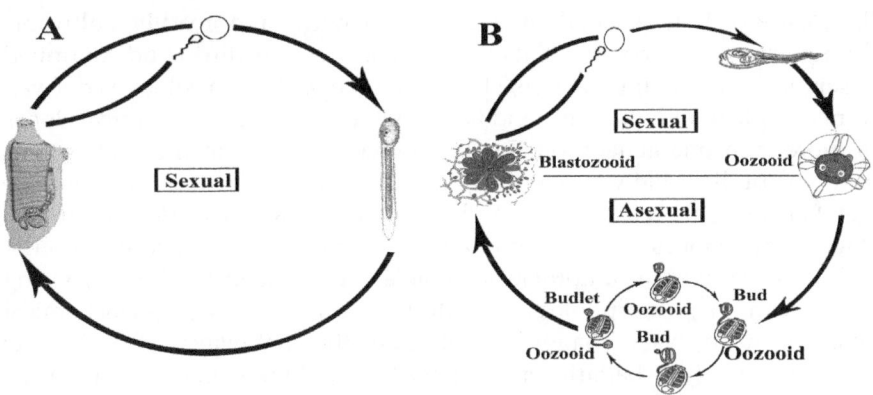

FIGURE 10.3

Life cycle of benthic ascidians. A. Solitary oviparous, indirect life cycle involving less differentiated tadpole (compiled from Google, modified). B. Colonial sexual ovoviviparous zooids with indirect life cycle involving differentiated tadpole. It is metagenic and alternates with asexual blastogenic cycle (compiled from Google, modified).

auto-regulative. In them, nearly complete twin larvae can develop from the halves of the same haploid or diploid egg. When one of the two blastomeres of *Styela plicata* is destroyed, the remaining half develops into a half larva (Conklin, 1906). Nakauchi and Takeshita (1983), who have extended the investigation of Conklin, have found that the hatched half larva can metamorphose and develop into a functional juvenile with completely normal morphology. Thus, the regulation occurs after metamorphosis but not in the middle phases of development as in ascidians. This ability to reconsititute parts or the whole of the body in the post-metamorphic stage is relevant to biomedical applications (Kurn et al., 2011). To escape from predators, many solitary ascidians have accumulated obnoxious substances. Being suspension-feeders, they are also known to synthesize anti-cancer substances, as sponges do (Shenkar and Swalla, 2011). Not surprisingly, solitary ascidians have emerged as model animals for the study of molecular control of embryogenesis and differentiation of specific cell lines (e.g. Satoh and Levine, 2005) and their genome has been partially or fully sequenced (e.g. Yokobori et al., 2003).

Before furthering this account on colonial ascidians, a word of appreciation must be stated. Thanks to the pioneering endeavors of Prof Armando Sabbadin at the University of Pavoda, Italy, the colonial ascidian *Botryllus schlosseri* has emerged as a model species for the study of clonal reproduction, regeneration and immunology (Sabbadin, 1958, 1978). Manni et al. (2007) and Kurn et al. (2011) have listed the following unique advantages offered by this colonial ascidian: 1. *B. schlosseri* is cosmopolitan in temperate waters (Ballarin and Manni, 2009). There is an urgent need to discover a tropical counterpart of

B. schlosseri. 2. It is small (0.5–1.5 mm length) and readily cultivable through many generations. 3. Its transparency renders direct and continued observations under microscope/stereomicroscope. 4. Each colony is derived from a single founder tadpole and is composed of genetically identical clones of more than one generation. 5. The bidirectional arrangement of zooids, buds and budlets allows a microsurgical tungsten needle to damage, remove (e.g. Lauzon et al., 2002) and graft a bud or zooid of a specific developmental stage among clones of many generations. 6. Intervascular microinjection of cells or tracer dyes via ampullae is possible (e.g. Laird et al., 2005a, b). Using this technique, morphological, ultrastructural, immuno-cytochemical analyses have been made to delineate the pathways of hemocyte differentiation (e.g. Ballarin et al., 1993) and pollutant induced alteration in their function (Cima and Ballarin, 2004). 7. Using RNA interference and morpholinos, the function of a specific gene(s) in a colony can be determined (e.g. De Tomaso et al., 2005). 8. The genetic basis of self/non-self discrimination of or colonial specificity is also possible (e.g. Rinkevich and Weissman, 1987).

In colonial ascidians, the tadpole-like larva (1.2–1.5 mm length) metamorphoses into a fully functional oozoid, the founder of a new colony. The zooid bears a single palleal bud on its right side. In a week's time at 19°C, the oozoid is resorbed by a process called take over and replaced by its bud, which develops into a mature suspension-feeding blastozooid. In its turn, the blastozooid generates two or more palleal buds on each side of its body capable of originating budlets (Fig. 10.3B) and replacing the adult zooids, when they die. Hence, an adult colony (1.5 mm length) consists of many suspension-feeding blastozooids grouped into a stellate system with a central cloacal siphon, buds and budlets. Because adult zooids are cyclically replaced by the newly developing buds, a healthy colony can grow in size, having up to thousands of zooids and buds maintained by synchronized arrival of a new generation of zooids and death of old ones. Thus, the blastozooid initiates a life long process of cyclic blastogenesis.

Each blastogenic cycle, "a highly coordinated event that is punctuated by the death and resorption of the entire generation of ongoing zooids coincident with succession of incoming generation of zooids" (Laird et al., 2005a), lasts for 6–7 days (Lauzon et al., 1992, 2002). It culminates by the take over process of a synchronized wave of programmed cell and zooid death lasting for a period of 26–36 hours. To show that this is an intrinsic process rather than a competitive elimination of old senescent zooids by newly developing primary buds, Watkins (1958) performed an elegant experiment. The surgical removal of all the buds within a colony did not extend the life span of old zooids. Hence, the take over process is an intrinsic one. From a series of experiments involving buddectomy, hemibuddectomy and zooidectomy, Lauzon et al. (2002) have shown that the buddectomy and hemibuddectomy extended the take over period to 36–48 and 48–60 hours, respectively. In zooidectomized colonies, the bud development was severely curtailed. These findings strongly suggest that components of dying zooids are reutilized for the newly developing primary buds.

In colonial ascidians, the epicardium, septum and hemoblasts harbor multipotent stem cells that enable the clonal reproduction (Kawamura et al., 2008). The epicardium is located at the pharyngeal floor and runs parallel to the gut (Fig. 10.4A). It may be a tube-like or v-shaped sac. It is capable of not only regenerating itself but also the pharynx, neural structures, gut, heart and gonad (Berril, 1950, Scott, 1957). Buds can also rise from epithelial diverticulum called stolon (Fig. 10.4D). A septum is located within the stolon and consists of a single row of layered multipotent epithelium. The sac-like vesicles located along epidermal cells of stolon develop into stolonial buds. The multipotent septal cells are of hemocoel origin, as they can be destroyed

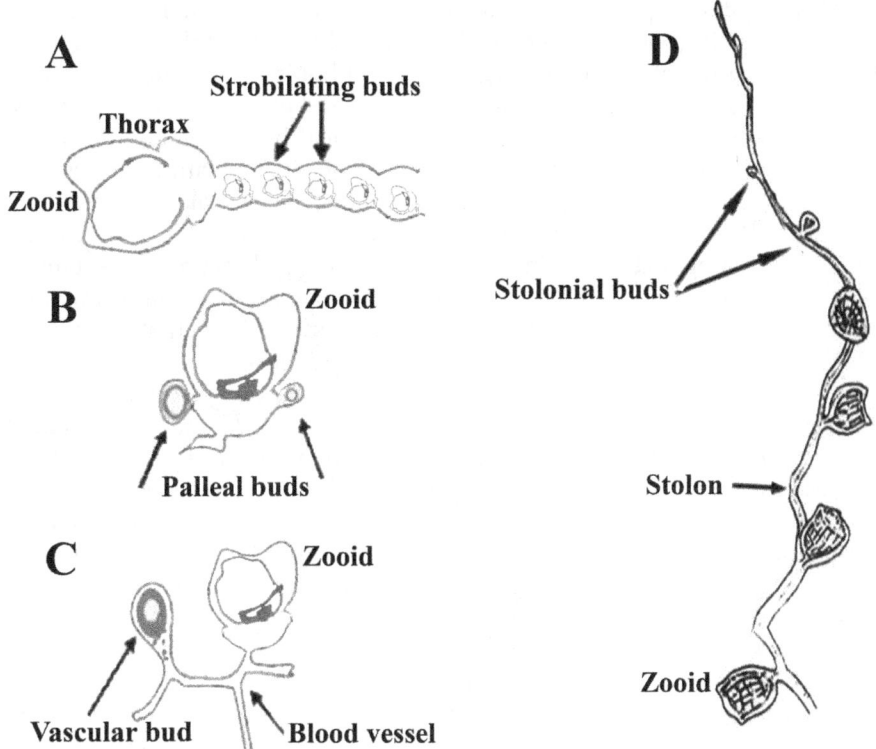

FIGURE 10.4

Sources of asexual reproduction in colonial ascidians. A. Abdominal budding (strobilization) in polycitorid and polycinid ascidians, B. Palleal budding, C. Vascular budding (common in botryllid ascidians), D. Stolonial budding in *Perophora* (compiled and redrawn from Nakuchi, 1982).

by gamma radiation and restored by the introduction of hemoblasts (Freeman, 1964). The atrial (peribranchial) epithelium is a specialized tissue encircling the pharynx (Fig. 10.4C). In *Botryllus* and *Symplegma*, the bud primordium is a thickened (cell proliferation) disk of the atrial epithelium. Whereas the bud primordium is limited to a specific region in *Botryllus* and *Symplegma*, it is formed anywhere along the zooidal body axis in *Polyandrocarpa*. Hence *Polyandrocarpa* zooids, when cut into pieces, the missing parts of zooidal fragments are regenerated by the atrial epithelium (Taneda, 1985).

Clonal reproduction occurs by (i) outgrowth development (budding) or self-division (strobilization) (Berril, 1950). Rarely, *Clavelina lepardiformis*, when divided into anterior and posterior segment, as in holothuroids, the segments regenerate missing body parts (Brien, 1968). In colonial tunicates, the formative tissue(s) of a nascent clone is known to arise from different sources: (i) the stoloneal septum in perophorids (Brien and Brien-Gavage, 1927), (ii) atrial epithelium in polyclinids and clavelinids (Nakauchi, 1984), (iii) palleal (polystyelids, Kawamura and Nakauchi, 1984) and (iv) vascular system (botryllids, Berril, 1941). Considering the dermal origin of the bud, a different picture emerges. In perophorids (e.g. *Perophora viridis*, Freeman, 1964, *P. orientalis*, Mukai et al., 1983) and clavelinids (e.g. *Clavelina lepadiforms*, Kawamura et al., 2008), the stoloneal buds originate from the mesenchyme-derived septum within the stolon. In polycinid species that bud by strobolization (e.g. *Alpidium palladium*, *Amoroucium constellatum*, Berril, 1951), the nascent bud is derived from the digestive or epicardial tissue of an elongated abdominal zone, i.e. endodermal origin. In the styelids (e.g. *Botryllus*, Berril, 1941, *Polyandrocarpa misakiensis*, Fujiwara and Kawamura, 1992), the nascent bud emerges from peribranchial epithelium of ectodermal origin. Besides, *Botryllus* and *Botrylloides* can also develop vascular buds (Rinkevich et al., 2007). Evidently, coloniality has emerged many times independently (Brown and Swalla, 2012) and has retained multipotent/pluripotent cells in different lineages resulting in their persistence in different locations of the colonies. However, it is not clear whether some of these differentiated cells are dedifferentiated/transdifferentiated stem cells to perform clonal reproduction. Some authors have suggested the possible transdifferentiation by activation of stem cells from the existing cells.

Fujiwara and Kawamura (1992) have brought histochemical and immunological evidences for the role played by atrial epithelium in the formation of nascent bud in *P. misakiensis*. The bud primordium is formed with changes in atrial epithelial cells from squamous to cuboidal, swelling nuclei, distinct appearance of nucleoli and accumulation of RNA. These cells, designated as 'activated stem cells', are identified by a marker *Pae 1* antigen. This tissue specific *Pae 1* antigen is expressed in atrial epithelial cells but selectively disappears, as organogenesis proceeds. Further studies of Kawamura et al. (1991) have brought evidence for the contribution by a small population of circulating blood cells in the clonal development of the buds. The evaginated nascent primordium consists of outer and inner layers of epithelium containing mesenchymal cells including hemoblasts. In the most posterior part of the bud, the inner activated epithelium invaginates into the mesenchymal space and forms the gut rudiment. The pharyngeal rudiment evaginates into the atrial cavity. Thus, the gut rudiment is formed by atrial epithelium through a transdifferentiation-like process.

The colonials *Perophora* and *Botryllus* have been subjected to a series of experiments. In *P. viridis*, endoblastic vesicles are formed along the stolon in colloboration with lymphocytes and mesenchymal septum. Ligation of a stolon fraction between adjacent zooids hindered elongation of the stolon (Goldin, 1948). Exposure of the stolon to thiourea (2×10^{-2} M) completely inhibited the elongation. On the other hand, thyroxine (10^{-8}–10^{-5} M) stimulated the elongation. Hence, the thyroid hormone facilitated the elongation of interzooidal stolon but thiourea, an inhibitor of thyroid hormone biosynthesis, induced stolonial budding. Recently, a thyroid hormone receptor gene was cloned from *Ciona intestinalis* (Yagi et al., 2003). In a pioneering study, Freeman (1964) demonstrated that gamma radiation inhibited/blocked stolonial budding in *P. viridis* but an injection of a population of lymphocyte-like cells (hemoblasts?) into the irradiated stolon restored the budding. Separate injections of eight morphologically distinct populations of blood cells revealed that lymphocytes alone reinstated the budding potential in irradiated colonies. Freeman (1971) further showed that activity of dorsal half of the thorax plays a major role in determination of the timing of strobolization. In all, the colloboration between the epidermal cells of stolonial septum and lymphocytes is responsible for bud formation; the stobolization process is controlled by the activities of thyroid cells. Due to differences in sources of stolonial and palleal budding, the injection of enriched population of putative stem cells into gamma irradiated *Botryllus* did not restore the budding potential (Laird and Weissman, 2004a).

Besides palleal budding, some species of *Botryllus* and *Botrylloides* are also capable of stress-induced vascular budding (Fig. 10.4C). In fact, vascular budding is reported in *Botryllus schlosseri* (Milkman, 1967), *B. primigenus* (Oka and Watanabe, 1957), *Botrylloides leachi* (Rinkevich et al., 2007) and *B. violaceus* (Brown et al., 2009a). In *Botryllus primigenus*, the budding is part of its life history (Oka and Watanabe, 1957). In *Botrylloides gascoi* and *B. leachi*, it occurs in the natural habitats, when the respective colonies recover from aestivation (Bancroft, 1903) or hibernation (Burighel et al., 1976). But it occurs in *Botrylloides violaceus* and *Botryllus schollsseri* only after the surgical removal of all the zooids and palleal buds (Voskoboynik et al., 2007). Rinkevich et al. (2007) have reported for the first time for any prochordate the Whole Body Regeneration (WBR) in *Botrylloides leachi* from an isolated miniscule (0.1 mm) of a blood vessel containing 100–300 blood cells (cf 200 archaeocytes required for WBR in sponge, see Pandian, 2016) within a period

TABLE 10.3

Temporal sequence of whole body regeneration by vascular budding in *Botrylloides leachi* (from Rinkevich et al., 2007, modified)

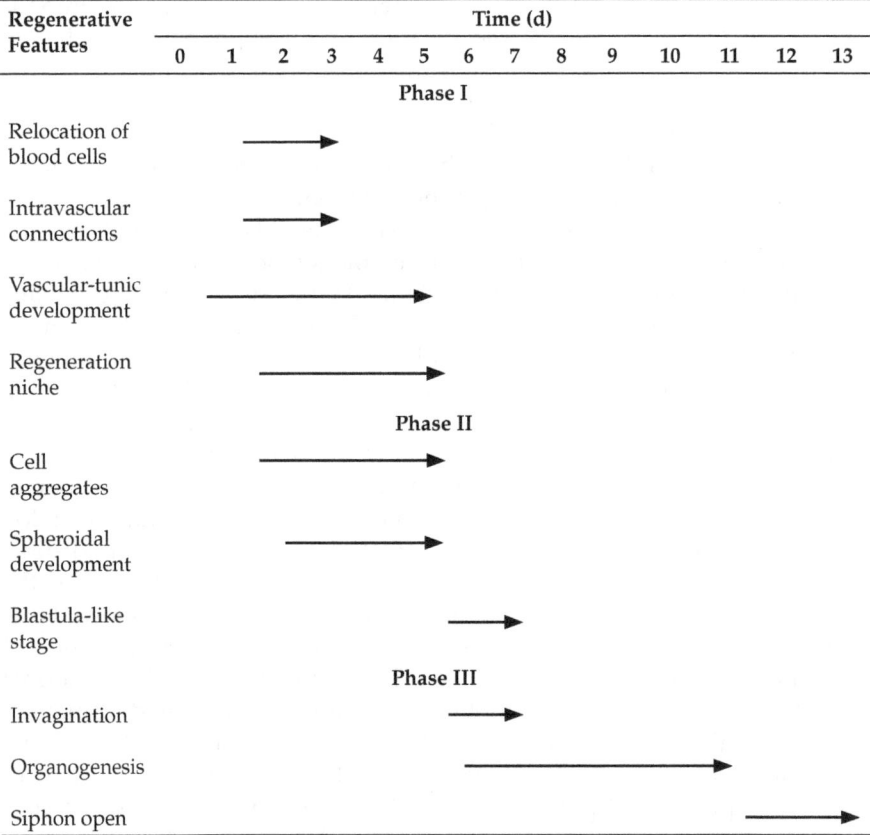

Regenerative Features	Time (d)													
	0	1	2	3	4	5	6	7	8	9	10	11	12	13

Phase I — Relocation of blood cells; Intravascular connections; Vascular-tunic development; Regeneration niche.

Phase II — Cell aggregates; Spheroidal development; Blastula-like stage.

Phase III — Invagination; Organogenesis; Siphon open.

of 13–14 days (Table 10.3) which involved epimorphosis but without blastema formation, as in asteroids, cf Table 5.12. In all, the WBR involved multiple regenerating niches *B. leachi* instead of a single blastema, as in non-asteroid echinoderms (see Table 5.12) and systemic induction in *B. leachi* instead of centralized induction in other regenerative animals.

Retinoid Acid (RA) is a low molecular weighing biologically active metabolite of Vitamin A. In vertebrates, a heterodimeric transcription factor, comprising the RA Receptor (RAR) and the Retinoid X Receptor (RXR), mediates the RA signal (Mangeldorf and Evans, 1955). Where several regeneration niches are established, buds of *B. leachi* develop simultaneously within the ampullae. RAR expression pattern is localized specifically in all these buds indicating the activation role played by RA (Rinkevich et al., 2007).

In *P. misakiensis* too, RA is considered to indirectly act on multipotent cells, for the coelomic cells treated with RA are activated to induce dedifferentiation in the atrial epithelium (Hara et al., 1992, see also Kawamura et al., 1993).

10.3 Germline Lineage

Primordial Germ Cells (PGCs) are the progenitors of germ cell lineage and possess the ability to differentiate into either oogonia or spermatogonia. Hence they carry heritable information from one generation to the next. In animals like fishes, undifferentiated germ cells are distinguished from somatic cell lineages by their large, round nucleus, single large nucleolus, a cytoplasm relatively clearer of organelles and basophilic bodies of granular cytoplasmic material called nuage-germ granules. Proteins and RNAs localized in the nuage are products of the genes like *vasa*. Hence, *vasa*, the first molecular marker, can be used to trace the origin and development of germ cell lineage (e.g. fishes Pandian, 2011). The thaliacean and colonial ascidian tunicates are metagenics, alternating between sexual and asexual reproduction. Hence, tracing the germ cell lineage through sexual and non-sexual stages in these metagenic tunicates may prove to be a fascinating field of research.

As in other animals like fishes, solitary ascidians contain maternal transcripts required for the formation of the germ cell (Brown et al., 2009). In the solitary ascidian *Ciona intestinalis*, early *vasa* mRNA is localized to a characteristic cytoplasm known as postplasm (Fujumura and Takamura, 2000). The mRNA transcripts are sequestered into a posterior germ cell lineage during early cleavage. *Vasa* transcripts are further sequestered into the B 7.6 cells. It is followed by final unequal cleavage forming the B 8.12 lineage of cells that eventually form the germ cells (Shirae-Kurabayshi et al., 2006). During gastrulation, the cells divide and most maternal vasa mRNA transcripts are separated from the rest of postplasmic compartments into a lineage of B 8.12 identified as the PGCs. The PGCs are then localized to the tip of endodermal strand of the tail and move into the gonad after metamorphosis (Takamura et al., 2002). From day 1 following metamorphosis, when the gonad rudiment becomes apparent amidst the degenerated tissues of the tail, it requires 11–12 days for the testis and ovary to become visible (Yamamoto and Okada, 1999). In another solitary *Boltenia villosa*, in which *bv-vasa* mRNA is expressed in germ cells in B 8.11 and B 8.12 as well as in adult gonads, more or less the same sequence of germ cell lineage is described by Brown and Swalla (2007).

Life span of colonial *Botryllus schlosseri* is one year in Scotland but 12–20 months in Italy, where most colonies reproduce for two reproductive seasons (Sabbadin, 1955a). Following metamorphosis, colonies undergo at least 8–12 blastogenic cycles prior to sexual maturity (Brown et al., 2009a). Oocytes are spawned, when primary buds approach adulthood. Sperms are released after one or two days of spawning. The weekly/fortnightly blastogenic cycles and life long turnover of zooids may require (i) the persistence of germ line progenitors through generations, (ii) the mobility and migration of progenitors from buds to zooids and (iii) their enrichment (Brown et al., 2009b). These requirements are met by the colonials.

As in solitary ascidians, the maternally derived *vasa* mRNAs segregate early in development to a posterior lineage of cells and thereby confirms that early germ line formation occurs also in colonial ascidians (Brown et al., 2009b). However, the colonials have a second source to generate germline cells during the adult stage. From their cytochemical and ultrastructure study, Akhmadieva et al. (2007) indicated the presence of germ cells in the buds of adult colonies of *B. tubaratus*. In the same year, Brown and Swalla (2007) reported the expression of bot-vasa mRNA in putative spermatogonia and oocytes of oozooids as well as some circulating cells in the zooids of mature colonies of *B. violaceus*. Continuing their good work, Brown et al. (2009b) confirmed that vasa expression was dynamic in the adult gonads as well as scattered throughout the circulating system. According to Sabbadin and Zaniolo (1979), the germline precursors appear first in the budlets. They are mobile and migrate to a niche, from which they differentiate into gonads. Oocytes can also migrate between successive generations until the colony sexually matures. Obviously, the motile and migrative germ cells are hemoblasts. Available evidence goes to show that the circulating pluripotent stem cells, derived from embryonic mesenchyme (cf p 131–132) are progenitors of blood cell line (Sabbadin, 1959) and germline (Sabbadin and Zaniolo, 1979). The potential of circulating hemoblasts has been also experimentally proven in zooidectomized and buddectomized (Manni et al., 2007).

Employing their newly cloned *Botryllus* vasa homolog, the *BpVas* of *Botryllus primigenus*, Sunanga et al. (2006) found its expression specifically in germline cells like the loose cell mass, oogonia and early oocytes in the ovary and primordial testis (compact cell mass) spermatogonia and early spermatocytes in the testis (Fig. 10.5). An ultrastructure study has shown that the loose mass of cells consist of undifferentiated hemoblasts. Following the zooidectomy and buddectomy, *BpVas* transcripts completely disappear. Nevertheless, *BpVas*-positive cells *de nova* appear again after a fortnight in these buddectomized and zooidectomized colonies. Figure 10.5 is a schematic representation of the origin and development of germline in different tissues/organs, as indicated by the expression of *BpVas* in adult colonies of *B. primigenus*. *Vasa* is expressed in cell populations with high level of aldehyde dehydrogenase (ALDH) activity in functional germline precursors. In a thoughtfully conceived study, Laird et al. (2005a) have transplanted single undifferentiated cells from a donor into a recipient colony of *B. schlosseri*. The single undifferentiated cells contribute to either somatic or germline chimeras but not to both. Clearly, the undifferentiated hemoblasts remain

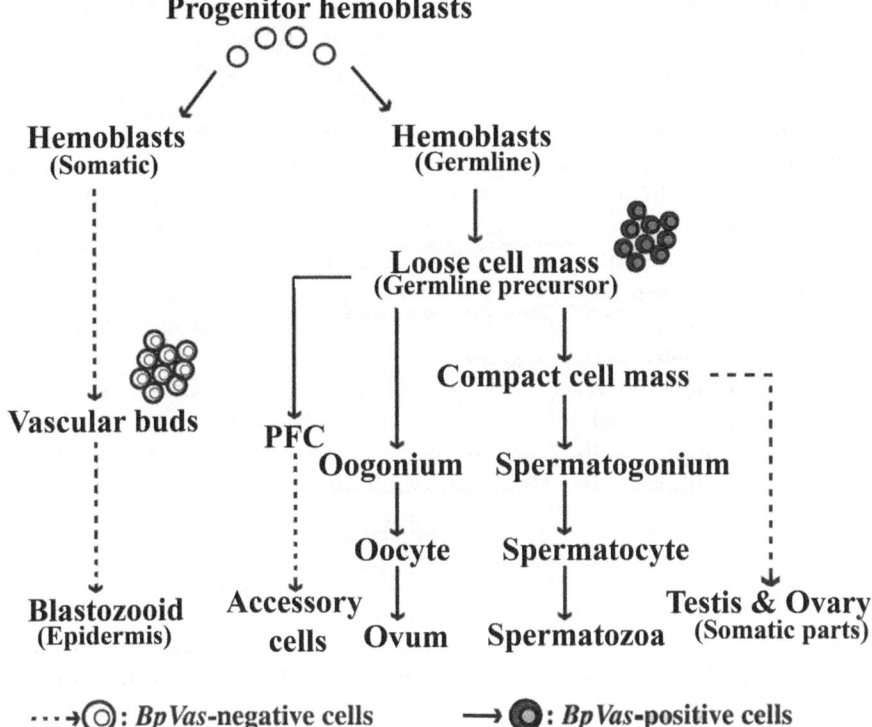

FIGURE 10.5

Tentative pathways, through which the expression of *BpVas* germline lineage in *Botryllus primigenus*. PFC = Primary Follicle Cells (modified and redrawn from Sunanga et al., 2006).

already differentiated into somatic and germ line progenitors and co-exist in adults. From their serial transplantation assay, Laird et al. (2005b) found that the transplanted progenitor cells are capable of reconstitution and self-renewal.

The contact between two or more conspecific colonies results in either their mixing to form chimera (e.g. Rinkevich and Weissman, 1987, Paz and Rinkevich, 2002) or rejection during which the interacting tissues are eliminated (Kurn et al., 2011). A single, highly polymorphic histocompatible locus called FU/HC is responsible for rejection or fusion. Gonads are seeded, gametogenesis occur in colonies well after fusion and involves circulating germline progenitors. Germline parasitism is common among fused chimeras (Stoner and Weissman, 1996). Stoner et al. (1999) have also shown that genetically distinct primitive germ cells (pgcs) and somatic cells (sgcs) compete for access to developing gonads and somatic organs. The resulting traits are hierarchial, reproducible and heritable. Fusion (parabiosis) between the colonial clones homozygous for two loci for pigmentation genes (AAbb) with opposite genotype (aaBB) can be experimentally induced. These genes control two simple Mendelian traits namely hemolymph-transferable orange pigmentation and non-hemolymph transferable intersiphonal double band (Sabbadin and Graziani, 1967). Hence, the presence/absence of the intersiphonal double band serves as a marker to identify the two original colonies in a chimera. A series of parabiosis experiments have brought unequivocal evidence for the capability of the germ line-committed hemoblasts to differentiate into either oogonia or spermatogonia. Other experimental crosses have shown that cross-fertlization results in higher yield of normal larvae and long-living colonies than those of self-fertilization (Gasparini et al., 2015).

Social or foraging intra-specific grouping is common in many animal species. In colonial ascidians, tissue fusion or natural chimerism is another form of grouping. Usually, colonial fusion is followed by resorption of one partner in botrylloids. However, freely circulating hemoblast stem cells from the 'loser' partner may parasitize the soma and germline of the 'winner'. As a result, a range of chimeras from bi- to -multichimeras are produced. Rinkevich and his associates have related the allogenic resorption in bi-, tri- and -multichimeras in *B. schlosseri* and bi-, tri-, and -hexachimeras in *Botrylloides* sp (Table 10.4). With injection of 'parasitic' hemoblasts into the germline of the host, a kind of heterosis occurs. With increasing number of partners, i.e. from bi- to -multiplechimersism, it is advantageous for the 'hybrid host', when life history parameters like zooid number per colony, zooid size and growth rate are considered. In terms of characters related to allogenic resorption, the responses of *B. schlosseri* and *Botrylloides* sp differ.

To escape from accumulation of harmful mutation with such high magnitude of asexual reproduction, the colonial ascidians have developed a few strategic adaptations. 1. Tracer experiments on bud formation and development in different morphological and chimeric phenotypes of *Polyandrocarpus stolonifera* have revealed that not all the parental epithelial tissues surrounding the bud primordium enter the growing bud (Kawamura and Watanabe, 1981). In this context, Kawamura and Fujiwara (2000) considered that budding involves the purge of parental somatic cells and tissues harboring harmful mutations. 2. In *P. misakiensis*, Kawamura and Fujiwara (2000) list at least three genetic variations in a clonal colony. 3. More importantly, when terminal ends of two or more conspecific colonies coalesce, a sort of genomic mixing occurs and eventually results in germline parasitism, a common feature amidst the colonial ascidians (e.g. Stoner and Weissman, 1996). This sort of genomic coalescence in the colonies has ample scope for elimination of harmful mutations and introduction of useful genes. 4. Colonial ascidians have also developed strategies to eliminate senescence in both germline and somatic progenitors. 4a. Telomerase activity affords protection to chromosomes of germline and stem cells from fatal shortening

TABLE 10.4

Life history characteristics and characters related to allogenic resorption in bi- (BC), tri- (TC) and -multiple chimeri (MC) *Botryllus schlosseri* and *Botrylloides* sp (compiled and condensed from Rinkevich and Shapira, 1999, Paz and Rinkevich, 2002)

Tested Characters	BC vs MC Comparison In	
	B. schlosseri	*Botrylloides* sp
Life history characteristics		
Life span (day)	Not affected, 156–182	Not affected, 273–284
Zooid (no./colony)	Increased in MC, 64 to 98	Increased in MC, 73 to 233
Zooid size (µm)	Increased by 2.2 times	Increased by 3.5 times
Age at maximum size (day)	Decreased by 0.8 times	Increased by 3.7 times
Growth rate (µm/day)	Increased by 2 times	Increased by 3.4 times
Characters related to allogenic resorption		
Disconnection (%)	Decreased from 50 to 0% with increasing partners	Increased from 27 to 100% with increasing partners
Surviving partners (%)	Increases with increasing partners	–
Time for resorption (day) First partner	Decreased from 76 to 23 with increasing partners	Decreased from 53 to 118 with increasing partners
Second partner	Decreased from 132 to 66 with increasing partners	Increased from 46 to 144 with increasing partners
Maximum size (µm) of First partner	50% larger in BC	> 2 times larger in MC
Second partner	2 times larger in TC	3 times larger in MC

due to replication. Laird and Weissman (2004a) have found the highest level of telomerase activity in the gonads, developing embryos and bud primordium of *Botryllus schlosseri* and thereby eliminates senescence in the colony, the most important requirement to sustain asexual reproduction. 4b. With weekly/fortnightly blastogenic cycle, the life span of zooids is kept at the minimal duration. In all, the colonial ascidians successfully employ these adaptive strategies to purge harmful mutations and escape from senescence.

11

New Findings and Highlights

Introduction

The inferences drawn from this comprehensive presentation have harvested many new findings related to reproduction and development in primitive deuterosomes namely echinoderms and prochordates. Accomplishing a major treatise series on invertebrates, L.H. Hyman (1955) saluted only "the echinoderms as a noble group designed to puzzle the zoologists". 1. Echinoderms are unique and distinctly differ from all other animal taxons in that their embryos/larvae display bilateral symmetry but the adult pentamerous radial symmetry. 2. They may not be economically an important taxon but research on them have fetched two Nobel Prizes one for cell cycle and the other for phagocytosis. 3. With extra-ordinary potential for regeneration, their genomes have 70% homology with humans. These unique features may inspire more and more people into the fold of echinoderm research.

11.1 Structure and Distribution

The pentamerous radius symmetry in adult echinoderms has eliminated the scope for the development of an anterior head and lateral appendages. As a result, they are either sessile/sedentary or slow motile. The obligate need for acquisition of food by facing the oral surface over the substratum in echinoids, asteroids and ophiuroids has unbreakably tied the distribution of > 74% of echinoderms to benthic habitats; others are sessile/sedentary. As an immense consequence, the echinoderms are unable to colonize pelagic realm as plankton and the entire column of sea water as nekton.

11.2 Fecundity, Size and Depth

In sea urchins, the algae which promote the growth of the gonads (commercially valuable roe) are identified. For example, *Laminaria japonica* facilitates a greater allocation for gonadal growth, while *Rhodemenia palmata* for somatic growth (p 22). At species level, fecundity linearly increases with increasing body size in all the classes of echinoderms, irrespective of direct or indirect mode of development in gonochorics, parthenogenics and protandrics (Fig. 1.8). Lecithotrophic egg size of ophiuroids and asteroids increases with increasing depth up to 2,000–5,000 m but remains constant in their planktotrophs, albeit at different levels (Fig. 1.10B). At interspecies level, a lecithotrophic egg size decreases with increasing body size. Yet, the relationship becomes linear in lecithotrophic and viviparous ophiuroids, when body size is considered in weight (Fig. 1.10D) instead of disk size. Notably, the lecithotrophic asteroids and ophiuroids produce larger and larger eggs with increasing depth. But the planktotrophs produce equal sized eggs, irrespective of increasing depth (Fig. 1.10A). A striking contrast is that ophiuroids larger than 18 mm size do not produce planktotrophic eggs but only asteroids larger than 110 mm size produce planktotrophic eggs (Fig. 1.10C).

11.3 Aquaculture: Sea Urchins and Cucumbers

Despite its small scale fishery, aquaculture of holothuroids has much scope for conservation and provision of alternate employment for millions of Asian fisherfolks. The following deserve to be mentioned: 1. Due to the emergence of viral diseases, the abandoned coastal shrimp farms may profitably be restored by switching to sea cucumber culture (p 70). 2. In the cucumbers, mucous secretion can effectively be used as an external marker to recognize the ripening gonad (Fig. 1.6B). 3. The scope for wider application of previsceral coelomic fluid to induce spawning has to be explored for each culturable holothuroid species. 4. In sea urchins and asteroids, thyroxine regulates the larval duration and offers considerable scope to reduce the risky larval duration. It is synthesized endogenously but in inadequate quantities; it can be acquired in adequate quantities from algal feed. Its use in holothuroid larvae remains to be tested. 5. The use of abundant benthic macrophytic algae and sea grasses as ingredient of synthetic feed must be explored (p 53). 6. The claimed medicinal property of edible holothuroids must experimentally be proved.

11.4 Intromittent Organ

Among aquatic invertebrates, echinoderms display the simplest reproductive system, in comparison to others, such as molluscs (Pandian, 2017). A known but not adequately recognized feature is the virtual absence of an intromittent organ in any of the echinoderm taxon. As a rule, fertilization in them is external. But brooding and viviparity occurs in some species of all the five classes (Table 1.22). More than 3% of ophiuroids brood their eggs/embryos (Stohr et al., 2005). It is not known how the externally fertilized eggs/embryos/larvae are attracted to one or other external brood chambers. For example, as many as 1,160 externally fertilized eggs/embryos reach the external brood chamber of *Leptasterias hexactis* (Hyman, 1955). In some holothurians, internal brooding occurs in organs like the marsupium, coelomic chamber and even within the intragonadal chamber. In internally brooding ophiuroids too, ovoviviparity/viviparity occurs in the bursal sacs or within the ovary. In the absence of an intromittent organ, it is not known how these internally brooding/viviparous holothuroids and ophiuroids ensure (internal) fertilization. Research to identify one or other chemical cues that attract conspecific sperm may prove rewarding. An important but not yet recognized consequence of the absence of an intromittent organ is the high frequency of unfertilized females and low frequency of Fertilization Success (FS). For example, in the protandric brooding holothuroid *Leptosynapta clarkii*, in which male ratio remains as high as 0.5 even in the largest size group, 44% of females remain unfertilized and FS in other females is as low as 48% (p 55).

11.5 Gonad and Hormonal Economy

In deuterostomes, both echinoids and larvaecea have explored the scope to economize the gonadal functions by harboring germ cells and nutritive cells within the gonad. The supporting role played by the Nutritive Phagocytes (NPs) in maturation of gametes of echinoids is well known (see Fig. 3.3). In larvacea, the role of nutritive syncytial or accessory cells (Galt and Fenuax, 1990) *per se* remains to be confirmed, though it is most likely (see p 198). As a consequence, the costs of hormones and translocation of nutrients are reduced in echinoids. For example, the E_2 level is high as 4–460 pg/g gonads in asteroids, in comparison to 5–160 pg/g in echinoids (Table 7.1). Endocrine disruptors alter the 'hormonal climate' and interfere with the regenerative process. Parasitic castration is not common among echinoderms, and it induces partial sterility alone, when it occurs.

11.6 Regenerative Potential

Despite their amazing autotomic and regenerative potential, the reported descriptions on regeneration of smaller or larger body part(s) is surprisingly limited to 204 species only, i.e. ~ 3% of echinoderms (Table 5.5). The rate of number of species reported to regenerate is low; for example, hardly 0.4 species is annually added to the list of regenerating holothuroids. Nevertheless, its prevalence within each of the reported species is fairly high. It ranges from > 50 to 100% in crinoids (Table 5.1), 38 to 69% in asteroids (Table 5.2) and 13 to 98% in ophiuroids (Table 5.6). A sum of 300 mt of ophiuroid arm biomass is suggested to be injected into the trophodynamics of the Swedish coastal aquatic system (p 143). Interestingly, cephalochordates also inject a few million tons of larvae into the trophodynamics of the Cape Blanc (northwest Africa, p 194). The present analysis has brought to light for the first time that the reduced food consumption by autotomized asteroids is more due to want of adequate digestive enzymes from the reduced pyloric caceca rather than their ability to handle the prey (p 118). Regeneration is accomplished more at the cost of reproduction (see Fig. 5.3D, 5.6B). Regenerating tissues/organs grow faster but differentiation of tissue types is a slower process. Regeneration is epimorphic in crinoids, holothuroids and ophiuroids but morphallactic in asteroids and echinoids. A few genes (*Anbmp, Afuni, EpenHg, ArHox 1, vasa, piwi*) and growth factors (neurotransmitters, neuropeptides) responsible for regeneration in echinoderms have been identified; some of these genes are also cloned (Table 5.11).

11.7 Gonads of Cloners

The presence of two groups within clonally reproducing holothuroids, ophiuroids and asteroids has been recognized. In Group 1, the temporally separated clonal and sexual reproduction coexists almost throughout the life time. In Group 2, the duration of clonal reproduction is limited to 10–30 or 66% of the life time. In them, sexual reproduction alone occurs beyond the specific point in their life time (Fig. 4.2). In solitary hemichordates, a series of paired gonads is located between the post-branchial and pre-hepatic trunk region. However, the number of gonads is drastically reduced to a single symmetric pair in the metagenic colonial cephalodiscids and to a single median gonad in rhapdopleurans (Fig. 8.2). In the pelagic tunicates too, the sexually reproducing larvacea produce a large number of eggs (~ 0.1 to 1.9 eggs per day, p 196). But the number of absolute (life time) fecundity is drastically reduced to just two–six eggs in metagenic colonial salpids and doliolids. Hence, clonal reproduction imposes temporal (Fig. 4.1) or body size-wise (Fig. 4.2) separation from sexual reproduction in echinoderms, drastic reduction in the number of gonads in colonial hemichordates and the number of eggs in pelagic colonial tunicates. Apparently, the clonal reproduction somehow regulates the expression of PGCs, reduces gonad formation or inhibits mitotic proliferation of oogonium in these deuterostomes. This can be a potentially rich area for research by endocrinologists, cytologists and molecular biologists.

11.8 Clonal Reproduction

The incisive analysis of clonal reproduction in echinoderms has brought to light many new findings for the first time. 1. Barring crinoids, ~ 2% of other echinoderms are capable of clonal reproduction during larval or adult stage by fission, clonal autotomy or larval cloning (Table 4.4). 2. Of 137 and odd clonal species, 87, 9 and 4% have opted to clonal propagation by fission, larval cloning and clonal autotomy, respectively. 3. Larval cloners do not continue clonal reproduction by fission or clonal autotomy in adults. Hence, these three clonal modes mutually eliminate each other. 4. Clonal and sexual reproduction is distinctly separated temporally as well as by body size in holothuroids, asteroids and ophiuroids. Within them, clonal reproduction is limited to 30–60% of their respective life span in Group 2, while in others, it continues until life time in Group 1 (Fig. 4.2). 5. Intervals between successive fissions are in the range of 90 days, > 90 days and > 1–2 years in ophiuroids, holothuroids and asteroids (Fig. 4.3). Hence, clonal reproductive potential decreases in the order of ophiuroids < holothuroids < asteroids. This has relevance to epimorphic regeneration in ophiuroids and holothuroids and morphallaxic regeneration in asteroids. 6. In asteroids and possibly in ophiuroids too (Fig. 4.6), the presence of more than five non-regenerating arms, inhibits clonal reproduction by fission. 7. Fission is triggered by an internal factor at an individual level and is not a synchronized event at the population or species level. Hence, the trigger depends on nutrients accumulated within the body, i.e. food availability rather than population density. 8. Evidence has shown the presence of Multiple Stem Cells (MSCs) and Primordial Germ Cells (PGCs). MSCs are dispersed in adequate numbers in the oral disk and arms of asteroids and ophiuroids, and the number of

MSCs decreases from proximate to distal end of the arm. In holothuroids, the PGCs are located in the dorsal mesentery.

11.9 Autoregulation and Stem Cells

The echinoids and solitary ascidians are known for their amazing autoregulation potential. Though no direct evidence is yet available, it is possible to derive a few inferences on stem cells of echinoderms and other aquatic invertebrates. About 99% of echinoderms reproduce sexually alone (see also Table 4.4), implying the disappearance of Embryonic Stem Cells (ESCs) or totipotency, once the cleavage is commenced. However, the blastomeres remain totipotent up to two-cell stage in echinoids and solitary ascidians as well as up to four-cell stage in lecithotrophic echinoids (see Table 1.18, p 45). Remarkably, a planktotrophic egg of *Arbacia punctulata* as small as 80 µm and a lecithotrophic egg of *Clypeaster rosaceus* as small as 270 µm can autoregulate at the blastula stage itself and produce viable twins and quadruplets, respectively. Contrastingly, one of the two-blastomeres of *Ciona intestinalis* develops into a half larva, which autoregulate at the metamorphosis to produce a viable juvenile. These observations have thus far remained isolated. This comparative account has indicated for the first time that the delayed auto-regulation ability of *C. intestinalis* is more relevant to biomedical application. Interestingly, echinoids and solitary ascidians are known for autoregulation of early development but not known to reproduce asexually. Clearly, autoregulation and clonal reproduction eliminate each other.

Incidentally, considerable efforts have been made to fuse eggs and to increase its size of asteroids and echinoids (p 167). In echinoids, it is a pre-requisite to induce androgenics and triploids. For, the echinoid eggs are released from the ovary after complete maturation, i.e. after releasing the polar body and so on. Following fusion, cytoplasm of two oocytes does not mix. As a result, the induced androgenic and triploid echinoids do not survive beyond the blastula/gastrula stage (Fig. 6.1). Incidentally, the terminal ends of conspecific individuals of colonial ascidians may fuse (p 212). But the colony, from which one or more zooids are removed, continues to survive but its integrity is lost, as evidenced by the loss of synchronized maturation of its zooids.

Clonal propagation involves fission/autotomy or budding. As mentioned elsewhere, it can be bidirectional, when the fissioned ramets are developed into fully functional individuals, as in echinoderms. But it is unidirectional, when only one of the two ramets is fully developed into a functional individual, as in enteropneust worms (Table 8.1). Using these features, an attempt is made to group the asexual clonal propagation in some aquatic invertebrates (cf Pandian, 2016). From Table 11.1, the following inferences can be drawn: 1. Barring botryllid ascidians, the presence of totipotent ESCs is reported from none of these taxa. Like the neoblasts of turbullareans, the hemoblasts of botryllids are totipotent. 2. Following amputation, the solitary enteropneusts undergo unidirectional cloning. 3. Accompanying fission, the claimed bidirectional cloning in the pteropod mollusc *Clio pyrimidata* remains to be confirmed. On the other hand, the bidirectional cloning is an established feature of ophiuroids, holothuroids and asteroids. In these echinoderms and *C. pyrimidata*, multipotent stem cells arise from different sources; for example, from septum and hepatopancreas in *C. pyrimidata* and the oral disk in asteroids (Fig. 11.1) and ophiuroids (Fig. 4.3). In holothuroids, the missing tissues/organs/systems are regenerated mostly by dedifferentiation and transdifferentiation. For example, an electron microscopic study has shown that the respiratory tree is regenerated from the cloaca of *Clalolabes schemeltzii*, which has undergone fission at a single plane (Table 5.8) but the trees are developed in the middle as well as anterior and posterior middles bearing no cloaca in *Stichopus herrmanni*, which has undergone fissions at double and treble planes (Table 4.11, Fig. 11.1). Hence, it is likely that the epimorphic (holothuroids, ophiuroids) and morphallaxic (asteroids) regenerative processes involve the multipotent, migratory coelomocytes and the likes (Table 5.11). Arising from different sources of tissues/organs, the multipotent cells play a major role in accomplishing bidirectional cloning in echinoderms. Figure 11.1 shows the bidirectional cloning in representative carnivorous asteroids with pyloric caeca as a storage organ and sediment-feeding holothuroids with no special storage organ. 4. Unlike uni- and bi-directional clonal propagation resulting in the generation of one or two progenies, budding is a process leading to the production of a large number of buds/budlets. These buds have also as many tissues/organs/systems, as their parent. However, the generation of buds/externae arising from the internae of the parasitic colonial rhizocephalan crustaceans is exhausted, implying the non-renewability of the cluster of cells injected by vermogen larva into the body of the host crab (Table 11.1). The vermogen injected cells can also be totally eliminated, when the host crab molts. Secondly, the number of tissue types differentiated by the injected cells is also limited to four–five types only. Considering these features, Pandian (2016) designated these cluster cells as oligopotent. Contrastingly, budding involves evagination of a fraction of the parental body and the buds possess almost all tissue types. The body source, from which the buds arise, can be stolonial as in pterobranchs, pelagic tunicates and some benthic ascidians, or palleal, as in botryllids or post-thoracic, as in some benthic ascidians or vascular as in botryllids (see Fig. 10.4).

TABLE 11.1

Groups of taxons that undergo asexual propagation

Taxon, Reference	Reported or Inferred Observations
	Unidirectional cloning
Hemichordates, solitary enteropneust worms, Tweedel (1966), see Chapter 8	Proboscis and collar harbor multipotent stem cells and on amputation, the remaining proboscis and collar can regenerate the missing parts. Pre-branchial trunk holds multipotent stem cells to regenerate the remaining trunk
	Bidirectional cloning by fission
Pteropod mollusca *Clio pyrimidata* (van der Spoel, 1979, Pandian, 2017)	Columnar muscles and septum, the mesodermal derivatives harbor multipotent stem cells responsible for regeneration of all missing somatic and gonadal tissues/organs, respectively. Hepatopancreas, an endodermal derivative harbors multipotent stem cells responsible for regenerating the missing parts of the gut
Echinoderms: A. Epimorphic, Holothuroids and Ophiuroids, B. Morphallaxic asteroids	Regeneration of all missing organs/systems involving dedifferentiation, redifferentiation and transdifferentiation. Many organs like the pharynx and cloaca hold multipotent stem cells. Coelomocytes plays major role (see Table 5.11, Fig. 11.1)
	Budding
*Parasitic colonial rhizocephalan crustaceans? (see Pandian, 2016)	The vermogen larva injects a cluster of cells into the body of host crab. 'Cultured' in hemocoel the cells proliferate to produce only a few cell types in its internae and externae (buds) namely epidermis and nerve cells derived from ectoderm as well as muscle cells and connective tissue from mesoderm
Hemichordates colonial pterobranchs (Chapter 8)	Stolonial budding clumped vertically or spread horizontally
Pelagic tunicates	Stolonial budding generates sexual progenies
Benthic colonial tunicates	Sexual and asexual stolonial, palleal, post-thorax and vascular buds (see Fig. 10.4)

*an analogous situation has been described for the 'sacculunized' males of entoconchid prosobranch (Pandian, 2017)

11.10 Cloning and Coloniality

Being the closest relatives of vertebrates (Delsuc et al., 2006), much effort has been made to trace the phylogeny of ascidians (e.g. Zeng et al., 2006). But this account has made an attempt to trace the origin of coloniality among the clonal primitive deuterostomes. Table 11.2 summarizes major features that characterize clonal reproduction and coloniality in these deuterostomes.

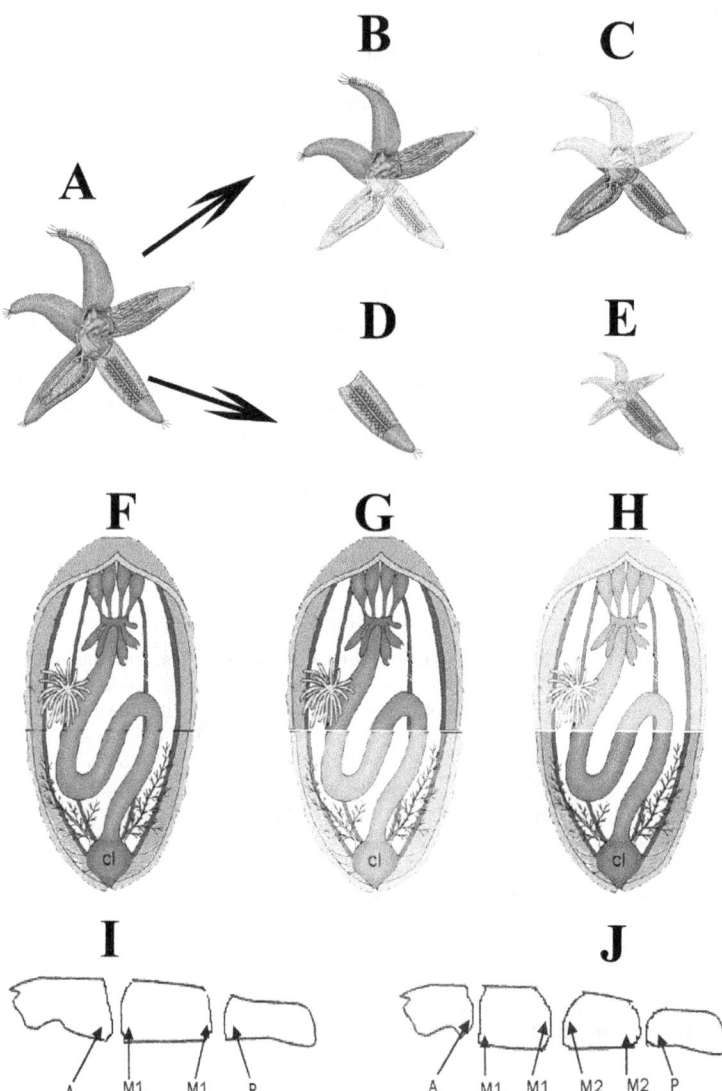

FIGURE 11.1

Clonal fission in a representative asteroid and holothuroid. A. A star fish (from untraceable Google source), B and C regenerating star fish, which underwent fission, D and E regenerating star fish, which underwent clonal autotomy. F. A holothuroid (redrawn from Kamenev and Dolmatov, 2016), G and H. Regenerating holothuroid. Note the lighter colored regenerative body parts of the star fish and holothuroid. *Stichopus herrmanni* undergoing fission at (I) double and (J) treble planes (redrawn from Hartati et al., 2016). A, M and B represent anterior, middle and posterior ramets; M1 and M2 represent anterior middle and posterior middle.

TABLE 11.2

Features that characterize solitary and colonials mode of reproductive strategies in primitive deuterostomes

Solitary	Colonial
Echinodermata	
Speciose, ~ 7,000 species. Motile. Fairly large 4–3,000 g. 2–5 years life span. Gonochorics, rarely protandrics. Oviparous, brooders rare. ~ 1.9, 2.6 and 3.7% of ophiuroids, asteroids and holothuroids reproduce clonally also. In them, *fission* is most common mode of clonal reproduction resulting in bidirectional cloning. Including echinoids, budding in them is limited to larval stage. Intervals between successive fissions is > 60 days to 1–2 years	–
Hemichordata	
Enteropneusts: ~ 90 species, Motile, 10–50 cm long vermiform gonochorics. Possess a large number of symmetrical gonadal pairs. Oviparous. Amputation of proboscis, collar or pre-branchial trunk results in *unidirectional cloning*	*Pterobranchs*: < 40 species, Sessile, 5–10 mm sized vertically or horizontally spreading colonies. Gonads limited to a single pair or only one. Ovoviviparous. Gonochorics individually but simultaneous hermaphrodites colonially. Reproduce clonally mostly by budding as well as sexually
Cephalochordata	
35 species, Motile, Gonochoric. Sexual reproduction alone. Oviparous. Prolific breeders producing tons and tons of larvae	–
Urochordata	
Pelagic tunicates	
Larvacea: ~ 70 species, Motile, Small 0.1–5.0 μm long. Short generation time and life span, semelparous, protandrics, oviparous > 300 eggs	*Thaliacea*: 75 species, Motile, Small 0.1–5.0 μm long, Short generation time and life span, semelparous, protandrics/protogynics, viviparous, Metagenics clonal reproduction by stolonial budding producing > 400 progenies, in comparison to 2–6 eggs produced by sexual reproduction. Population doubles in 2.5 days.
Benthic ascidians	
~ 1,690 species. Sessile. ~ 12 no./m². Grows to 5–15 cm height. Life span 1–3 years. Simultaneous hermaphrodites. Sexual reproduction alone. oviparous. 1 of the 2 blastomeres survives and autoregulates at metamorphosis	< 1,125 species, Sessile, 0.5–1.5 μm, life span 12–20 months, Gonochorics individually but simultaneous hermaphrodites colonially. Ovoviviparous or viviparous. Clonal reproduction by cyclic budding only but fragmentation of colonies limited to didemonids. Stolonial, palleal, post-thoracic and/or vascular budding. Buds are regenerated once every few days. Seasonal sexual reproduction release ~ 50 eggs/mm² colony area

From it, the following may be inferred: 1. Fission produces bidirectional cloning in echinoderms but unidirectional cloning in solitary enteropneusts and ascidians, in which cloning was experimentally induced. 2. Budding usually leads to the formation of coloniality in pterobranchs, thaliaceans and ascidians. Remarkably, the coloniality has imposed reductions in species number and body size, gonads and fecundity as well as switching from gonochorism to simultaneous hermaphroditism, oviparity to ovoviviparity/ viviparity. 3. Clonal budding reduces genetic diversity and speciation, especially in benthic sessile (but not in motile pelagic tunicates) pterobranchs and ascidians. Consequently, the colonials are less speciose with 40 and 1,125 species, in comparison to 90 and 1,690 species in hemichordates and benthic tunicates, respectively. 4. The repeated and rapid process of budding (in comparison to fission interval of 90 days to 1–2 years in echinoderms) results in size reduction. For example, the size ranges from 10 to 50 cm in enteropneusts in comparison to 5 to 10 mm in pterobranchs and 5–15 cm height in solitary ascidians, in comparison to 0.5 to 1.5 μm width in colonial ascidians. 5. As a consequence, the life span is also shortened, for example, from 1–3 years in sessile ascidians to 12–20 months in *Botryllus* (see also Rinkevich and Shapira, 1999, Paz and Rinkevich, 2002). An extreme seems to be semelparity, as in thaliaceans. 6. Solitary enteropneusts and larvaceans are motile but pterobranchs and colonial ascidians are sessile. 7. Sexuality is switched from gonochorism in solitary echinoderms, enteropneusts and larvacean to colonial simultaneous hermaphroditism in pterobranchs and colonial ascidians. 8. With inclusion of clonal propagation, the importance of sexual reproduction is dramatically reduced at levels of gonad number (pterobranchs) and fecundity (thaliaceans). 9. Mode of development is also switched from oviparity in solitaries to ovoviviparity/viviparity in colonial pterobranchs, thaliaceans and ascidians. Briefly, the repeated and rapid budding in pterobranchs, thaliaceans and ascidians leads to colonial formation. Coloniality imposes reductions in species number and body size, generation time and life span, gonad number and fecundity as well as switching from gonochorism to simultaneous hermaphorditism and oviparity to ovoviviparity/viviparity.

12

References

Abbott, D.P. and Newberry, A.T. 1980. Urochordates: the tunicates. In: *Intertidal Invertebrates of California*. (eds) Morris, R.H., Abbott, D.P. and Haderlie, E.C., Stanford University Press, Stanford, pp 179–229.

Achituv, Y. and Sher, E. 1991. Sexual reproduction and fission in the sea star *Asterina burtoni* from the Mediterranean coast of Israel. Bull Mar Sci, 48: 670–678.

Acuna, J. 2001. Pelagic tunicates: why gelatinous? Am Nat, 158: 100–107.

Agatsuma, Y. 2001a. Ecology of *Strongylocentrotus intermedius*. In: *Edible Sea Urchins: Biology and Ecology*. (ed) Lawrence, J.M., Elsevier, Amsterdam, pp 333–346.

Agatsuma, Y. 2001b. Ecology of *Hemicentrotus pulcherrimus*, *Pseudocentrotus depressus* and *Anthocidaris crassispina* in southern Japan. In: *Edible Sea Urchins: Biology and Ecology*. (ed) Lawrence, J.M., Elsevier, Amsterdam, pp 363–374.

Akhmadieva, A.V., Shukalyuk, A.I., Alexandrova, Y.N. and Isaeva, V.V. 2007. Stem cells in asexual reproduction of the colonial ascidian *Botryllus tubaratus* (Tunicata: Ascidiacea). Rus J Mar Biol, 33: 181–186.

Alcorn, N.J. and Allen, J.D. 2009. How do changes in parental investment influence development in echinoid echinoderms? Evol Dev, 11: 719–727.

Allen, J.D., Zakas, C. and Podolsky, R.D. 2006. Effects of egg size reduction in larval feeding on juvenile quality for a species with facultative-feeding development. J Exp Mar Biol Ecol, 331: 186–197.

Allen, J.D. 2012. Effects of egg size reductions on development time and juvenile size in three species of echinoid echinoderms: implications for life history theory. J Exp Mar Biol Ecol, 422-423: 72–80.

Al-Rashdi, K.M., Claereboudt, M.R. and Al-Busaidi, S.S. 2007. Density and size distribution of the sea cucumber *Holothuria scabra* (Jaeger, 1935) at size exploited sites in Mahout Bay, Sultanate of Oman. Agri Mar Sci, 12: 43–51.

Alves, L.S.S., Pereira, A. and Ventura, C. 2002. Sexual and asexual reproduction of *Coscinasterias tenuispina* (Echinodermata: Asteroidea) from Rio de Janeiro, Brazil. Mar Biol, 140: 95–101.

Amemiya, S. and Oji, T. 1992. Regeneration in sea lilies. Nature, 357: 546–547.

Anderson, S.C., Flemming, J.M., Watson, R. and Lotze, H.K. 2010. Serial exploitation of global sea cucumber fisheries. Fish Fisher, 12: 317–339.

Andrew, N. and Byrne, M. 2001. The ecology of *Centrostephanus rodgersii*. In: *Edible Sea Urchins: Biology and Ecology*. (ed) Lawrence, J.M., Elsevier, Amsterdam, pp 149–160.

Angerer, R.C. and Angerer, L.M. 2001. Sea urchin embryo: Specification of cell fate. Encyclopedia of Life Sciences, John Wiley, doi: 10.1002/9780470015902.a0001513.pub2.

Arnone, M.I., Byrne, M. and Martinez, P. 2015. Echinodermata. In: *Evolutionary Developmental Biology of Invertebrates: Deuterostomia*. (ed) Wanninger, A., Springer Verlag, Vienna, 6: 1–58.

Asha, P.S. and Muthiah, P. 2002. Spawning and larval rearing of sea cucumber *Holothuria (Theelothuria) spinifera* Theel. SPC Beache-de-mer Inf Bull, 16: 1–14.

Asha, P.S. and Muthiah, P. 2005. Effects of temperature, salinity and pH on larval growth, development and survival of the sea cucumber *Holothuria spinifera* Theel. Aquaculture, 250: 823–829.

Asha, P.S. and Muthiah, P. 2006. Effects of single and combined microalgae on larval growth, development and survival of the commercial sea cucumber *Holothuria spinifera* Theel. Aquacult Res, 37: 113–118.

Asha, P.S. and Muthiah, P. 2007. Growth of the hatchery-produced juveniles of commercial sea cucumber *Holothuria (Theelothuria) spinifera* Theel. Aquacult Res, 38: 1082–1087.

Asha, P.S. and Muthiah, P. 2008. Reproductive biology of the commercial sea cucumber *Holothuria spinifera* (Echinodermata: Holothuroidea) from Tuticorin, Tamil Nadu, India. Aquacult Int, 16: 231–242.

Asha, P.S., Rajagopalan, M. and Diwakar, K. 2011. Influence of salinity on hatching rate, larval and early juvenile rearing of sea cucumber *Holothuria scabra* Jaeger. J Mar Biol Ass Ind, 53: 218–224.

Asha, P.S. and Diwakar, K. 2013. Effect of stocking density on the hatching rate, larval and early juvenile rearing of edible sea cucumber *Holothuria scabra* (Jaeger, 1883). Ind J Geo-Mar Sci, 42: 191–195.

Asha, P.S. and Diwakar, K. 2015. Field observation of asexual reproduction by fission in sea cucumber *Holothuria atra*. Mar Fish Infor Serv, 223–224.

Asha, P.S., Johnson, B., Ranjith, L. et al. 2015. Status of sea cucumber resources and impact of fishing ban on the livelihood of fishers in Gulf of Mannar and Palk Bay. Mar Fish Inf Ser, 226: 3–9.

Azariah, J. 1969. Physiological and histochemical studies on some prochordates (Cephalochordates). PhD Thesis, University of Madras.

Azariah, J. 1994. Cephalochordata. In: *Reproductive Biology of Invertebrates*. (eds) Adiyodi, K.G. and Adiyodi, R.G., Oxford & IBH Publishing, New Delhi, 6B: 395–401.

Bak, R.P.M., Carpay, M.J.E. and De Ruyter van Steveninck, E.D. 1984. Densities of the sea urchin *Diadema antillarum* before and after mass mortalities on the coral reefs of Curacao. Mar Ecol Prog Ser, 17: 105–108.

Ballarin, L., Cima, F. and Sabbadin, A. 1993. Histoenzymatic staining and characterization of the clonal ascidian *Botryllus schlosseri* (Urochordata: Ascidiacea). Curr Pharm Des, 14: 138–147.

Ballarin, L. and Manni, L. 2009. Stem cells in sexual and asexual reproduction of *Botryllus schlosseri* (Ascidiacea: Tunicata): An overview. In: *Marine Stem Cells*. (eds) Rinkevich, B. and Mataranga, V., Springer Verlag, New York, pp 267–280.

Balser, E.J. 1998. Cloning by ophiuroid echinoderm larvae. Biol Bull, 194: 187–193.

Bancroft, F.W. 1903. Aestivation of *Botryllus gascoi* Della Valle. In: *Mark Anniversary*. Henry Holt, New York, 8: 147–166.

Bannister, R., McGonnell, I.M., Graham, A. et al. 2005a. *Afuni*, a novel transforming growth factor-beta gene is involved in arm regeneration of the brittle star *Amphiura filiformis*. Dev Genes Evol, 215: 393–401.

Bannister, R., McGonnell, I.M., Graham, A. et al. 2005b. Coelomic expression of a novel bone morphogenetic protein in regenerating arm of the brittle star *Amphiura filiformis*. Dev Genes Evol, 218: 33–38.

Barbaglio, A., Mozzi, D., Sugni, M. et al. 2006. Effects of exposure to ED contaminants (TPT-Cl and fenarimol) on crinoid echinoderms: comparative analysis of regenerative development and correlated steroid levels. Mar Biol, 149: 65–77.

Barbaglio, A., Sugni, M., Di Benedetto, C. et al. 2007. Gametogenesis correlated with steroid levels during the gonadal cycle of the sea urchin *Paracentrotus lividus* (Echinodermata: Echinoidea). Comp Biochem Physiol, 147A: 466–474.

Barker, M.F. and Xu, R.A. 1993. Effects of estrogens on gametogenesis and steroid levels in the ovaries and pyloric caeca of *Sclerasterias mollis* (Echinodermata: Asteroidea). Invert Reprod Dev, 24: 53–58.

Barker, M.F. 2001. The ecology of *Evechinus chloroticus*. In: *Edible Sea Urchins: Biology and Ecology*. (ed) Lawrence, J.M., Elsevier, Amsterdam, pp 245–260.

Barker, M.F. and Scheibling, R.E. 2008. Rates of fission, somatic growth and gonadal development of a fissiparous sea star, *Allostichaster insignis* in New Zealand. Mar Biol, 153: 815–824.

Barrios, J.V., Gaymer, C.F., Vásquez, J.A. and Brokordt, K.B. 2008. Effect of the degree of autotomy on feeding, growth, and reproductive capacity in the multi-armed sea star *Heliaster helianthus*. J Exp Mar Biol Ecol, 361: 21–27.

Battaglene, S.C., Seymour, T.E., Ramofafia, C. and Lane, I. 2002. Spawning induction of three tropical sea cucumbers, *Holothuria scabra*, *Holothuria fuscogilva* and *Actinopyga mauritiana*. Aquaculture, 207: 29–47.

Baumiller, T.K. and Messing, C.G. 2007. Stalked crinoid locomotion, and its ecological and evolutionary implications. Palaeontol Electron, 10: 1–10.

Bay-Schmith, E. and Pearse, J.S. 1987. Effect of fixed day lengths on photoperiodic regulation of gametogenesis in the sea urchin *Strongylocentrotus purpuratus*. Int J Invert Reprod Dev, 11: 287–294.

Benitez-Villalobos, F., Avila-Poveda, O.H. and Gutierrez-Mendez, I.S. 2013. Reproductive biology of *Holothuria fuscocinerea* (Echinodermata: Holothuroidea) from Oaxaca, Mexico. Sex Early Dev Aquat Org, 1: 13–24.

Berrill, N.J. 1941. The development of the bud in *Botryllus*. Biol Bull, 80: 169–184.

Berrill, N.J. 1950. *The Tunicata with an Account of the British Species*. Ray Society, London, p 354.

Berrill, N.J. 1951. Regeneration and budding in tunicates. Biol Rev, 26: 456–475.

Bertrand, S. and Escriva, H. 2011. Evolutionary crossroads in developmental biology: amphioxus. Development, 138: 4819–4830.

Bingham, B.L., Burr, J. and Wounded Head, H. 2000. Causes and consequences of arm damage in the sea star *Leptasterias hexactis*. Can J Zool, 78: 596–605.

Birkeland, C., Chia, F.-S. and Strathmann, R.R. 1971. Development, substratum selection, delay of metamorphosis and growth in the seastar *Mediaster aequalis* Stimpson. Biol Bull, 41: 99–108.

BOBLME, 2015. Report of FAO-BOBLME project on Evaluation of the current conservation measures on sea cucumber stocks in Palk Bay and Gulf of Mannar of India, Chennai, p 97.

Bodnar, A.G. 2015. Cellular and molecular mechanisms of negligible senescence: Insight from the sea urchin. Invert Reprod Dev, 59: 23–27.

Boffi, E. 1972. Ecological aspects of ophiuroids from the phytal of S.W. Atlantic Ocean warm waters. Mar Biol, 15: 316–328.

Bolton, T.F., Thomas, F.I.M. and Leonard, C.N. 2000. Maternal energy investment in eggs and jelly coats surrounding eggs of the echinoid *Arbacia punctulata*. Biol Bull, 199: 1–5.

Bone, Q. 1998. *The Biology of Pelagic Tunicates*. Oxford University Press, New York.

Bonham, K. and Held, E.E. 1963. Ecological observations on the sea cucumbers *Holothuria atra* and *H. leucospilota* at Rongelap Atoll, Marshall Islands. Pacific Sci, 17: 305–314.

Bookbinder, L.H. and Shick, J.M. 1986. Anaerobic and aerobic energy metabolism in ovaries of the sea urchin *Strongylocentrotus droebachiensis*. Mar Biol, 93: 103–110.

Boolootian, R.A. and Moore, A.R. 1956. Hermaphroditism in echinoids. Biol Bull, 111: 328–335.

Borges, M., Yokoyama, L.Q. and Amaral, C.Z. 2009. Gametogenic cycle of *Ophioderma januarii*, a common Ophiodermatidae (Echinodermata: Ophiuroidea) in southeastern Brazil. Zoologia, 26: 118–126.

Bosch, I., Rivkin, R.B. and Alexander, S.P. 1989. Asexual reproduction by oceanic planktotrophic echinoderm larvae. Nature, 337: 169–170.

Bottger, S.A., Eno, C.C. and Walker, C.W. 2011. Methods for generating triploid green sea urchin embryos: An initial step in producing triploid adults for land based and near shore aquaculture. Aquaculture, 318: 199–206.

Boudouresque, C.F. and Verlaque, M. 2001. Ecology of *Paracentrotus lividus*. In: *Edible Sea Urchins: Biology and Ecology*. (ed) Lawrence, J.M., Elsevier, Amsterdam, pp 177–216.

Bourgoin, A. and Guillou, M. 1994. Arm regeneration in two populations of *Acrocnida brachiata* (Montagu) (Echinodermata: Ophiuroidea) in Douarnenez Bay (Brittany: France): An ecological significance. J Exp Mar Biol Ecol, 184: 123–39.

Bowmer, T. and Keegan, B.F. 1983. Field survey of the occurrence and significance of the regeneration of *Amphiura filiformis* (Echinodermata: Ophiuroidea) from Galway, west coast of Ireland. Mar Biol, 74: 65–71.

Braconnot, J.C. 1963. Study of the annual cycle of salps and doliolids in Rade de Ville frenche-sur-mer. ICES Int Counc Explor Sea, 28: 21–36.

Braconnot, J.C. 1970. Contribution al'etude des stades successifis dans le cycle des tuniciers pelagiques Doliolides. 1 Les stades larvaire, oozoide, nourrice et gastrozoide. Arch Zool Exp Genet, 111: 629–668.

Brandriff, B., Hinegardner, R.T. and Steinhardt, R. 1975. Development and life cycle of parthenogentically activated sea urchin embryos. J Exp Zool, 192: 13–24.

Brien, P. 1932. L'heteromorphose chez les Tuniciers. La regeneration bithoracique et monothoracique des fragments oesophagiens de *Clavelina lepadeformis* (Muller). Bull Acad Belg Cl Sci, 18: 975–1005.

Brien, P. 1968. Blastogenesis and morphogenesis. In: *Advances in Morphogenesis*. (eds) Abercrombie, M., Brachet, J. and King, T.J., Academic Press, New York, 7: 151–203.

Britt, D.T., Yeomans, D., Housen, K. and Consolmagno, G. 2002. Asteroid density, porosity and structure. Asteroids, 3: 485–500.

Brooks, J.M. and Wessel, G.M. 2002. The major yolk protein in sea urchins is a transferrin-like, iron binding protein. Dev Biol, 245: 1–12.

Brown, F.D. and Swalla, B.J. 2007. Vasa expression in a colonial ascidian *Botrylloides violaceus*. Evol Dev, 9: 165–177.

Brown, F.D., Keeling, E.L., Le, A.D. and Swalla, B.J. 2009a. Whole body regeneration in a colonial ascidian, *Botrylloides violaceus*. J Exp Zool, 312B: 885–900.

Brown, F.D., Tiozzo, S., Roux, M.M. et al. 2009b. Early lineage specification of long-lived germline precursors in the colonial ascidian *Botryllus schlosseri*. Development, 136: 3485–3494.

Brown, F.D. and Swalla, B.J. 2012. Evolution and development of budding by stem cells: Ascidian coloniality as a case study. Dev Biol, 369: 151–162.

Bureau, D., Campbell, A. and Hartwick, E.B. 1997. Roe enhancement in the red sea urchin *Strongylocentrotus franciscanus*, fed the bull kelp, *Nereocystis luetkeana*. Bull Aquacul Ass, 1: 26–30.

Burighel, P., Burnetti, R. and Zaniola, G. 1976. Hibernation of the colonial ascidian *Botrylloides leachi* (Savigny): Histological observation. Bull Zool, 43: 293–301.

Byrne, M. 1986. Induction of evisceration in the holothurians *Eupentacta quinquesemita* and evidence for the existence of an endogenous evisceration factor. J Exp Biol, 120: 25–39.

Byrne, M. 1990. Annual reproductive cycles of the commercial sea urchin *Paracentrotus lividus* from an exposed intertidal and a sheltered subtidal habitats on the west coast of Ireland. Mar Biol, 104: 275–289.

Byrne, M. 1991. Reproduction, development and population biology of the Caribbean ophiuroid *Ophionereis olivacea*, a protandric hermaphrodite that broods its young. Mar Biol, 111: 387–399.

Byrne, M. and Cerra, A. 1996. Evolution of intragonadal development in the diminutive asterinid sea stars *Patiriella vivipara* and *P. parvivipara* with an overview of development in the asterinidae. Biol Bull, 191: 17–26.

Byrne, M. 2001. The morphology of autotomy structures in the sea cucumber *Eupentacta quinquesemita* before and during evisceration. J Exp Biol, 204: 849–863.

Byrne, M. 2006. Life history diversity and evolution in the Asterinidae. Integ Comp Biol, 46: 243–254.

Byrne, M., Sewell, M.A. and Prowse, T.A.A. 2008. Nutritional ecology of sea urchin larvae: influence of endogenous and exogenous nutrition on echinopluteal growth and phenotypic plasticity in *Tripneustes gratilla*. Func Ecol, 22: 643–648.

Caballes, C.F., Pratchett, M.S., Kerr, A.M. and Rivera-Posada, J.A. 2016. The role of maternal nutrition on oocyte size and quality, with respect to early larval development in the coral eating starfish *Acanthaster planci*. PLoS One, 11(6): e0158007.

Cameron, A., Leachy, P.S. and Davidson, E.H. 1996. Twins raised from separated blastomeres develop into sexually mature. Dev Biol, 178: 514–519.

Cameron, R.A., Hough-Evans, B.R., Britten, R.J. and Davidson, E.H. 1987. Lineage and fate of each blastomere of the eight-cell sea urchin embryo. Genes Dev, 1: 75–84.

Cameron, R.A., Mahairas, G., Ras, J.P. et al. 2000. A sea urchin genome project: Sequence, scan, virtual map and additional resources. Proc Natl Acad Sci, USA, 97: 9514–9518.

Candelaria, A.G., Murray, G., File, S.K. and Garcia-Arraras, J.E. 2006. Contribution of mesenterial muscle dedifferentiation to intestine regeneration in the sea cucumber *Holothuria glaberrima*. Cell Tissue Res, 325: 55–65.

Candia-Carnevali, M.D. 2006. Regeneration in echinoderms: Repair, regrowth, cloning. Invert Surv J, 3: 64–76.

Candia-Carnevali, M.D., Bonasoro, F., Patruno, M. et al. 2001. PCB exposure and regeneration in crinoids (Echinodermata). Mar Ecol Prog Ser, 215: 155–167.

Candia-Carnevali, M.D., Thorndyke, M.C. and Matranga, V. 2009. Regenerating echinoderms: A promise to understand stem cell potential. In: *Stem Cells in Marine Organisms*. (eds) Rinkevich, B. and Matranga, V., Springer Verlag, Heidelberg, pp 165–186.

Carrier, T.J., King, B.L. and Coffman, J.A. 2015. Gene expression changes associated with the developmental plasticity of sea urchin larva in response to food availability. Biol Bull, 228: 171–180.

Casareto, B.E. and Nemoto, T. 1986. Salps of the Southern Ocean (Australian sector) during the 1983–84 summer, with special reference to the species *Salpa thompsoni*, Foxton 1961. Mem Natl Inst Polar Res, 40: 221–239.

Chaet, A.B. and McConnaughy, R.A. 1955. Physiologic activity of nerve cell extracts. Biol Bull, 117: 407–408.

Chao, S.M. and Tsai, C.C. 1995. Reproduction and population dynamics of the fissiparous brittle star *Ophiactis savignyi* (Echinodermata: Ophiuroidea). Mar Biol, 124: 77–78.

Chao, S.N., Chen, C.P. and Alexander, P.S. 1993. Fission and its effect on population structure of *Holothuria atra* (Echinodermata: Holothuroidea) in Taiwan. Mar Biol, 116: 109–115.

Chao, S.N., Chen, C.P. and Alexander, P.S. 1994. Reproduction and growth of *Holothuria atra* (Echinodermata: Holothuroidea) at two contrasting sites in southern Taiwan. Mar Biol, 119: 565–570.

Chat, A.B. 1962. A toxin in the coelomic fluid of scalded starfish (*Asterias forbesi*). Proc Soc Exp Biol Med, 109: 791–794.

Chen, T.Y. 1931. On a hermaphrodite specimen of the Chinese amphioxux. Peaking Nat Hist Bull, 5: 11–16.

Chen, Y., Shin, P.K.S. and Cheung, S.G. 2008. Growth, secondary production and gonad development of two co-existing amphioxus species (*Branchiostoma belcheri* and *B. malayanum*) in subtropical Hong Kong. J Exp Mar Biol Ecol, 357: 64–74.

Chinn, S. 2006. Habitat distribution and comparison of brittle star (Echinodermata: Ophiuroidea) arm regeneration on Moorea, French Polynesia. eSchol Univ California, Berkeley.

Chino, Y., Saito, M., Yamasu, K. et al. 1994. Formation of the adult rudiment of sea urchins is influenced by thyroid hormone. Dev Biol, 161: 1–11.

Cima, F. and Ballarin, L. 2004. Tributyltin, sulfhydryl interaction as a cause of immunotoxicity in phagocytes of tunicates. Ecotoxicol Env Saf, 58: 386–395.

Clements, L.A., Bell, S.S. and Kurdziel, J.P. 1994. Abundance and arm loss of the infaunal brittlestar *Ophiophragmus filograneus* (Echinodermata: Ophiuroidea) with an experimental determination of regeneration rates in natural and planted seagrass beds. Mar Biol, 121: 97–104.

Cloney, R.A. 1990. Urochordata-Ascidiacea. In: *Reproductive Biology of Invertebrates*, (eds) Adiyodi, K.G. and Adiyodi, R.G., Oxford IBH Publishing, New Delhi, Vol 4: Part B, pp 391–451.

Cochran, R.C. and Engelmann, F. 1972. Echinoid spawning induced by a radial nerve factor. Science, 178: 423–424.

Colombera, D. and Venier, G. 1981. Il numero dei chromosomi di cinque specie di echinodermi. Caryologia, 33: 503–507.

Conand, C. 1981. Sexual cycle of three commercially important holothurian species (Echinodermata) from the Lagoon of New Caledonia. Bull Mar Sci, 31: 523–543.

Conand, C. and De Ridder, C. 1990. Reproduction asexuce par scission chez *Holothuria atra* (Holothuroidea) dans des populations de platiers recifaux. In: *Echinoderm Research*. (eds) De Ridder, C. et al., Balkema, Rotterdam, 71–76.

Conand, C. and Byrne, M. 1993. A review of recent developments in the world sea cucumber fisheries. Mar Fish Rev, 55: 1–13.

Conand, C. 1993. Reproductive biology of the holothurians from the major communities of the New Caledonian Lagoon. Mar Biol, 216: 439–450.

Conand, C. 1996. Asexual reproduction by fission in *Holothuria atra*: variability of some parameters in populations from tropical Indo-Pacific. Oceanol Acta, 19: 209–216.

Conand, C., Morel., C. and Mussard, R. 1997. A new study on asexual reproduction in holothurians. Fission in *Holothuria leucospilota* populations on Reunion Island in the Indian Ocean. SPC Beche-de-Mer Info Bull, 9: 5–11.

Conand, C., Armand, J., Dijoux, N. and Garryer, J. 1998. Fission in a population of *Stichopus chloronotus* on Reunion Island, Indian Ocean. SPC Beche-de-Mer Inf Bull, 10: 15–23.

Conand, C. 2004. Monitoring fissiparous population of *Holothuria atra* in a fringing reef on Reunion Island (Indian Ocean). SPC Beche-de-Mer Inf Bull, 20: 22–25.

Conklin, E.G. 1906. Does half an ascidan egg give rise to a whole larva? Arch Entwiknungsmech, 2: 727–753.

Crozier, W.J. 1917. Multiplication by fission in holothurians. Am Nat, 51: 560–566.

Crump, R.G. and Barker, M.F. 1985. Sexual and asexual reproduction in geographically separated populations of the fissiparous asteroid *Coscinasterias calamaria* (Gray). J Exp Mar Biol Ecol, 88: 109–127.

Dan, K. and Kubota, H. 1960. Data on the spawning of *Comanihus japonica* between 1937 and 1955. Embryologia, 5: 21–37.

Daponte, M.C., Capitanio, F.L. and Esnal, G.B. 2001. A mechanism for swarming in the tunicate *Salpa thompsoni* (Foxton, 1961). Antarct Sci, 13: 240–245.

Daponte, M., Palmieri, M.A., Casareto, B. and Esnal, G.B. 2013. Reproduction and population structure of the salp *Iasis zonaria* (Pallas, 1774) in the southwestern Atlantic Ocean (34° 30'S to 39° 30'S) during three successive winters (1999–2001). J Plankton Res, 35: 813–830.

Davidson, E.H., Cameron, R.A. and Ransick, A. 1998. Specification of cell fate in the sea urchin embryo: Summary and some proposed mechanisms. Development, 125: 3269–3290.

Davidson, E.H., Rast, J.P. Oliveri, P. et al. 2002. Genomic regulatory network for development. Science, 295: 1669–1778.

Davis, L.V. 1967. The suppression of autotomy in *Linckia multiflora* (Lamark) by a parasitic gastropod *Stylifer linckiae*. Veliger, 9: 343–346.

Dawbin, W.H. 1949. Autoevisceration and the regeneration of viscera in the holothurians *Stichopus mollis* (Hulton). Trans Soc, New Zealand, 77: 497–523.

De Tomaso, A.W., Nyholm, S.V., Palmeri, K.J. et al. 2005. Isolation and characterization of a protochordate histocompatibility locus. Nature, 438: 454–459.

Deibel, D. and Lowen, B. 2012. A review of life cycle and life-history adaptations of pelagic tunicates to environmental conditions. ICES J Mar Sci, 69: 358–369.

Deichmann, E. 1922. On some cases of multiplication by fission and coalescence in holothurians. Vid Medd Dansk Naturh Foren, 73: 199–213.

Delavault, R. 1975. Hermaphroditism in echinoderms: Studies on asteroids. In: *Intersexuality in the Animal Kingdom*. (ed) Reinboth, R., Springer Verlag, Berlin, pp 188–194.

Delroisse, J., Fourgon, D. and Eeckhaut, I. 2013. Reproductive cycles and recruitment in *Ophiomastix venosa* and *Ophiocoma scolopendrina*, two co-existing tropical ophiuroids from the barrier reef of Toliara (Madagascar). Cah Bio Mar, 54: 593–603.

Delsuc, F., Brinkman, H., Chorrout, D. et al. 2006. Tunicates and not cephalochordates are the closest living relative of vertebrates. Nature, 439: 965–968.

de Jong Westman, M., Qian, P.Y., March, B.Y. and Carefoot, T.H. 1995. Artificial diets in sea urchin culture: effects of dietary proteins level and other additives on egg quality, larval morphometrics and larval survival in the green sea urchin *Strogylocentrotus droebachiensis*. Can J Zool, 73: 2080–2090.

den Besten, P.J., Herwig, H.J., Zandee, D.I. and Voogt, P.A. 1989. Effects of cadmium and PCBs on reproduction of the sea star *Asterias rubens*: aberrations in the early development. Ecotoxicol Environ Saf, 18: 178–180.

den Besten, P.J., Herwig, H.J., Smaal, A.C. et al. 1990. Interference of polychlorinated biphenyls (Clophen A50) with gametogenesis in the sea star *Asterias rubens* L. Aquat Toxicol, 18: 231–246.

Desdevises, Y., Maillet, V., Fuentes, M. and Escriva, H. 2011. A snapshot of the population structure of *Branchiostoma lanceolatum* in the Racou Beach, France, during its spawning season. PLoS ONE, 6: e18520.

Diaz-Guisado, D., Gaymer, C.F., Brokordt, K.B. and Lawrence, J.M. 2006. Autotomy reduces feeding, energy storage and growth of the sea star *Stichaster striatus*. J Exp Mar Biol Ecol, 338: 73–80.

Dix, T.G. 1977. Reproduction in Tasmanian population of Heliocidaris erythrogramma (Echinodermata: Echinometridae). Aust J Mar Freshwat Res, 28: 509–520.

Dolmatov, I.Y. 2014. New data on asexual reproduction, autotomy, and regeneration in holothurians of the order Dendrochirotida. Rus J Mar Biol, 40: 228–232.

Dolmatov, I.Y., Frolova, L.T., Zhakharova, E.A. and Ginanova, T.T. 2011. Development of respiratory trees in the holothurians *Apostichopus japonicus* (Aspidochirotida: Holothuroidea). Cell Tissue Res, 346: 327–338.

Dolmatov, I.Y., Ginanova, T.T. and Frolova, L.T. 2016. Metamorphosis and definitive organogenesis in the holothurians *Apostichopus japonicus*. Zoomorphology, 135: 173–188.

Dolmatov, Y., Khang, N.A. and Kamenev, Y.O. 2012. Asexual reproduction, evisceration, and regeneration in holothurians (Holothuroidea) from Nha Trang Bay of the South China Sea. Rus J Mar Biol, 38: 243–252.

Donahue, J.K. 1940. Occurrence of estrogens in the ovaries of certain marine invertebrates. Endocrinoogy, 27: 149–152.

Doty, J.E. 1977. Fission in *Holothuria atra* and holothurian population growth. M.S. Thesis, Univ Guam, p 54.

Downey, M.E. 1973. *Stafishes from the Caribbean and the Gulf of Mexico*. Smithsonian Contribution to Zoology, 126: 1–168.

Duffy, L., Sewell, M.A. and Murray, B.G. 2007. Chromosome number and chromosome variation in embryos of *Evechinus chloroticus* (Echinoidea: Echinometridae): Is there conservation of chromosome number in the Phylum Echinodermata? New findings and a brief review. Invert Reprod Dev, 50: 219–231.

Dumont, C., Pearce, C.M., Stazicker, C. et al. 2006. Can photoperiod manipulation affect gonad development of a boreo-arctic echinoid (*Strongylocentrotus droebachiensis*) following exposure in the wild after the autumnal equinox? Mar Biol, 149: 365–378.

Dupont, S. and Thorndyke, M.C. 2006. Growth or differentiation? Adaptive regeneration in the brittle stars *Amphiura filiformis*. J Exp Biol, 209: 3879–3881.

Eaves, A.A. and Palmer, A.R. 2003. Wide spread cloning in echinoderm larvae. Nature, 425: 146.

Ebert, T.A. 2008. Longevity and lack of senescence in the red sea urchin *Strongylocentrotus franciscanus*. Exp Gerontol, 43: 734–738.

Ebert, T.A., Russel, M.P., Gamblia, G. and Bodmar, A. 2008. Growth, survival and longevity estimates for rock-boring sea urchin *Echinometra lucunter lucunter* (Echinodermata: Echinoidea) in Bermuda. Bull Mar Sci, 82: 381–403.

Eckelbarger, K.J., Young, C.M. and Cameron, J.L. 1989a. Ultrastructure and development of dimorphic sperm in the abyssal echinoid *Phrissocystis multispina* (Echinodermata: Echinoidea): Implications for deep sea reproductive biology. Biol Bull, 176: 257–271.

Eckelbarger, K.J., Young, C.M. and Cameron, J.L. 1989b. Modified sperm ultrastructure in four species of soft-bodied echinoids (Echinodermata: Echinothuriidae) from the bathyal zone of the deep sea. Biol Bull, 177: 230–236.

Edmondson, C.H. 1935. Autotomy and regeneration in Hawaiian starfishes. Oceas Pap Bernice Pauachi Bishop Mus, Honolulu, 11: 1–29.

Emlet, R.B. 1995. Developmental mode and species geographic names in regular sea urchins (Echinodermata: Echinoidea). Evolution, 49: 476–489.

Emlet, R.E. and Hoegh-Guldberg, O. 1997. Effects of egg size on post larval performance: experimental evidence from a sea urchin. Evolution, 51: 141–152.

Emson, R.H. and Mladenov, P.V. 1987. Studies of the fissiparous holothurian *Holothuria parvula* (Selenka) (Echinodermata: Holothuroidea). J Exp Mar Biol Ecol, 111: 195–211.

Emson, R.H. and Wilkie, I.C. 1980. Fission and autotomy in echinoderms. Oceanogr Mar Biol Ann Rev, 18: 155–250.

Emson, R.H. and Wilkie, I.C. 1984. An apparent instance of recruitment following sexual reproduction in the fissiparous brittlestar *Ophiactis savignyi* (Muller & Troschel). J Exp Mar Biol Ecol, 77: 23–38.

Eno, C.C., Bottger, S.A. and Walker, C.W. 2009. Methods for karyotyping and for localization of developmentally relevant genes on the chromosomes of the purple sea urchin, *Strongylocentrotus purpuratus*. Biol Bull, 217: 306–312.

Evans, T., Rosenthal, E.T., Youngblom, J. et al. 1983. Cyclin: a protein specified by maternal mRNA in sea urchin eggs that is destroyed at each cleavage division. Cell, 33: 389–396.

Falkner, I., Byrne, M. and Sewell, M.A. 2006. Maternal provisioning in *Ophionereis fasciata* and *O. schayeri*: Brittle stars with contrasting modes of development. Biol Bull, 211: 204–207.

Falkner, I., Sewell, M.A. and Byrne, M. 2015. Evolution of maternal provisioning in ophiuroid echinoderms: Characterization of egg composition in planktotrophic and lecithotrophic developers. Mar Ecol Prog Ser, 525: 1–13.

FAO. 1995. Yearbook, fishery statistics, catches, and landings, FAO Rome.

Fell, H.B. 1946. The embryology of the viviparous ophiuroid *Amphipholis squamata* Delle Chiaje. Trans R Soc New Zealand, 75: 419–464.

Flood, P. and Deibel, D. 1998. The appendicularian house. In: *The Biology of Pelagic Tunicates*. (ed) Bone, Q., Oxford University Press, New York, pp 105–124.

Flood, P.R., Braun, J.G. and De Leon, A.R. 1976. On the annual production of Amphioxus larvae (*Branchiostoma senegalense* Webb) off Cap Blanc, North West Africa. Sarsia, 61: 63–70.

Foxton, P. 1966. The distribution and life history of *Salpa thompsoni* Foxton, with observations on a related species, *Salpa gerlachei* Foxton. Discovery Rep, 34: 1–116.

Francis, N., Gregg, T., Owen, R. et al. 2006. Lack of age associated telomere shortening in long- and short-lived species of sea urchins. FEBS Letters, 580: 4713–4717.

Freeman, G. 1964. The role of blood cells in the process of asexual reproduction in the tunicate *Perophora viridis*. J Exp Zool, 156: 157–183.

Freeman, G. 1971. A study of the intrinsic factors which control the initiation of asexual reproduction in the tunicate *Amaroecium constellatum*. J Exp Zool, 178: 433–456.

Fuentes, M., Benito, E., Bertrand, S. et al. 2007. Insights into spawning behavior and development of the European amphioxus (*Branchiostoma lanceolatum*). J Exp Zool B Mol Dev Evol, 308: 484–493.

Fuji, A. 1960. Studies on the biology of the sea urchin. II Size at first maturity and sexuality of two sea urchins *Strongylocentrotus nudus* and *S. intermedius*. Bull Fac Fish Hokkaido Univ, 11: 43–48.

Fujimura, M. and Takamura, K. 2000. Characterization of an ascidian DEAD-box gene, *Ci-DEAD1*: specific expression in the germ cells and its mRNA localization in posteriormost blastomeres in early embryos. Dev Genes Evol, 210: 64–72.

Fujita, T. and Ohta, S. 1990. Size structures of dense population of the brittle star *Ophiura sarsii* (Ophiuroidea: Echinodermata) in the bathyal zone around Japan. Mar Ecol Prog Ser, 64: 113–122.

Fujiwara, S. and Kuwamura, K. 1992. Ascidian budding as a transdifferentiation-like system: Multipotent epithelium is not undifferentiated. Dev Growth Diff, 34: 463–472.

Futch, C.R. and Diwineil, S.E. 1977. Nearshore marine ecology at Hutchinson Island, Florida: 1971–1974: 4. Lancelets and Fishes. Fla Mer Res Publ, 24: 1–23.

Gaffney, J. and Goad, L.J. 1974. Progesterone metabolism by echinoderm *Asterias rubens* and *Marthasterias gracilis*. Biochem J, 138: 309–311.

Gage, J.D. and Tyler, P.A. 1982. Growth and reproduction of the deep sea brittlestar *Ophiomuscium lymani* Wyvile Thompson. Oceanol Acta, 5: 73–83.

Galt, C.P. and Fenaux, R. 1990. Urochordata-thaliacea. In: *Reproductive Biology of Invertebrates*. (eds) Adiyodi, K.G. and Adiyodi, R.G., Oxford IBH Publishing, New Delhi, Vol 4: Part B, pp 471–527.

Ganot, P., Bouquet, J.M., Kallesoe, T. and Thompson, E.M. 2007. The Oikopleura coenocyst, a unique chordate germ cells permitting rapid, extensive modulation of oocyte production. Dev Biol, 302: 591–600.

Garcia-Arraras, J.E., Estrada-Rodgers, L., Santago, R. et al. 1998. Cellular mechanisms of intestine regeneration in the sea cucumber *Holothuria glaberrima* Slenka (Holothuroidea: Echinodermata). J Exp Zool, 281: 288–304.

Garcia-Arraras, J.E. and Greenberg, M.J. 2001. Visceral regeneration in holothurians. Microsc Res Tech, 55: 438–451.

Garcia-Arraras, J.E. and Dolmatov, I.Y. 2010. Echinoderms: Potential model systems for studies on muscle regeneration. Curr Pharm Des, 16: 942–955.

Garrett, F.K., Mladenov, P.V. and Wallis, G.P. 1997. Evidence of amictic reproduction in the brittle-star *Ophiomyxa brevirima*. Mar Biol, 129: 169–174.

Gasparini, F., Manni, L., Cima, F. et al. 2015. Sexual and asexual reproduction in the colonial ascidan *Botryllus schlosseri*. Genesis, 53: 105–120.

Geneviere, A.-M., Aze, A., Even, Y. et al. 2009. Cell dynamics in early embryogenesis and pluripotent embryonic cell lines: From sea urchin to mammals. In: *Stem Cells in Marine Organisms*. (eds) Rinkevich, B. and Matranga, V., Springer Verlag, Dordrecht, pp 215–244.

George, S.B., Cellario, C. and Fenaux, L. 1990. Population differences in egg quality of *Arbacia lixula* (Echinodermata: Echinoidea): proximate composition of eggs and larval development. J Exp Mar Biol Ecol, 141: 107–1.

George, S.B., Lawrence, J.M. and Fenaux, L. 1991. The effect of food ration on the quality of eggs of *Luidia clathrata* (Say) (Echinodermata: Asteroidea). Invert Reprod Dev, 20: 237–242.

George, S.B. 1994. Population differences in maternal size and offspring quality of *Leptasterias epiculora* (Brandt) (Echinodermata: Asteroidea). J Exp Mar Biol Ecol, 175: 121–131.

George, S.B. 1996. Echinoderms egg and larval quality as a function of adult nutritional state. Oceanol Acta, 19: 297–308.

Goldin, A. 1948. Regeneration in *Perophora viridis*. Biol Bull, 94: 184–193.

Goldstone, J.V., Goldstone, H.M.H., Morrison, A.M. et al. 2007. Cytochrome P450 1 genes in early deuterostomes (tunicates and sea urchins) and vertebrates (chicken and frog): Origin and diversification of the CYP1 gene family. Mol Biol Ecol, 24: 2619–2631.

Gordeaux, J.E.A. 1990. Urochordata-thaliacea. In: *Reproductive Biology of Invertebrates*. (eds) Adiyodi, K.G. and Adiyodi, R.G., Oxford IBH Publishing, New Delhi, Vol 4: Part B, pp 453–470.

Gosselck, F. 1975. The distribution of *Branchiostoma senegalense* (Acrania Branchiostomidae) in the off-shore shelf region off North West Africa. Int Rev Ges Hydrobiol Hydrogr, 60: 199–207.

Gosselck, F. and Kuehner, E. 1973. Investigation on the biology of *Branchiostoma senegalense* larvae off-shore shelf region off North West Africa. Mar Biol, 22: 67–73.

Guzman, H.M., Guevara, C.A. and Hernandez, I.C. 2003. Reproductive cycle of two commercial species of sea cucumber (Echinodermata: Holothuroidea) from Caribbean Panama. Mar Biol, 142: 271–279.

Hadel, V.F., Monteiro, A.M.G. and Ditadi, A.S.F. 1999. Filo Echinodermata. *In: Biodiversidade do Estado de São Paulo: Síntese do conhecimento no final do século*. (eds) Migotto, A.E. and Tiago, C.G., Invertebrados marinhos. FAPESP, São Paulo, XX Parte, 3: 259–271.

Hagen, N.T. 1987. Sea urchin outbreaks and nematode epizootics in Vestfjorden, northern Norway. Sarsia, 72: 213–229.

Hagen, N.T. 1992. Macroparasitic epizootic disease: a potential mechanism for the termination of sea urchin outbreaks in northern Norway. Mar Biol, 114: 469–478.

Hagen, N.T. 1996. Parasitic castration of the green sea urchin *Strongylocentrotus droebrachiensis* by the nematode endoparasite *Echinomermella matsi*: reduced reproductive potential and reproductive death. Dis Aquat Org, 24: 215–226.

Hagen, N.T. 1997. Out-of-season maturation of echinoid broodstock in fixed light regimes. Bull Aquacult Ass, Canada, 97: 1–61.

Hair, C., Mills, D.J., McIntyre, R. and Southgate, P.C. 2016. Optimising methods for community-based sea cucumber ranching. Experimental releases of cultured juvenile *Holothuria scabra* in to seagrass meadows in Papua New Guinea. Aquacult Rep, 3: 198–208.

Hamel, J.F. and Mercier, A. 1996. Evidence of chemical communication during the gametogenesis of holothuroids. Ecology, 77: 1600–1616.

Hamel, J.F. and Mercier, A. 1999. Mucus as a mediator of gametogenic synchrony in the sea cucumber *Cucumaria frondosa* (Holothuroidea: Echinodermata). J Mar Biol Ass UK, 79: 212–129.

Hamel, J.-F., Conand, C., Pawson, D.L. and Mercier, A. 2001. The sea cucumber *Holothuria scabra* (Holothuroidea: Echinodermata). Its biology and exploitation as Beache-de-Mer. Adv Mar Biol, 4: 130–223.

Hamel, J.F. and Mercier, A. 2004. Synchronous gamete maturation and reliable spawning induction method in holothurians. In: *Advances in Sea Cucumber Aquaculture and Management*. (eds) Lovatelli, A., Conand, C., Purcell, S. et al., FAO Fish Tech Paper No. 463. FAO, Rome, pp 359–372.

Han, Q., Keesing, J.K. and Liu, D. 2016. A review of sea cucumber aquaculture, ranching, and stock enhancement in China. Rev Fisher Sci Aquacul, 24: 326–341.

Hara, K., Fujiwara, S. and Kawamura, K. 1992. Retinoic acid can induce a secondary axis in developing buds of a colonial ascidian, *Polyandrocarpa misakiensis*. Dev Growth Differ, 34, 437–445.

Haramoto, S., Komatsu, M. and Yamazaki, Y. 2007. Patterns of asexual reproduction in the fissiparous seastar *Coscinasterias acutispina* (Asteroidea: Echinodermata) in Japan. Zool Sci, 24: 1075–1081.

Harrington, F. and Ozaki, H. 1986. The effect of estrogen on protein synthesis in echinoid coelomocytes. Comp Biochem Physiol, 84B: 417–421.

Harriot, V.J. 1982. Sexual and asexual reproduction of *Halothuria atra* Jaeger at Heron Island Reef, Great Barrier Reef. Aust Mus Mem, 16: 53–66.

Harriot, V.J. 1985. Reproductive biology of three congeneric sea cucumber species *Holothuria atra, H. impatients* and *H. edulis* at Heron Reef, Great Barrier Reef. Aust J Mar Freshw Res, 36: 51–57.

Harrold, C. and Pearse, J.S. 1980. Allocation of pyloric caecum results in fed and starved sea stars *Pisaster giganteus* (Stimpson): somatic maintenance comes before reproduction. J Exp Mar Biol Ecol, 48: 169–183.

Hart, M.W. 1995. What are the costs of small egg size for a marine invertebrate with feeding planktonic larvae? Am Nat, 146: 415–426.

Hartati, R., Widianingish, Endrawati, H. 2016. The growth of sea cucumber *Stichopus herrmanni* after transverse induced fission in two and three fission plane. Ilmu Kelautan, 21: 93–100.

Harvey, E.B. 1936. Parthenogenetic merogony or cleavage without nuclei in *Arbacia punctulata*. Biol Bull, 71: 101.

Harvey, E.B. 1940. A new method of producing twins, triplets, and quadruplets in *Arbacia punctulata*, and their development. Biol Bull, 78: 202–216.

He, D., Yang, G., Fang, S. et al. 1988. Study of zooplankton ecology in Zhejiang coastal upwelling system-Zooplankton biomass and abundance in major groups. Acta Oceanol Sin, 7: 607–620.

Hendler, G. 1979. Sex reversal and viviparity in *Ophiolepis kieri*, n. sp, with notes on viviparous brittlestar from the Caribbean (Echinodermata: Ophiuroidea). Proc Biol Soc Wash, 92: 783–795.

Hendler, G. and Littman, B.S. 1986. The ploys of sex: Relationships among the mode of reproduction, body size and habitats of coral-reef brittle stars. Coral Reefs, 5: 31–42.

Henschke, N., Smith, J.A., Everett, J.D. and Suthers, I.M. 2015. Population drives of a *Thalia democratica* swarm: insights from population modeling. J Plankton Res, 37: 1074–1087.

Herbst, C. 1982. Experimentalle untersuchungen uber den Einfluss der verandenten chemischen Zussamensetzung des ungebeden Mediums auf die Entwicklung der Tiere. 1. Teil versuche an Seeigleiern. Z Wiss Zool, 55: 446–518.

Heron, A.C. 1972. Population ecology of a colonizing species: the pelagic tunicate *Thalia democratica*. 1. Individual growth rate and generation time. Oecologia, 10: 269–293.

Heron, A.C. and Benham, E.E. 1985. Life history parameters as indicators of growth rate in three salp populations. J Plankton Res, 7: 365–379.

Herrera, J.C., McWeeney, S.K. and McEdward, L.R. 1996. Diversity of energetic strategies among echinoid larvae and the transition from feeding to non feeding development. Oceanol Acta, 19: 313–321.

Hertwig, O. 1876. Beiträge zur Kenntnis der Bildung, Befruchtung und Theilung des thierischen Eis. Morph Jahrbuch, 1: 347–434.

Heyland, A. and Hodin, J. 2004. Heterochronic developmental shift caused by thyroid hormone in larval sand dollars and its implications for phenotypic plasticity and the evolution of nonfeeding development. Evolution, 58: 524–538.

Heyland, A., Reitzel, A.M. and Hodin, J. 2004. Thyroid hormones determine developmental mode in sand dollars (Echinodermata: Echinoidea). Evol Dev, 6: 382–392.

Hines, G.A. and Watts, S.A. 1992. Sex steroid levels in the testes, ovaries and pyloric caeca during gametogenesis in the sea star *Asterias vulgaris*. Gen Comp Endocrinol, 87: 451–460.

Hines, G.A., McClintock, J.B. and Watts, S.A. 1992. Levels of estrogen and progesterone in male and female *Eucidaris tribuloides* (Echinodermata: Echinoidea) over an annual gametogenetic cycle. Fla Sci, 55: 85–91.

Hirai, S. and Kanatani, H. 1971. Site of production of meiosis-inducing substance in ovary of starfish. Exp Cell Res, 67: 224–227.

Hodgin, J. 1990. Sex determinations compared in *Drosophila* and *Caenorhabditis*. Nature, 344: 35–47.

Hodin, J., Hoffman, J., Miner, B.J. and Davidson, B.J. 2001. Thyroxine and the evolution of lecithotrophic development in echinoids. In: *Echinoderms*. (ed) Barker, M.F., Balkema Lisse, The Netherlands, pp 447–452.

Holland, N.D. and Dan, K. 1975. Ovulation in an echinoderm (*Comamhus japonica*). Experientia, 21: 1078–1080.

Holland, N.D. and Holland, L.Z. 1989. Fine structural study of the cortical reaction and formation of the egg coats in a lancelet (Amphioxus), *Branchiostoma floridae* (Phylum Chordata: Subphylum Cephalochordata Acrania). Biol Bull, 176: 111–122.

Holy, J. 1998. Chlorpropham [isopropyl *n*-(3-chlorophenyl) carbamate] disrupts microtubule organization, cell division, and early development of sea urchin embryos. J Toxicol Environ Heal, 54A: 319–333.

Hooper, R.G., Cuthbert, F.M. and McKeever, T. 1996. Sea urchin aquaculture—phase II: ration size seasonal growth rates, mixed kelt diets, fish diet, old urchin growth and baiting confinement experiments. Can Cent Fish Innovation, St. John's, New Foundland.

Hopper, D.R., Hunter, C.L. and Richmond, R.H. 1998. Sexual reproduction of the tropical sea cucumber *Actinopyga maurtiana* (Echinodermata: Holothuroidea) in Guam. Bull Mar Sci, 63: 1–9.

Hotchkiss, F.H. 2000. On the number of rays in starfish. Am Zool, 40: 340–354.

Hyman, L.H. 1955. *The Invertebrates*: *Echinodermata*. McGraw-Hill Book Co, New York, Vol IV, p 763.

Hyman, L.H. 1959. *The Invertebrates*: *Smaller Coelomate Groups*: *Phylum Hemichordata*. McGraw-Hill Book Co, New York, Vol V, Chapter 17, pp 72–207.

Ishikawa, M. 1975. Parthenogenetic activation and development. In: *The Sea Urchin Embryo-Biochemistry and Morphogenesis*. (ed) Czihak, G., Springer Verlag, Berlin, pp 148–169.

Jackle, W.B. 1994. Multiple modes of asexual reproduction by tropical and subtropical sea star larvae: An unusual adaptation for gene dispersal and survival. Biol Bull, 186: 62–71.

Jaeckle, W.B. 1995. Variation in the size, energy content and biochemical composition of invertebrate eggs: correlates to the mode of larval development. In: *Ecology of Marine Invertebrate Larvae*. (ed) McEdward, L., CRC Press, Boca Raton, Fl.

James, B.D. and Pearse, J.S. 1969. Echinoderms from the Gulf of Suez and the northern Red Sea. J Mar Biol Ass India, 11: 78–125.

James, B.D., Rajapandian, M.E., Basker, P.K. and Gopinathan, C.P. 1988. Successful induced spawning and rearing of the holothurians *Holothuria* (*Metriayla*) *scabra* Jaeger at Tuticorin. Mar Fish Inf Ser, Trend Env Ser, 87: 30–33.

James, B.D. 2004. Captive breeding of the sea cucumber, *Holothuria scabra* from India. In: *Advances in Sea Cucumber Aquaculture and Management*. (eds) Lovatelli, A., Conand, C., Purcell, S. et al., FAO Fish Tech Paper No. 463. FAO, Rome, pp 385–398.

Jangoux, M. 1990. Diseases of Echinodermata. In: *Diseases of Marine Animals*. (ed) Kinne, O., Biologische Anstalt Helgoland, Hamburg, pp 439–560.

Johnson, L.G. and Cartwright, C.M. 1996. Thyroxine-accelerated larval development in the crown-of-thorns starfish *Acanthaster planci*. Biol Bull, 190: 299–301.

Johnson, L.G. 1998. Stage-dependent thyroxine effects on sea urchin development. N Zool J Mar Freshw Res, 32: 531–536.

Johnson, M.S. and Threlfall, T.J. 1987. Fissiparity and population genetics of *Coscinasterias calamaria*. Mar Biol, 93: 517–525.

Jones, G.P. and Andrew, N.L. 1990. Herbivory and patch dynamics on rocky reefs in temperate Australia: the roles of fish and sea urchins. Aust J Ecol, 15: 505–520.

Kamenev, Y.O. 2015. Ultrastructure of internal organs, asexual reproduction and regeneration in holothurian *Cladolabes schmeltzii*, A.V. Zhirmunsky Institute of Marine Biology FEB RAS, Vladivostok, Russia.

Kamenev, Y.O. and Dolmatov, I.Y. 2015. Posterior regeneration following fission in the holothurians *Clalolabes schmeltzii* (Dendrochirotidae: Holothuroidea). Microsc Res Tech, 78: 540–552.

Kamenev, Y.O. and Dolmatov, I.Y. 2016. Anterior regeneration after fission in the holothurians *Clalolabes schmeltzii* (Dendrochirotidae: Holothuroidea). Microsc Res Tech, 79: 1–22.

Kanatani, H. 1974. Presence of L-methyladenine in the sea urchin gonad and its relation to oocyte maturation. Dev Growth Diff, 16: 157–170.

Kanatani, H. 1975. Maturation-inducing substances in asteroid and echinoid oocytes. Am Zool, 15: 493–505.

Karako, S., Achituv, Y., Peri-Treves, R. and Katcoff, D. 2002. *Asterina burtoni* (Asteroidea: Echinodermata) in the Mediterranean and the Red Sea. Does asexual reproduction facilitate colonization? Mar Ecol Prog Ser, 234: 139–145.

Kaul-Strehlow, S.K. and Rottinger, E. 2015. Hemichordata. In: *Evolutionary Developmental Biology of Invertebrates: Deuterostomia*. (ed) Wanninger, A., Springer Verlag, Vienna, 6: 59–89.

Kawamura, K. and Fujiwara, S. 1955. Cellular and molecular characterization of transdifferentiation in the process of morphallaxis of budding tunicates. Semin Cell Biol, 6: 117–126.

Kawamura, K. and Watanabe, H. 1981. Studies of Japanese compound styelid ascidians. III. A new possibly asexual *Polyandrocarpa* from Shimoda Bay. Publ Seto Mar Biol Lab, 26: 425–436.

Kawamura, K. and Nakauchi, M. 1984. Changes of the atrial epithelium during the development of anteroposterior organization in palleal buds of the polystyelid ascidian, *Polyandrocarpa misakiensis*. Mem Fac Sci Kochi Univ, 5D: 15–35.

Kawamura, K., Hara, K. and Fujiwara, S. 1993. Developmental role of endogenous retinoids in the determination of morphallactic field in budding tunicate. Development, 117: 835–845.

Kawamura, K. and Fujiwara, S. 2000. Advantage or disadvantage: Is sexual reproduction beneficial to survival of the tunicate *Polyandrocarpa misakiensis*. Zool Sci, 17: 281–291.

Kawamura, K., Sugino, Y., Sunanaga, Y. and Fujiwara, S. 2008. Multipotent epithelial cells in the process of regeneration and asexual reproduction in colonial tunicates. Dev Growth Diff, 50: 1–11.

Kawamura, K., Tiozzo, S., Manni, L. et al. 2011. Germline cell formation and gonad regeneration in solitary and colonial ascidians. Dev Dyn, 240: 299–308.

Keats, D.W., Steele, D.H. and South, G.R. 1984. Depth-dependent reproductive output of the green sea urchin *Strongylocentrotus droebachiensis* (O.F. Muller), in relation to the nature and availability of food. J Exp Mar Biol Ecol, 80: 77–91.

Keesing, J. 2001. The ecology of *Heliocidaris erythrogramma*. In: *Edible Sea Urchins: Biology and Ecology*. (ed) Lawrence, J.M., Elsevier, Amsterdam, pp 261–270.

Keesing, J.K. and Hall, K.C. 1998. Review of harvests and status of world sea urchin fisheries points to opportunities for aquaculture. J Shellfish Res, 17: 1597–1604.

Kelly, M. and Cook, E. 2001. The ecology of *Psammechinus miliaris*. In: *Edible Sea Urchins: Biology and Ecology*. (ed) Lawrence, J.M., Elsevier, Amsterdam, pp 217–224.

Kelly, M.S., Hunter, A.J., Scholfield, C. and McKenzie, J.D. 2000. Morphology and survivalship of larval *Psammechinus miliaris* (Gmelin) (Echinoidea: Echinodermata) in response to varying food quality and quantity. Aquaculture, 183: 223–240.

Kenny, R. 1969. Growth and asexual reproduction of the starfish *Nepanthia belcheri* (Perrier). Pacific Sci, 23: 51–55.

Kille, F. 1939. Regeneration of gonad tubules following in the sea cucumber *Thyone briareus* (Lesuerer). Biol Bull, 76: 70–79.

Kille, F.R. 1942. Regeneration of the reproductive system following binary fission in the sea cucumber *Holothuria parvula*. Biol Bull, 83: 55–66.

Kirchhoff, N.T., Eddy, S., Harris, L. and Brown, N.P. 2008. Nursery-phase culture of green sea urchin *Strongylocentrotus droebachiensis* using 'on-bottom' cages. J Shellfish Res, 27: 921–927.

Kirchhoff, N.T., Eddy, S. and Brown, N.P. 2010. Out-of-season gamete production in *Strongylocentrotus droebachiensis*: Photoperiod and temperature manipulation. Aquaculture, 303: 77–85.

Kitazawa, C. and Komatsu, M. 2000. Larval development and asexual development of the sea star *Distolasterias nipon*. Proc 10th Internatl Echinoderm Conf, Dunedin, New Zealand, Abstract, 123.

Kitazawa, C. and Amemiya, S. 2001. Regulating potential in development of a direct developing echinoid, *Peronella japonica*. Dev Growth Differ, 43: 73–82.

Kjesbu, O.S., Murua, F., Saborida-Rey, F. and Whithames, P.R. 2010. Method development and evaluation of stock reproductive potential of marine fish. Fisher Res, 104: 1–7.

Klinger, T.S., Lawrence, J.M. and Lawrence, A.L. 1997. Gonad and somatic production of *Strongylocentrotus droebachiensis* fed manufactured feeds. Bull Aquacul Ass, 1: 35–37.

Knot, K.E., Balser, E.J. and Jackle, W.B. et al. 2003. Identification of asteroid genera with species capable of larval cloning. Biol Bull, 204: 246–255.

Koguchi, S., Sugino, Y.M. and Kuwamora, K. 1993. Dynamics of epithelial stem cells in the process of stolonal budding of the colonial ascidian *Perophora japonica*. Mem Fac Sci Kochi Univ (SERD), 14: 7–14.

Kon, T., Nohara, M., Yamanoue, Y. et al. 2007. Phylogenetic position of a whale-fall lancelet (Cephalochordata) inferred from whole mitochondrial genome sequences. BMC Evol Biol, 7: 127.

Kondo, M. and Akasaka, K. 2012. Current status of echinoderm genome analysis—What do we know? Curr Genome, 13: 134–143.

Krishnan, S. 1986. Histochemical studies on reproductive and nutritional cycles of the holothurians *Holothuria scabra*. Mar Biol, 2: 54–55.

Krishnaswamy, S. and Krishnan, S. 1967. A report on reproductive cycle of holothurian *Holothuria scabra* Jaeger. Curr Sci, 6: 155–156.

Kurn, U., Rendulic, S., Tiozzo, S. and Lauzon, R.J. 2011. Asexual propogation and regeneration in colonial ascidians. Biol Bull, 221: 43–61.

Laird, D.J. and Weissman, I.L. 2004a. Continuous development precludes radioprotection in a colonial ascidian. Dev Comp Immunol, 28: 201–209.

Laird, D.J. and Weissman, I.L. 2004b. Telomerase maintained in self-renewing tissues during serial regeneration of the urochordate *Botryllus schlosseri*. Dev Biol, 273: 185–194.

Laird, D.J., Chang, W.T., Weissman, I.L. and Lauzon, R.J. 2005a. Identification of a novel gene involved in asexual organogenesis in the budding ascidian *Botryllus schlosseri*. Dev Dyn, 235: 997–1005.

Laird, D.J., De Tomaso, A.W. and Weissman, I.L. 2005b. Stem cells are units of natural selection in colonial ascidian. Cell, 123: 1351–1360.

Lambeth, L. 2000. The subsistence use of *Stichopus variegatus* (now *S. herrmanni*) in the Pacific Islands. SPC Beche-de-Mer Inf Bull, 13: 18–21.

Larson, B.R., Vadas, R.L. and Keser, M. 1980. Feeding and nutritional ecology of the sea urchin *Strongylocentrotus droebachiensis* in Maine, USA. Mar Biol, 59: 49–62.

Lauzon, R.J., Ishizuka, K.J. and Weissman, I.L. 1992. A cyclical, developmentally regulated death phenomenon in a colonial urochordate. Dev Dyn, 194: 71–83.

Lauzon, R.J., Patton, C.W. and Weissman, I.L. 1993. A morphological and immunohistochemical study of programmed cell death in *Botryllus schlosseri* (Tunicata, Ascidiaceae). Cell Tissue Res, 272: 115–127.

Lauzon, R.J., Ishizuka, K.J. and Weissman, I.L. 2002. Cyclical generation and degeneration of organs in a colonial urochordate involves crosstalk between old and new: A model for development and regeneration. Dev Biol, 249: 348.

Lavado, R., Barbaglio, A., Candia Carnevali, M.D. and Porte, C. 2006a. Steroid levels in crinoid echinoderms are altered by exposure to model endocrine disruptors. Steroids, 71: 489–497.

Lavado, R., Sugni, M., Candia Carnevali, M.D. and Porte, C. 2006b. Triphenyltin alters androgen metabolism in the sea urchin *Paracentrotus lividus*. Aquat Toxicol, 79: 247–256.

Lawrence, J.M. and Lane, J.M. 1982. The utilization of nutrients by post-metamorphic echinoderms. In: *Echinoderm Nutrition*. (eds) Jangoux, M. and Lawrence, J.M., Balkema, Rotterdam, pp 331–371.

Lawrence, J.M., Klinger, T.S., McClintock, J.B. et al. 1986. Allocation of nutrient resources to body components by regenerating *Luidia clathrata* (Say) (Echinodermata: Asteroidea). J Exp Mar Biol Ecol, 102: 47–53.

Lawrence, J.M. 1987. *A Functional Biology of Echinoderms.* Croom Held, London, p 340.

Lawrence, J.M. and Ellwood, A. 1991. Simultaneous allocation of resources to arm regeneration and to somatic and gonadal production in *Luidia clathrata* (Say) (Echinodermata: Asteroidea). In: *Biology of Echinodermata.* (eds) Yanagisawa, T., Yasumasu, I., Oguro, C. et al., A.A. Balkema, Rotterdam, pp 543–48.

Lawrence, J.M. and Larrain, A. 1994. The cost of arm autotomy in the starfish *Stichaster striatus.* Mar Ecol Prog Ser, 109: 311–3.

Lawrence, J.M. and Vásquez, J. 1996. The effect of sublethal predation on the biology of echinoderms. Oceanol Acta, 19: 431–440.

Lawrence, J.M. 2001a. *Edible Sea Urchins: Biology and Ecology.* Elsevier, Amsterdam, p 419.

Lawrence, J.M. 2001b. The edible sea urchins. In: *Edible Sea Urchins: Biology and Ecology.* (ed) Lawrence, J.M., Elsevier, Amsterdam, pp 1–4.

Lawrence, J.M. 2001c. Sea urchin roe cuisine. In: *Edible Sea Urchins: Biology and Ecology.* (ed) Lawrence, J.M., Elsevier, Amsterdam, pp 415–416.

Lawrence, J.M. and Pomory, C.M. 2008. Position of arm loss and rate of arm regeneration by *Luidia clathrata* (Echinodermata: Asteroidea). Cah Biol Mar, 49: 369–373.

Lawrence, J.M. 2010. Energetic costs of loss and regeneration of arms in stellate echinoderms. Integ Comp Biol, 50: 506–514.

Lawrence, J.M. and Gaymer, C.F. 2012. Autotomy of rays of *Heliaster helianthus* (Asteroidea: Echinodermata). Zoosymposia, 7: 173–176.

Laxminarayana, A. 2006. Asexual reproduction by induced transverse fission in the sea cucumbers *Bohadschia marmorata* and *Holothuria atra.* SPC Beche-de-Mer Inf Bull, 23: 35–37.

Lee, J., Byrne, M. and Uthicke, S. 2008a. The influence of population density on fission and growth of *Holothuria atra* in natural mesocosms. J Exp Mar Biol Ecol, 365: 126–135.

Lee, J., Byrne, M. and Uthicke, S. 2008b. Asexual reproduction by fission of a population of *Holothuria hilla* at One Tree Reef Great Barrier Reef. SPC Beche-de-Mer Inf Bull, 27: 17–23.

Lee, J., Uthicke, S. and Byrne, M. 2009. Asexual reproduction and observations of sexual reproduction in the aspidochirotid sea cucumber *Holothuria difficilis.* J Invert Reprod Dev, 53: 87–92.

Lessios, H.A. 2007. Reproductive isolation between species of sea urchins. Bull Mar Sci, 81: 191–208.

Levitan, D.R. 1989. Density-dependent size regulation in *Diadema antillarum*: effects on fecundity and survivorship. Ecology, 70: 1414–1424.

Levitan, D.R. 1991. Influence of body size and population density on fertilization success and reproductive output in a free spawning invertebrate. Biol Bull, 181: 261–268.

Levitan, D.R., Sewell, M.A. and Chia, F.-S. 1992. How distribution and abundance influence fertilization success in the sea urchin *Strongylocentrotus franciscanus.* Ecology, 73: 248–254.

Levitan, D.R. 1993. The importance of sperm limitation to the evolution of egg size in marine invertebrates. Am Nat, 141: 517–536.

Lipani, C., Vitturi, R., Sconzo, G. and Barbata, G. 1996. Karyotype analysis of the sea urchin *Paracentrotus lividus* (Echinodermata): Evidence for a heteromorphic chromosome sex mechanism. Mar Biol, 127: 67–72.

Liu, X., Zhu, G., Zhao, Q. et al. 2004. Studies on hatchery techniques of the sea cucumber, *Apostichopus japonicus.* In: *Advances in Sea Cucumber Aquaculture and Management.* (eds) Lovatelli, A., Conand, C., Purcell, S. et al., FAO Fish Tech Paper No. 463. FAO, Rome, pp 287–296.

Localli, T.C. 2000. Larval budding, metamorphosis and the evolution of life history patterns in the echinoderms. Invert Biol, 119: 234–241.

Loeb, V. and Santora, J.A. 2012. Population dynamics of *Salpa thompsoni* near the Antarctic Peninsula: growth rates and interannual variations in reproductive activity (1993–2009). Prog Oceanogr, 96: 93–107.

Lokani, P. 1996. Illegal fishing of sea cucumber (Beache-de-Mer) by Papua New Guinea fisherman in the Torres strait protected zone. South Pacific Commission Beache-de-Mer Inf Bull, 8: 2–6.

Lombard, F., Renaud, F., Sainsbury, C. et al. 2009a. Apendicularian ecophysiology. 1. Food concentration dependent clearance rate, assimilation efficiency, growth and reproduction of *Oikopleura dioica*. J Mar Syst, 78: 606–616.

Lombard, F., Sciandra, A. and Gorsky, G. 2009b. Appendicularian ecophysiology. 2. Modeling nutrition, metabolism, growth and reproduction of the appendicularian *Oikopleura dioica*. J Mar Sys, 78: 617–629.

Lucas, J.S. 1982. Quantitaive studies of feeding and nutrition during larval development of the coral reef asteroid *Acanthaster planci* (L.). J Exp Mar Biol Ecol, 65: 173–193.

Luis, O., Delgado, F. and Gago, J. 2005. Year-round captive spawning performance of the sea urchin *Paracentrotus lividus*: relevance for the use of its larvae as live feed. Aquat Liv Res, 18: 45–54.

Maderspacher, F. 2008. Theodor Boveri and the natural experiment. Curr Biol, 18: R279–R286.

Mangelsdorf, D.J. and Evans, R.M. 1955. The RXR heterodimers and orphan receptors. Cell, 83: 841–850.

Manni, L., Zaniolo, G., Cima, F. et al. 2007. *Botryllus schlosseri*: A model ascidian for the study of asexual reproduction. Dev Dyn, 236: 335–352.

Mariante, F.L.F., Lemos, G.B. and Eutrópio, F.J. 2010. Reproductive biology in the starfish *Echinaster (Othilia) guyanensis* (Echinodermata: Asteroidea) in southeastern Brazil. Zoologia, 27: 897–901.

Marin, M.G., Moschino, V., Cima, F. and Celli, C. 2000. Embryotoxicity of butyltin compounds to the sea urchin *Paracentrotus lividus*. Mar Environ Res, 50: 231–235.

Marsh, A.G. and Watts, S.A. 2001. Energy metabolism and gonad development. In: *Edible Sea Urchins: Biology and Ecology*. (ed) Lawrence, J.M., Elsevier, Amsterdam, pp 27–42.

Marsh, L.M. 1977. Coral reef asteroids of Palau, Caroline Islands. Micronesia, 13: 251–281.

Marshall, D.J. and Bolton, T.F. 2007. Effects of egg size on the development time of non-feeding larvae. Biol Bull, 212: 6–11.

Mary Bai, M.M. 1994. Studies on generation of the holothurians *Holothuria (Metriatyla) scabra* Jaegar. CMFRI Bull, 46: 44–50.

Mashanov, V.S., Dolmatov, I.Y. and Heinzeller, T. 2005. Transdifferentiation in holothurian gut regeneration. Biol Bull, 209: 184–193.

Mashanov, V.S., Zueva, O.R. and Garcia-Arraras, J.E. 2013. Radial glial cells play a key role in echinoderm regeneration. BMC Biology, 11: 418–449.

Matsuno, T. and Tsushima, M. 2001. Carotenoids in sea urchins. In: *Edible Sea Urchins: Biology and Ecology*. (ed) Lawrence, J.M., Elsevier, Amsterdam, pp 115–138.

McAlister, J.S. and Moran, A.L. 2013. Effects of variation in egg energy and exogenous food on larval development in congeneric sea urchins. Mar Ecol Prog Ser, 490: 155–167.

McBride, S.C., Lawrence, J.M., Lawrence, A.L. and Mulligan, T.M. 1999. Ingestion, absorption and gonad production of adult *Strongylocentrotus franciscanus* fed different ration of a prepared diet. J World Aquacult Soc, 30: 364–370.

McClanahan, T.R. and Muthiga, N.A. 2001. The ecology of *Echinometra*. In: *Edible Sea Urchins: Biology and Ecology*. (ed) Lawrence, J.M., Elsevier, Amsterdam, pp 225–244.

McClintock, J.B. and Pearse, J.S. 1987. Reproductive biology of the common Antarctic crinoid *Promachocrinus kerguelensis* (Echinodermata: Crinoidea). Mar Biol, 96: 375–383.

McClintock, J.B. and Watts, S.A. 1990. The effects of photoperiod on gametogenesis in the tropical sea urchin *Eucidaris tribuloides* (Lamarck) (Echinodermata: Echinoidea). J Exp Mar Biol Ecol, 139: 175–184.

McDonald, K.A. and Vaughn, D. 2010. Abrupt change in food environment induces cloning in plutei of *Dendraster excentricus*. Biol Bull, 219: 38–49.

McEdward, L.R. and Chia, F.C. 1991. Size and energy content of eggs of echinoderms with pelagic lecithotrophic development. J Exp Mar Biol Ecol, 147: 95–102.

McEdward, L.R. 1996. Experimental manipulation of parental investment in echinoid echinoderms. Am Zool, 36: 169–179.

McEdward, L.R. and Morgan, K.H. 2001. Interspecific relationships between egg size and the level of parental investment per offspring in echinoderms. Biol Bull, 200: 33–50.

McEdward, L.R. and Miner, B.G. 2001a. Echinoid larval ecology. In: *Edible Sea Urchins: Biology and Ecology*. (ed) Lawrence, J.M., Elsevier, Amsterdam, pp 59–78.

McEdward, L.R. and Miner, B.G. 2001b. Larval and life cycle patterns in echinoderms. Can J Zool, 79: 1125–1170.

McGovern, T.M. 2003. Plastic reproductive strategies in a clonal marine invertebrate. Proc Roy Soc, 270 B: 2517–2522.

Meidel, S.K. and Scheibling, R.E. 1999. Effects of food type and ration on reproductive maturation and growth of the sea urchin *Strongylocentrotus droebachiensis*. Mar Biol, 134: 155–166.

Menard, F., Fromentin, J.M., Goy, J. and Dallot, S. 1998. Timing of sperm shedding and release of aggregates in the salp *Thalia democratica* (Urochordata: Thaliacea). Mar Biol, 129: 607–614.

Menge, B.A. 1975. Brood or broadcast? The adaptive significance of different reproductive strategies in the two intertidal sea stars *Leptasterias hexactis* and *Pisaster ochraceus*. Mar Biol, 31: 87–100.

Mercier, A., Battaglene, S.C. and Hamel, J.F. 1999. Daily burrowing cycle and feeding activity of juvenile sea cucumbers *Holothuria scabra* in response to environmental factors. J Exp Mar Biol Ecol, 239: 125–156.

Mercier, A., Battaglene, S.C. and Hamel, J.F. 2000. Periodic movement, recruitment and size-related distribution of the sea cucumbers *Holothuria scabra* in Solomon Islands. Hydrobiologia, 440: 81–100.

Mercier, A. and Hamel, J.-F. 2002. Perivisceral coelomic fluid as a mediator of spawning induction in tropical holothurians. Invert Reprod Dev, 41: 223–234.

Mercier, A., Hidalgo, R.Y. and Hamel, J.F. 2004. Aquaculture of the Galapagos sea cucumber, *Isostichopus fuscus*. In: *Advances in Sea Cucumber Aquaculture and Management*. (eds) Lovatelli, A., Conand, C., Purcell, S. et al., FAO Fish Tech Paper No. 463. FAO, Rome, pp 347–358.

Metz, E.C., Yanakimaschi, H. and Palumbi, S.R. 1991. Gamete compatibility and reproductive isolation of closely related Indo-Pacific sea urchin *Echinometra*. In: *Biology of Echinodermata*. (eds) Yanakisawa et al., A.A. Balkema, Rotterdam, pp 131–137.

Metz, E.C., Kane, R.E., Yanagimachi, H. and Palumbi, S.R. 1994. Fertilization between closely related sea urchins is blocked by incompatibilities during sperm-egg attachment and early stages of fusion. Biol Bull, 187: 23–34.

Meyer, D.L., Dame, E.A. and Lask, P.B. 2008. Decline of crinoids on the reefs of Curaco and Bonaire Netherlands Antilles. Proc 11th Int Coral Reef Symp, Ft. Lauderdale, Florida, pp 875–878.

Milkman, R. 1967. Genetic and developmental studies on *Botryllus schlosseri*. Biol Bull, 132: 229–243.

Miller, R.L. and Cosson, J. 1997. Timing of sperm shedding and release of aggregate in the salp *Thalia democratic* (Urochordata: Thaliacea). Mar Biol, 129: 607–614.

Minor, M.A. and Scheibling, R.E. 1997. Effects of food ration and feeding regime on growth and reproduction of the sea urchin *Strongylocentrotus droebachiensis*. Mar Biol, 129: 159–167.

Mladenov, P.V., Emson, R.H., Colpitts, L.V. and Wilkie, I.C. 1983. Asexual reproduction in the West Indian brittle star *Ophiocomella ophiactoides* (H.L. Clark) (Echinodermata: Ophiuroidea). J Exp Mar Biol Ecol, 72: 1–23.

Mladenov, P.V. 1986. Reproductive biology of the feather star *Florometra serratissima*: gonadal structure, breeding pattern, and periodicity of ovulation. Can J Zool, 64: 1642–1651.

Mladenov, P.V., Carson, S.F. and Walker, C.W. 1986. Reproductive ecology of an obligately fissiparous population of the sea star *Stephanasterias albula* (Stimpson). J Exp Mar Biol Ecol, 96: 155–175.

Mladenov, P.V. and Emson, R.H. 1988. Density, size structure and reproductive characteristics of fissiparous brittle stars in algae and sponges: evidence for interpopulational variation in levels of sexual and asexual reproduction. Mar Ecol Prog Ser, 42: 181–194.

Mladenov, P.V., Igdoura, S. and Asotra, S. 1989. Purification and partial characterization of an autotomy-promoting factor from the sea star *Pycnopodia helianthoides*. Biol Bull, 176: 169–175.

Mladenov, P.V. and Emson, R.H. 1990. Genetic structure of population of two closely related brittle stars with contrasting sexual and asexual life histories with observation on the genetic structure of a second sexual species. Mar Biol, 104: 265–274.

Mladenov, P.V. and Burke, P. 1994. Echinodermata: Asexual reproduction. In: *Reproductive Biology of Invertebrates*. (eds) Adiyodi, K.G. and Adiyodi, R.G., Oxford IBH Publishing, New Delhi, Vol 6: Part B, pp 339–383.

Mladenov, P.V. 1996. Environmental factors influencing asexual reproductive processes in echinoderms. Oceanol Acta, 19: 227–235.

Monks, S.P. 1904. Variablity and autotomy of Phataria. Acad Nat Sci Phil, 56: 596–600.

Montgomery, E.M. and Palmer, A.R. 2012. Effect of body size and shape on locomotion in the bat star (*Patiria miniata*). Biol Bull, 222: 222–232.

Moore, H.B. 1932. A hermaphrodite sea-urchin. Nature, 130: 59–59.

Moore, K.A. and Lemischka, I.R. 2006. Stem cells and their niches. Science, 311: 1880–1885.

Moran, A.L. and McAlister, J.S. 2009. Egg size as a life history character of marine invertebrates: Is it all its cracked up to be? Biol Bull, 216: 226–242.

Moran, A.L., McAlister, J.S. and Whitehill, E.A. 2013. Egg as energy: Revisiting the scaling of egg size and energetic content among echinoderms. Biol Bull, 224: 184–191.

Morgan, R. and Jangoux, M. 2004. Assessing arm regeneration and its effect during the reproductive cycle in the gregarious brittle star *Ophiothrix fragilis*. Cah Biol Mar, 45: 277–80.

Mortensen, T. 1921. Studies of the development and larval forms of echinoderms. G.E.C. Gad, Copenhagen, p 261.

Mortensen, T. 1928–1951. *A Monograph of the Echinoidea-I*. V.C.A. Reitzel, Copenhagen.

Moschino, V. and Marin, M.G. 2002. Spermiotoxicity and embryotoxicity of triphenyltin in the sea urchin *Paracentrotus lividus* Lmk. Environ Biol Toxicol, 16-175–181.

Motnikar, S., Bryl, P. and Boyer, J. 1997. Conditioning green sea urchins in tanks: the effect of semi-moist diets on gonad quality. Bull Aquacul Ass, 1: 21–25.

Motokawa, T. and Tsuchi, A. 2003. Dynamic mechanical properties of body-wall dermis in various mechanical states and their implications for the behavior of sea cucumbers. Biol Bull, 205: 261–275.

Motokawa, T., Sato, E. and Umeyama, K. 2012. Energy expenditure associated with softening and stiffening of echinoderm connective tissue. Biol Bull, 222: 150–157.

Mozzi, D., Dolmatov, I., Bonasoro, F. and Candia-Carnevali, M.D. 2006. Visceral regeneration in the crinoids *Antedon mediterranea*: Basic mechanisms, tissues and cells involved in gut regeneration. Cent Eur J Biol, 1: 609–635.

Mueller, B., Bos, A.R., Graf, G. and Gumanao, G.S. 2011. Size-specific locomotion rate and movement pattern of four common Indo-Pacific sea stars (Echinodermata, Asteroidea). Aquat Biol, 12: 157–164.

Munuswamy, N. and Subramonian, T. 1985. Oogenesis and shell gland activity in a freshwater fairy shrimp *Streptocephalus dichotomus* Baird (Crustacea: Anostraca). Cytobios, 44: 137–147.

Mukai, H., Koyama, H. and Watanabe, H. 1983. Studies on the reproduction of three species of *Perophora* (Ascidiacea). Biol Bull, 164: 251–266.

Muthiga, N.A. 2006. The reproductive biology of a new species of sea cucumber, *Holothuria (Mertensiothuria) arenacava* in a Kenyan marine protected area: the possible role of light and temperature on gametogenesis and spawning. Mar Biol, 149: 585–593.

Nakauchi, M. 1982. Asexual development of ascidians: its biological significance, diversity, and morphogenesis. Am Zool, 22: 753–763.

Nakauchi, M. and Takeshita, T. 1983. Ascidian one-half embryos can develop into functional adults. J Exp Zool, 227: 155–158.

Navarro, P.G., Garcia-Sanz, S. and Tuya, F. 2012. Reproductive biology of the sea cucumber *Holothuria santori* (Echinodermata: Holothuroidea). Sci Mar, 76: 741–752.

Newman, H.H. 1921. On the development of spontaneously parthenogenetic eggs of *Asterina* (*Patiria*) *miniata*. Biol Bull, 40: 186–204.

Nichols, D. 1994. Reproductive seasonality in the comatulid crinoid *Antedon bifida* (Pennant) from the English Channel. Phil Trans R Soc Lond, 343B: 113–134.

Nilsson, H.C. 1999. Effects of hypoxia and organic enrichment on growth of the brittlestars *Amphiura filiformis* (OF Mu¨ller) and *Amphiura chiajei* Forbes. J Exp Mar Biol Ecol, 237: 11–30.

Novelli, A.A., Argese, E., Tagliapietra, D. et al. 2002. Toxicity of tributyltin and triphenyltin to early life-stages of *Paracentrotus lividus* (Echinodermata: Echinoidea). Environ Toxicol Chem, 21: 859–864.

Oetken, M., Bachmann, J., Schulte-Oehlmann, U. and Oehlmann, J. 2004. Evidence for endocrine disruption in invertebrates. Int Rev Cytol, 236: 1–44.

Ohlstein, B., Kai, T., Decotto, E. and Sprading, A. 2004. The stem cell niche: Theme and variations. Curr Opin Cell Biol, 16: 693–699.

Oka, H. and Watanabe, H. 1957. Vascular budding, a new type of budding in *Botryllus*. Biol Bull, 112: 225–240.

Okazaki, K. and Dan, K. 1954. The metamorphosis of partial larvae of *Peronella japonica* Mortensen, a sand dollar. Biol Bull, 106: 83–99.

Okazaki, K. 1975. Normal development to metamorphosis. In: *The Sea Urchin Embryo. Biochemistry and Morphogenesis*. (ed) Czihak, G., Springer-Verlag, Heidelberg, pp 177–232.

Okumura, S.I., Kimura, K., Sakai, M. et al. 2009. Chromosome number and telomere sequence mapping of the Japanese sea cucumber *Apostichopus japonicus*. Fish Sci, 75: 249–251.

O'Loughlin, P.M., Eichler, J., Altoff, L. et al. 2009. Observations on reproductive strategies for some dendrochirotid holothuroids (Echinodermata: Holothuroidea: Dendrochirotidae). Mem Mus Victoria, 66: 215–220.

Omar, H.A., Razek, F.A.A., Rahman, S.H.A. and El Shimy, N.A. 2013. Reproductive periodicity of sea cucumber *Bohadschia vitiensis* (Echinodermata: Holothuroidea) in Hurghada area, Red Sea, Egypt. Egy J Aquat Res, 39: 115–123.

Ortani, G. 1958. Cleavage and development of egg fragments in ascidians. Acta Embryol Morph Exp, 1: 247–272.

Ottesen, P.O. and Lucas, J.S. 1982. Divide or broadcast: Interrelation of asexual and sexual reproduction in a population of the fissiparous hermaphroditic seastar *Nepanthia belcheri* (Asteroidea: Asterinidae). Mar Biol, 69: 223–233.

Overdyk, L.M., Scheibling, R.E. and Barker, M.F. 2016. Asexual reproduction and somatic growth of the fissiparous sea star *Allostichaster polyplax* in New Zealand. Mar Biol Res, 12: 85–95.

Packard, A. 1968. Asexual reproduction in *Balanoglossus* (Stomochordata). Proc R Soc, 171B: 261–272.

Paffenhofer, G.A. 1973. The cultivation of appendicularian through numerous generations. Mar Biol, 22: 183–185.

Paffenhofer, G.A. 1976. On the biology of Appendicularia of the southeastern North Sea. Proc 10th Eur Sym Mar Biol, 2: 437–455.

Paffenhofer, G.A. and Gibson, D.M. 1999. Determination of generation time and asexual fecundity of doliolids (Tunicata: Thaliacea). J Plankton Res, 21: 1183–1189.

Paffenhoffer, G.A. and Koster, M. 2011. From one to many: on life cycles of *Dolioletta genenbauri*. Uljanin (Tunicata: Thaliacea). J Plankton Res, 33: 1139–1145.

Pandian, T.J. and Fluchter, J. 1968. Rate and efficiency of yolk utilization in developing eggs of the sole *Solea solea*. Helgolander wissens Meeresuntersuch, 18: 53–60.

Pandian, T.J. 1975. Mechanism of heterotrophy. In: *Marine Ecology*. (ed) Kinne, O., John Wiley, London Vol 3 Part 1 pp 61–249.

Pandian, T.J. and Marian, M.P. 1985a. Nitrogen content of food as an index of absorption efficiency in fishes. Mar Biol, 85: 301–311.

Pandian, T.J. and Marian, M.P. 1985b. Estimation of absorption efficiency in polychaetes using nitrogen content of food. J Exp Mar Biol Ecol, 90 : 289–295.

Pandian, T.J. 1987. Fish. In: *Animal Energetics*. (eds) Pandian, T.J. and Vernberg, F.J., Academic Press, San Diego, 2: 357–465.

Pandian, T.J. 2010. *Sexuality in Fishes*. Science Publishers/CRC Press, USA. p 208.

Pandian, T.J. 2011. *Sex Determination in Fish*. Science Publishers/CRC Press, USA. p 270.

Pandian, T.J. 2012. *Genetic Sex Differentiation in Fish*. CRC Press, USA. p 214.

Pandian, T.J. 2013. *Endocrine Sex Differentiation in Fish*. CRC Press, USA. p 303.

Pandian, T.J. 2015. *Environmental Sex Determination in Fish*. CRC Press, USA, p 299.

Pandian, T.J. 2016. *Reproduction and Development in Crustacea*. CRC Press, USA, p 301.

Pandian, T.J. 2017. *Reproduction and Development in Mollusca*. CRC Press, USA, p 299.

Patricolo, E., Ortolani, G. and Cascio, A. 1984. The effect of L-thyroxine in the metamorphosis of *Ascidia malaca*. Cell Tissue Res, 214: 289–301.

Patruno, M., Beesley, P., Thorndyke, M.C. et al. 2000. Changes in ubiquitin conjugates and Hsp72 levels during arm regeneration in echinoderms. Mar Biotech, 3: 4–15.

Patruno, M., McGonnell, I.M., Beesely, P.W. et al. 2003. *AnBMP2/4* is a new member of the TGF-β super family isolated from a crinoids and involved in regeneration. Proc Roy Soc Lond, 270B: 1341–1347.

Pawson, D.L. and Miller, J.E. 2012. *Invertebrate Zoology.* Smithsonian National Museum of Natural History.

Paz, G. and Rinkevich, B. 2002. Morphological consequences for multi-partner chimerism in *Botrylloides*, a colonial urochordate. Dev Comp Immunol, 26: 615–622.

Pearce, C.M., Daggett, T.L. and Robinson, S.M.C. 2002. Effect of binder type and concentration on prepared feed stability and gonad yield and quality of the green sea urchin, *Strongylocentrotus droebachiensis.* Aquaculture, 205: 301–323.

Pearse, J.E. and Timm, R.W. 1971. Juvenile nematodes (*Echinocephalus pesudouncinatus*) in the gonads of sea urchin (*Centrostephanus coronotus*) and their effect on host gametogenesis. Biol Bull, 140: 95–103.

Pearse, J.S. 1968. Patterns of reproductive periodicities in four species of Indo-Pacific echinoderms. Proc Indian Acad Sci, 68B: 247–279.

Pearse, J.S. 1969a. Reproductive periodicities of the Indo-Pacific invertebrates in the Gulf of Suez. 1. The echinoids *Prionocidaris baculosa* (Lamarck) and *Lovenia elongata* (Gray). Bull Mar Sci, 19: 323–350.

Pearse, J.S. 1969b. Reproductive periodicities of the Indo-Pacific invertebrates in the Gulf of Suez. 2. The echinoid *Echinometra mathaei* (de Blainville). Bull Mar Sci, 19: 580–613.

Pearse, J.S. 1970. Reproductive periodicities of the Indo-Pacific invertebrates in the Gulf of Suez. 3. The echinoid *Diadema setosum* (Leske). Bull Mar Sci, 20: 697–720.

Pearse, J.S. and Cameron, R.A. 1991. Echinodermata. In: *Reproduction of Marine Invertebrates.* (eds) Giese, A.C., Pearse, J.S. and Pearse, V.B., Boxwood Press, Pacific Grove, CA, 6: 574–664.

Pearse, J.S. and Eernisse, D.J. 1982. Photoperiodic regulation of gametogenesis and gonadal growth in the sea star *Pisaster ochraceus.* Mar Biol, 67: 121–125.

Pearse, J.S., Pearse, V.B. and Newberry, A.T. 1989. Telling sex from growth: Dissolving Maynard Smith's paradox. Bull Mar Sci, 45: 433–456.

Pehrson, J.R. and Cohen, L.H. 1986. The fate of the small micromeres in sea urchin development. Dev Biol, 113: 522–526.

Pennington, J.T. 1985. The ecology of fertilization of echinoid eggs: the consequences of sperm dilution, adult aggregation and synchronous spawning. Biol Bull, 169: 417–430.

Perron, F.E. and Kohn, A.J. 1985. Larval dispersal and geographic distribution in coral reef gastropods of the genus *Conus.* (eds) Gabrie, C. and Salvat, B., Proc 5th Int Coral Reeg Cong, Tahiti, pp 67–72.

Pesando, D., Robert, S., Huitorel, P. et al. 2004. Effects of methoxychlor, dieldrin and lindane on sea urchin fertilization and early development. Aquat Toxicol, 66: 225–239.

Petersen, J.A. and Ditadi, A.S.F. 1971. Asexual reproduction in *Glossobalanus crozieri* (Ptychoderidae, Enteropneusta, Hemichordata). Mar Biol, 9: 78.

Picard, A., Palavan, G., Robert, S. et al. 2003. Effect of organochlorine pesticides on maturation of starfish and mouse oocytes. Toxicol Sci, 73: 141–148.

Pinder, L.C.V., Pottinger, T.G., Billinghurst, Z. and Depledge, M.H. 1999. Endocrine function in aquatic invertebrates and evidence for disruption by environmental pollutants. Endocrine Modulators Steering GROUP, Brussels, R&D Technical Report, p 153.

Pitt, R. and Duy, N.D.Q. 2004. Breeding and rearing of the sea cucumber *Holothuria scabra* in Viet Nam. In: *Advances in Sea Cucumber Aquaculture and Management.* (eds) Lovatelli, A., Conand, C., Purcell, S. et al., FAO Fish Tech Paper No. 463. FAO, Rome, pp 333–346.

Podolsky, R.D. and Strathmann, R.R. 1996. Evolution of egg size in free-spawners: consequences of the fertilization-fecundity trade-off. Am Nat, 148: 160–173.

Podolsky, R.D. 2001. Evolution of egg target size: An analysis of selection on correlated characters. Evolution, 55: 2470–2478.

Podolsky, R.D. 2002. Fertilization ecology of egg coats: physical versus chemical contributions to fertilization success of free-spawned eggs. J Exp Biol, 205: 1657–1668.

Pomory, C.M. and Lares, M.T. 2000. Rate of regeneration of two arms in the field and its effect on body components in *Luidia clathrata* (Echinodermata: Asteroidea). J Exp Mar Biol Ecol, 254: 211–20.

Pomory, C.M. and Lawrence, J.M. 2001. Arm regeneration in the field in *Ophiocoma echinata* (Echinodermata: Ophiuroidea): Effects on body composition and its potential role in a reef food web. Mar Biol, 139: 661–70.

Poulin, E. and Feral, J.-P. 1996. Why are there so many species of brooding Antartic echinoids? Evolution, 50: 820–830.

Pradheeba, M., Dilipan, E., Nobi, E.P. et al. 2011. Evaluation of sea grasses for nutritional value. Ind J Geo-Mar Sci, 40: 105–111.

Prowse, T.A. and Byrne, M. 2012. Evolution of yolk protein genes in the Echinodermata. Evol Dev, 14: 139–151.

Prowse, T.A.A., Sewell, M.A. and Byrne, M. 2008. Fuels for development: evolution of maternal provisioning in asternid sea stars. Mar Biol, 153: 337–349.

Purcell, S.W., Mercier, A., Conand, C. et al. 2013. Sea cucumber fisheries: global analysis of stocks, management measures and drugs of overfishing. Fish Fisher, 14: 34–59.

Purwati, P. and Luong-van, J.T. 2003. Sexual reproduction in a fissiparous holothurian species, *Holothuria leucospilota* Clark 1920 (Echinodermata: Holothuroidea). SPC Beech-de-Mer Inf Bull, 18: 33–38.

Purwati, P. 2004. Fissiparity in *Holothuria leucospilota* from Darwin tropical warm waters, northern Australia. SPC Beche-de-Mer Info. Bull, 20: 26–33.

Purwati, P. and Dwiono, S.A.P. 2005. Fission inducement in Indonesian holothuroids. SPC Beche-de-Mer Inf Bull, 22: 11–15.

Purwati, P., Dwiono, S.A., Indriana, L.F. and Fahmi, V. 2009. Shifting the natural fission plane of *Holothuria atra* (Aspidochirotida, Holothuroidea, Echinodermata). SPC Beche-de-Mer Inf Bull, 29: 16–19.

Radhika-Rajasree, R., Karthih, M.G., Gopalakrishnan, M. et al. 2016. Extraction and isolation of polyhydroxylated naphthoquinone pigments from sea urchin *Salmacis virugulata* L. Agassiz and its anti-cancer activity against human lung cancer cell lines. Proc Natl Conf, Sathyabama University, Chennai.

Raff, E.C., Popodi, E.M. and Sly, B.J. 1999. A novel ontogenetic pathway in hybrid embryos between species with different modes of development. Development, 126: 1937–1945.

Raff, R.A. and Byrne, M. 2006. The active evolutionary lives of echinoderm larvae. Heredity 97: 244–252.

Rahman, A. 2014. Sea cucumbers (Echinodermata: Holothuroidea): Their culture potentials, bioactive compounds and effective utilizations. Int Conf Env Agri Med Sci, doi:org/10.17758/IAAST.AIII 4056.

Rahman, M.A., Uehara, T. and Rahman, S.M. 2002. Effects of egg size on fertilization, fecundity and offspring performance: a comparative study between two sibling species of tropical sea urchins (Genus *Echinometra*). Pakistan J Biol Sci, 5114–5121.

Rahman, M.A. and Uehara, T. 2004. Interspecific hybridization and backcrosses between two sibling species of Pacific sea urchins (Genus *Echinometra*) on Okinawan intertidal reefs. Zool Stud, 43: 93–111.

Rahman, M., Yusoff, F.M., Arshad, A. et al. 2013. Population characteristics and fecundity estimates of short-spined white sea urchin *Salmacis sphaeroides* (Linnaeus, 1758) from the coastal waters of Johor, Malaysia. Asian J Ani Veteri Adv, 8: 301–308.

Ramsay, K., Kaiser, M.J. and Richardson, C.A. 2001. Invest in arms: behavioural and energetic implications of multiple autotomy in starfish (*Asterias rubens*). Behav Ecol Sociobiol, 50: 360–365.

Ransick, A., Cameron, R.A. and Davidson, E.H. 1996. Postembryonic segregation of the germ line in sea urchins in relation to indirect development. Proc Natl Acad Sci USA, 93: 6759–6763.

Rao, K.P. 1954. Bionomics of *Ptychotera flava*. J Madras Univ, 24B: 1–5.

Rao, K.P. 1955. Morphogenesis during regeneration in an enteropneust. J Ani Morphol Physiol, 1: 1–7.

Rao, P.S., Rao, K.H. and Shyamasundari, K. 1993. A rare condition of budding in bipinnaria larva (Asteroidea). Curr Sci, 65: 792–793.

Rasmussen, E. 1973. Systematics and ecology of the Isefjord marine fauna (Denmark). Ophelia, 11: 1–495.

Razek, F.A., Rahman, S.A., Mona, M.H. et al. 2007. An observation on the effect of environmental conditions on induced fission of the Mediterranean sand sea cucumber, *Holothuria arenicola* (Semper, 1868) in Egypt. Beche-de-Mer Inf Bull, 24: 45–48.

Reichenbach, N., Nishar, Y. and Saeed, A. 1996. Species and size-related trends in asexual propagation of commercially important species of tropical sea cucumbers (Holothuroidea). J World Aquacult Soc, 27: 475–482.

Reimer, C.L. and Crawford, B.J. 1995. Identification and partial characterization of yolk and cortical granule proteins in eggs and embryos of the starfish, *Pisaster ochraceus*. Dev Biol, 167: 439–457.

Reinardy, H.C., Emerson, C.E., Manley, J.M. and Bodnar, A.G. 2015. Tissue regeneration and biomineralization in sea urchins. Role of not signalling and presence of stem cells marker. PLoS One, 10, e0133860.

Renbo, W. and Yuan, C. 2004. Breeding and culture of the sea cucumber, *Apostichopus japonicus* Liao. In: *Advances in Sea Cucumber Aquaculture and Management*. (eds) Lovatelli, A., Conand, C., Purcell, S. et al., FAO Fish Tech Paper No. 463. FAO, Rome, pp 277–286.

Reunov, A.A., Crawford, B.J. and Reunova, Y.A. 2010. An investigation of yolk-protein localization in the testes of the starfish *Pisaster ochraceus*. Can J Zool, 88: 914–921.

Reverberi, G. and Ortolani, G. 1962. Twin larvae from halves of the same egg in ascidian. Dev Biol, 5: 84–100.

Richard, S.J.W. 1994. Age determination and life history characteristics of *Acanthaster planci* (L.). (Echinodermata: Asteroidea). PhD Thesis, James Cook University.

Rideout, R.S. 1978. Asexual reproduction as a means of population maintenance in the coral reef asteroid *Linckia multifora* on Guam. Mar Biol, 47: 287–295.

Riggenbach, E. 1903. Die selfest verstimmung der Tiere. Ergeb Anat Entwicklungesch, 12: 782–903.

Rinkevich, B. and Weissman, I.L. 1987. A long term study on fused subclones in the ascidian *Botryllus schlosseri*. The resorption phenomenon (Protochordata: Tunicata). J Zool (Lond), 213: 717–733.

Rinkevich, B. and Shapira, M. 1999. Multi-partner urochordate chimeras outperform two-partner chimera entities. Oikos, 87: 315–320.

Rinkevich, Y., Paz, G., Rinkevich, B. and Reshef, R. 2007. Systemic bud induction and retinoic acid signature underlie whole body regeneration in the urochordate *Botrylloides leachi*. PLoS Biol, 5: e71.

Rinkevich, Y., Matranga, V. and Rinkevich, B. 2009. Stem cells in aquatic invertebrates: Common premises and emerging unique themes. In: *Stem Cells in Marine Organisms*. (eds) Rinkevich, B. and Matranga, V., Springer Verlag, Dordrecht, pp 61–103.

Ritzmann, N.F., da Rocha, R.M. and Roper, J.J. 2009. Sexual and asexual reproduction in Diademnum rodriguesi (Ascidiacea, Didemnidae). Iheringia Ser Zool Porto Alegre, 99: 106–110.

Robinson, S.M.C. and Colborne, L. 1997. Enhancing roe of the green sea urchin using an artificial food source. Bull Aquacult Ass Can, 97: 14–20.

Robinson, S.M.C. and MacIntyre, A.D. 1997. Aging and growth of the green sea urchin. Bull Aquacult Ass, Canada, 97: 56–60.

Rubilar, T., Pastor de Ward, C.T. and Díaz de Vivar, M.E. 2005. Sexual and asexual reproduction of *Allostichaster capensis* (Echinodermata: Asteroidea) in Golfo Nuevo. Mar Biol, 146: 1083–1090.

Rubilar, T., Villares, G.L., Diaz de Vivar, M.E. and Pastor de Ward, C.T. 2011. Fission, regeneration, gonad production and lipid storage in the fissiparous sea star *Allostichaster capensis*. Revi Biol Trop, 53: 299–303.

Rubilar, T., Meretta, P.E. and Cledon, M. 2015. Regeneration rate after fission in the fissiparous sea star *Allostichaster capensis* (Asteroidea). Revi Biol Trop, 63: 321–328.

Rutherford, J.C. 1973. Reproduction, growth and mortality of the holothurian *Cucumaria pseudocurata*. Mar Biol, 22: 167–176.

Rutman, J., and Fishelson, L. 1985. Comparisons of reproduction in the Red Sea feather stars *Lamprometra klunzingeri* (Hartlamb). *Heieromelra savignii* (J. Miiller) and *Capilasler muhiradiaius* (L.). In: *Echinodermata*. (eds) Keegan, B.F. and O'Connor, B.D.-S., Balkema Press, Rotterdam, pp 195–201.

Rychel, A.L. and Swalla, B. 2009. Regeneration in hemichordates and echinoderms. In: *Stem Cells in Marine Organisms*. (eds) Rinkevich, B. and Matranga, V., Springer Verlag, Dordrecht, pp 245–266.

Sabbadin, A. 1955. Il ciclo biologico di *Botryllus schlosseri* (Pallas) [Ascidiacea] nella laguna di Venezia. Arch Oceanogr Limnol, 10 : 219–231.

Sabbadin, A. 1958. Analisi sperimentale dello sviluppo delle colonie di *Botryllus schlosseri* (Pallas) [Ascidiacea]. Arch It Anat Embriol, 63: 178–221.

Sabbadin, A. 1959. Analisi genetic del policromatismo di *Botryllus schlosseri* (Ascidicea). Arch Ocean Limnol, 12: 131–143.

Sabbadin, A. 1978. Genetics of the colonial ascidian, *Botryllus schlosseri*. In: *Marine Organisms: Genetics, Ecology and Evolution*. (eds) Battaglia, B. and Beardmore, J., Plenum Press, New York pp 195–209.

Sabbadin, A. and Zaniolo, G. 1979. Sexual differentiation and germ cell transfer in the colonial ascidian *Botryllus schlosseri*. J Exp Zool, 207: 289–304.

Saito, M., Seki, M., Amemiya, S. et al. 1998. Induction of metamorphosis in the sand dollar *Peronella japonica* by thyroid hormones. Dev Growth Differ, 40: 307–312.

Saotome, K. 1999. Chromosome number of sea urchin andromerogones during early development. Zool Sci, 16: 87–92.

Saotome, K. and Komatsu, M. 2002. Chromosomes of Japanese starfishes. Zool Sci, 19: 1095–1103.

Sarhadizadeh, N., Afkhani, M. and Ehsanpour, M. 2014. Evaluation of bioactivity of a sea cucumber *Stichopus herrmanni* from Persian Gulf. Eur J Exp Biol, 4: 254–258.

Sato, R., Tanaka, Y. and Ishimaru, T. 2001. House production by *Oikopleura dioica* (Tunicata, Appendicularia) under laboratory conditions. J Plankton Res, 23: 415–423.

Satoh, N. and Levine, M. 2005. Surfing with the tunicates into the post-genome era. Genes Dev, 19: 2407–2411.

Schatt, P. and Feral, J.P. 1996. Completely direct development of *Abatus cordatus*, a brood schizasteroid (Echinodermata: Echinoidea) from Kergulen, with a description of perigastrulation, a hypothetic new mode of gastrulation. Biol Bull, 190: 24–44.

Scheibling, R.E. and Lawrence, J.M. 1982. Differences in reproductive strategies of morphs of the genus *Echinaster* (Echinodermata: Asteroidea) from the eastern Gulf of Mexico. Mar Biol, 70: 51–62.

Scheibling, R.E. and Hatcher, B.G. 2001. The ecology of *Strongylocentrotus droebachiensis*. In: *Edible Sea Urchins: Biology and Ecology*. (ed) Lawrence, J.M., Elsevier, Amsterdam, pp 271–306.

Scheltema, R.S. 1966. Trans-Atlantic dispersal of veliger larvae from shoal water benthic mollusca. Sec Int Oceanogr Cong (Mosco) Abstract No, 375: 320.

Schoener, A. 1972. Fecundity and possible mode of development of some deep sea ophiuroids. Limnol Oceanogr, 17: 193–199.

Shoemarkers, H.J.N., Colenbrander, P.H.J.M. and Peute, J. 1977. Ultrastructural evidence for the existence of steroid synthesizing cells in the ovary of the starfish *Asterias rubens* (Echinodermata). Cell Tissue Res, 182: 275–279.

Shoemarkers, H.J.N., Dieleman, S.J., Van Bohemen, C.G. and Voogt, P.A. 1978. In: *Comparative Endocrinology*. (eds) Garllard, P.J. and Boer, H.H., Elsevier/North-Holland Biomedical Press, Amsterdam, pp 33–36.

Shoemarkers, H.J.N. 1979a. Steroids and reproduction of the female *Asterias rubens*. Ph.D Thesis, University of Utrecht.

Shoemarkers, H.J.N. 1979b. *In vitro* biosynthesis of steroids from cholesterol by the ovaries and pyloric caeca of the star fish *Asterias rubens*. Comp Biochem Physiol, 62B: 179–184.

Shoemarkers, H.J.N. and Voogt, P.A. 1980. *In vitro* biosynthesis of steroids from androstenedione by the ovaries and pyloric caeca of the starfish *Asterias rubens*. Gen Comp Endocrinol, 45: 242–248.

Shoemarkers, H.J.N. and Dieleman, S.J. 1981. Progesterone and estrone levels in the ovaries, pyloric caeca, and perivisceral fluid during the annual reproductive cycle of starfish, *Asterias rubens*. Gen Comp Endocrinol, 43: 63–70.

Shoemarkers, H.J.N. and Voogt, P.A. 1981. *In vitro* biosynthesis of steroids from androstenedione by the ovaries and pyloric caeca of the star fish *Asterias rubens*. Gen Comp Endocrinol, 45: 242–248.

Shoemarkers, H.J.N., van Bohemen, Ch.G. and Dieleman, S.J. 1981. Effects of estradiol-17 on the ovaries of the starfish *Asteria rubens*. Dev Growth Diff, 23: 125–135.

Scott, S.F.M. 1957. Regeneration and differentiation in buds of *Amaroecium* treated with nitrogen mustard. J Exp Zool, 135: 557–586.

Sea Urchin Genome Sequencing Consortium. 2006. The genome of sea urchin *Strongylocentrotus purpuratus*. Science, 314: 941–952.

Selvakumaraswamy, P. and Byrne, M. 2000. Reproduction, spawning and development of 5 ophiuroids from Australia and New Zealand. Invert Biol, 119: 394–402.

Seto, Y., Moriyama, Y., Fujita, D. and Komatsu, M. 2000. Sexual and asexual reproduction in two populations of the fissiparous asteroid *Cosinasterias acutispina* in Toyama Bay Jam. Benthos Res, 55: 85–93.

Seto, Y., Komatsu, M., Wakabayashi, K. and Fujita, D. 2013. Asexual reproduction of *Cosinasterias acutispina* (Stimpson, 1862) in tank culture. Cah Biol Mar, 54: 641–647.

Sewell, M.A. 1994. Small size, brooding, and protandry in the apodid sea cucumber *Leptosynapta clarki*. Biol Bull, 187: 112–123.

Sewell, M.A. and Chia, F.S. 1994. Reproduction of the intraovarian brooding apodid *Leptosynapta clarki* (Echinodermata: Holothuroidea) in British Columbia. Mar Biol, 121: 285–300.

Sewell, M.A. and Young, C.M. 1997. Are echinoderm egg size distribution bimodal? Bio Bull, 193: 297–305.

Sewell, M.A. 2005. Utilization of lipids during early development of the sea urchin *Evechinus chloroticus*. Mar Ecol Prog Ser, 304: 133–142.

Shaeffer, C.M. 2016. The effects of autotomy and regeneration on the locomotion and behavior or brittle stars (Echinodermata: Ophiuroidea) of Moorea, French Polynesia. Pre print.

Shane, B.S. 1994. *Introduction to Ecotoxicology*. CRC Press, Boca Raton, Fl.

Sharlaimova, N.S., Pinaev, G.P. and Petukhova, O.A. 2010. Comparative analysis of behavior and proliferative activity in culture of cells of coelomic fluid and of cells of various tissues of the sea star *Asterias rubens* L. isolated from normal and injured animals. Cell Tissue Biol, 4: 280–288.

Shenkar, N. and Swalla, B.J. 2011. Global Diversity of Ascidiacea. PLoS ONE, 6(6): e20657. doi.org/10.1371/journal.pone.0020657.

Shenkar, N. and Swalla, B.J. 2017. Hemichordata World Database. Website and databases developed and hosted by VLIZ. Contact: Billie Swalla.

Shibata, D., Hirano, Y. and Komatsu, M. 2011. Life cycle of the multiarmed sea star *Coscinasterias acutispina* (Stimpson, 1862) in laboratory culture: Sexual and asexual reproductive pathways. Zool Sci, 28: 313–317.

Shirae-Kurabayashi, M., Nishikata, T., Takamura, K. et al. 2006. Dynamic redistribution of vasa homolog and exclusion of somatic cell determinants during germ cell specification in *Ciona intestinalis*. Development, 133: 2683–2693.

Shyu, A., Raff, R.A. and Blumenthal, T. 1986. Expression of the vitellogenin gene in female and male sea urchins. Proc Nat Acad Sci, USA, 83: 3865–3869.

Shyu, A., Raff, R.A. and Blumenthal, T. 1987. A single gene encoding vitellogenin in the sea urchin Strongylocentrotus purpuratus: sequence of the 5′ end. Nucleic Acids Res, 15: 10405–10417.

Sinervo, B. and McEdward, L.R. 1988. Developmental consequences of an evolutionary change in egg size: An experimental test. Evolution, 42: 885–899.

Singletary, R. 1980. The biology and ecology of *Amphioplus coniortodes, Ophionephthys limicola,* and *Microphiopholis gracillima* (Ophiuroidea: Amphiuridae). Car J Sci, 16: 39–55.

Skold, H.N., Obst, M., Skold, M. and Akesson, B. 2009. Stem cells in asexual reproduction of marine invertebrates. In: *Stem Cells in Marine Organisms*. (eds) Rinkevich, B. and Matranga, V., Springer Verlag, Dordrecht, pp 105–138.

Skold, M., Loo, L.-O. and Rosenberg, R. 1994. Production, dynamics and demography of an *Amphiura filiformis* population. Mar Ecol Prog Ser, 103: 81–90.

Skold, M. and Rosenberg, R. 1996. Arm regeneration frequency in eight species of Ophiuroidea (Echinodermata) from European sea areas. J Sea Res, 35: 353–362.

Skold, M., Barker, M.F. and Mladenov, P.V. 2002. Spatial variability in sexual and asexual reproduction of the fissiparous sea star *Coscinasterias muricata*: the role of food and fluctuating temperature. Mar Ecol Prog Ser, 233: 143–155.

Sloan, N.A. 1985. Echinoderm fisheries of the world: a review. In: *Proc Fifth Int Echinoderm Conf, Galway*. (eds) Keegan, B.F. and Conner, B.D.S., Balkema, Rotterdam, pp 109–124.

Spirlet, C., Grosjean, P. and Jangoux, M. 2000. Optimization of gonad growth by manipulation of temperature and photoperiod in cultivated sea urchins *Paracentrotus livudus* (Lamarck) (Echinodermata). Aquaculture, 185: 85–99.

Stancyk, S.E., Golde, H.M., Pape-Lindstrom, P.A. and Dobson, W.E. 1994. Born to lose. I. Measures of tissue loss and regeneration by the brittlestar *Microphiopholis gracillima* (Echinodermata: Ophiuroidea). Mar Biol, 118: 451–462.

Starr, M., Himmelman, J.H. and Therriault, J.C. 1992. Isolation and properties of a substance from the diatom *Phaeodactylum tricornutum* which induces spawning in the sea urchin *Strongylocentrotus droebachiensis*. Mar Ecol Prog Ser, 79: 275–287.

Stebbing, A.R.D. 1970. Aspects of the reproduction and life cycle of *Rhabdopleura compacta* (Hemichordata). Mar Biol, 5: 205–212.

Sterling, K.A. and Shuster, S.M. 2011. Rates of fission in *Aquilonastra corallicola* Marsh (Echinodermata: Asteroidea) as affected by population density. Invert Reprod Dev, 55: 1–5.

Stien, A. 1993. The ecology and epidemiology of *Echinomermella matsi* (Nematoda), a parasite of the sea urchin *Strongylocentrotus droebachiensis*. PhD Thesis, University of Oslo.

Stohr, S. 2005. Who's who among baby brittle stars (Echinodermata: Ophiuroidea): postmetamorphic development of some North Atlantic forms. Zool J Linn Soc, 143: 543–576.

Stohr, S. and Segonzac, M. 2005. Deep-sea ophiuroids (Echinodermata) from reducing and non-reducing environments in the North Atlantic Ocean. J Mar Bioll Asso UK, 85: 383–402.

Stohr, S. and Segonzac, M. 2006. Two new genera and species of ophiuroid (Echinodermata) from hydrothermal vents in the East Pacific. Species Diversity 11: 7–32.

Stohr, S., O'Hara, T.D. and Thuy, B. 2012. Global diversity of brittle stars (Echinodermata: Ophiuroidea). PLoS ONE, 7(3): E31940.

Stoner, D.S. and Weissman, I.L. 1996. Somatic and germ cell parasitism in a colonial ascidian: Possible role for a highly polymorphic allorecognition system. Proc Natl Acad Sci USA, 93: 15254–15259.

Stoner, D.S., Rinkevich, B. and Weissman, I.L. 1999. Heritable germ and somatic cell lineage competitions in chimeric colonial protochordates. Proc Natl Acad Sci USA, 96: 9148–9153.

Strathmann, R.R. 1971. The feeding behavior of planktotrophic echinoderm larvae. Mechanisms, regulation and rates of suspension feeding. J Exp Mar Biol Ecol, 6: 109–160.

Strathmann, R.R., Jahn, T.L. and Fonesca, J.R.C. 1972. Suspension feeding by marine invertebrate larvae. Clearance of particles by ciliated bands of a rotifer, pluteus and trochophore. Biol Bull, 142: 505–519.

Strathmann, R.R. 1981. On barriers to hybridization between *Strongylocentrotus droebachiensis* (O.F. Muller) and *S. pallidus* (G.O. Sars). J Exp Mar Biol Ecol, 55: 39–47.

Strathmann, R.R. and Strathmann, M.F. 1982. The relationship between adult size and brooding in marine invertebrates. Am Nat, 119: 91–101.

Strathmann, R.R. 1985. Feeding and non-feeding larval development and life history evolution in marine invertebrates. Ann Rev Ecol System, 16: 339–361.

Strenger, A. 1973. *Sphaerechinus granularis* Violtter Seeigel. Grosses Zool Prak Gustav Fischer Verlad, Stuttgat, 18: 1–68.

Stuart, V. and Field, J.G. 1981. Respiration and ecological energetics of the sea urchin *Paracentrotus angulosus*. South Afr J Zool, 16: 90–95.

Sui, J., Zhang, Z.F., Shao, M.Y. and Hu, J.J. 2008. Cloning and characterization of a vasa-like gene in *Apostichopus japonicus* and its expression in tissues. J Fish Sci China, 03.

Sun, H., Liang, M., Yan, J. and Chen, B. 2004. Nutrient requirements and growth of the sea cucumber, *Apostichopus japonicus*. In: *Advances in Sea Cucumber Aquaculture and Management*. (eds) Lovatelli, A., Conand, C., Purcell, S. et al., FAO Fish Tech Paper No. 463. FAO, Rome, pp 297–310.

Sunanaga, T., Saito, Y. and Kuwamura, K. 2006. Postembryonic epigenesis of vasa-positive germ cells from aggregated homoblasts in the colonial ascidian *Botryllus primigenus*. Dev Growth Diff, 48: 87–100.

Suzuki, N. 1989. Sperm-activating peptides from sea urchin egg jelly. Bioorg Mar Chem, 3: 43–70.

Sweat, L.H. 2014. Indian river lagoon species inventory. Smithsonian Marine Station at Fort Pierce, irl_webmaster@si.edu.

Takahashi, N. and Kanatani, H. 1981. Effect of 17β-estradiol on growth of oocytes in cultures ovarian fragments of the starfish *Asterias pectinifera*. Dev Gorwth Diff, 23: 565–569.

Takahashi, N. 1982. The relation between injection of steroids and ovarian protein amounts in the starfish *Asterina pectinifera*. Bull Jap Soc Sci Fish, 48: 509–511.

Takamura, K. 2001. The origin of germ cells in *Ciona intestinalis*. In: *The Biology of Ascidians*. (eds) Sawada, H., Yokosawa, H. and Lambert, C., Springer Verlag, New York, pp 109–116.

Takamura, K., Fujimura, M. and Yamaguchi, Y. 2002. Primordial germ cells originate from the endodermal strand cells in the ascidian *Ciona intestinalis*. Dev Genes Evol, 212: 11–18.

Tan, S.Y. and Dee, M.K. 2009. Elie Metchnikoff (1845–1916): discoverer of phagocytosis. Singapore Med J, 50: 456–457.

Taneda, Y. 1985. Size regulation of regenerated organs in the compound ascidian, *Polyandrocarpa misakiensis*. J Exp Zool, 233: 331–334.

Thirunavukarassu, T. and Subhashini, P. 2016. Biological activities of seagrass metabolome and scope for *in vitro* synthesis. Modu Chem Mol Sci Chem Eng, doi.org/0.1016/B978-0-12-409547.117 13-5.

Thompson, B.A.W. and Riddle, M.J. 2005. Bioturbation behaviour of the spatangoid urchin *Abatus ingens* in Antarctic marine sediments. Mar Ecol Prog Ser, 290: 135–143.

Thompson, R.J. 1979. Fecundity and reproductive effort in the blue mussel (*Mytilus edulis*), the sea urchin (*Strongylocentrotus droebachiensis*) and the snow crab (*Chironecetes opilio*) from the populations in Nova Scotia and Newfoundland. J Fish Res Bd Can, 36: 955–964.

Thompson, R.J. 1982. The relationship between food ration and reproductive effort in the green sea urchin *Strongylocentrotus droebachiensis*. Oecologia, 56: 50–57.

Thompson, R.J. 1984. Partitioning of energy between growth and reproduction in three populations of the sea urchin *Strongylocentrotus droebachiensis*. In: *Advances in Invertebrate Reproduction*. (ed) Engels, W., Elsevier, Amsterdam, pp 425–432.

Thorndyke, M.C. and Candia-Carnevali, M.D. 2001. Regeneration of microhormones and growth factors in echinoderms. Can J Zool, 79: 1171–1208.

Thorne, B.V., Eriksson, H. and Byrne, M. 2013. Long term trends in population dynamics and reproduction in *Holothuria atra* (Aspidochirotida) in the southern Great Barrier Reef; the importance of sexual and asexual reproduction. J Mar Biol Ass UK, 93: 1067–1072.

Tomota, M., Ikeda, T. and Shiga, N. 1999. Production of *Oikopleura longicauda* (Tunicata: Appendicularia) in Toyama Bay, southern Japan Sea. J Plankton Res, 21: 2421–2430.

Townsely, S.J. and Townsely, M.P. 1973. A preliminary investigation on biology and ecology of holothurians at Fanning Island. Hawaii Inst of Geophysics, University of Hawaii.

Troedsson, C., Bouquet, J.M., Aksnes, D.L. and Thompson, E.M. 2002. Resource allocation between somatic growth and reproductive output in the pelagic chordate *Oikopleura dioica* allows opportunistic response to nutritional variation. Mar Ecol Prog Ser, 243: 83–91.

Tsukamoto, K., Yamada, Y., Okamura, A. et al. 2009. Positive buoyancy in eel leptocephali: an adaptation for life in the ocean surface layer. Mar Biol, 156: 835–846.

Tsushima, M., Byrne, B., Amemiya, S. and Matsuno, T. 1995. Comparative biochemical studies of carotenoids in sea urchins—III. Relationship between developmental mode and carotenoids in the Australian echinoids *Heliocidaris erythrogramma* and *H. tuberculata* and a comparison with Japanese species. Comp Biochem Physiol, 110 B: 719–723.

Tweedell, K.S. 1961. Regeneration of the enteropneust *Saccoglossus kowaleskii*. Biol Bull, 120: 118–127.

Tyler, P.A. 1977. Seasonal variation and ecology of gametogenesis in the genus *Ophiura* (Ophluroldea : Echinodermata) from the Bristol Channel. J Exp Mar Biol Ecol, 30: 185–197.

Tyler, P.A., Paterson, G.L.J., Sibuet, M. et al. 1995. A new genus of ophiuroid (Echinodermata: Ophiuroidea) from hydrothermal mounds along the Mid-Atlantic Ridge. J Mar Bioll Asso UK, 75: 977–986.

Uehara, T., Shingaki, M., Taira, K. et al. 1991. Chromosome studies in eleven Okinawan species of sea urchins with special reference to four species of the Indo-Pacific *Echinometra*. In: *Biology of Echinodermata*. (eds) Yanagisawa, T., Yasumasu, C., Ogura, N. et al., A.A. Balkema, Rotterdam, pp 119–129.

Unuma, T., Yamanoto, T. and Akiyama, T. 1996. Effects of oral administration of steroids on the growth of juvenile sea urchin *Pseudocentrotus depresses*. Suisanzoshoku, 44: 79–83.

Unuma, T., Yamamoto, T. and Akiyama, T. 1999. Effects of oral administration of steroids on the growth and gametogenesis in the juvenile red sea urchins *Pseudocentrotus depressus*. Biol Bull, 196: 199–204.

Unuma, T., Nakamura, A., Yamano, K. and Yokota, Y. 2010. The sea urchin major yolk protein is synthesized mainly in the gut inner epithelium and the gonadal nutritive phagocytes before and during gametogenesis. Mol Reprod Dev, 77: 59–68.

Urata, M. and Yamaguchi, M. 2004. The development of the enteropneust hemichordata *Balanoglossus misakiensis* Kuwano. Zool Sci, 21: 533–540.

Uthicke, S. 1997. Seasonality of asexual reproduction in *Holothuria* (*Halodeima*) *atra*, *H. edulis* and *Stichopus chloronotus* (Holothuroidea: Aspidochirodita) on the Great Barrier Reef. Mar Biol, 129: 435–441.

Uthicke, S. 1998. Respiration of *Holothuria* (*Halodeima*) *atra*, *Holothuria* (*Halodeima*) *edulis* and *Stichopus chloronotus*: Intact individuals and products of asexual reproduction. In: *Echinoderms: San Franscisco*. (eds) Mooi, R. and Telford, M., Balkema, Rotterdam, pp 531–536.

Uthicke, S., Benzie, J.A.H. and Ballment, E. 1999. Population genetics of the fissiparous holothurian *Stichopus chloronotus* (Aspidochirotida) on the Great Barrier Reef, Australia. Coral Reefs, 18: 123–132.

Uthicke, S. 2001. The process of asexual reproduction by transverse fission in *Stichopus chloronotus* (Greenfish). SPC Bêche-de-mer Inf Bull, 14: 23–25.

Uthicke, S. and Conand, C. 2005. Amplified fragment length polymorphism (AFLP) analysis indicates the importance of both asexual and sexual reproduction in the fissiparous holothurian *Stichopus chloronotus* (Aspidochirotida) in the Indian and Pacific Ocean. Coral Reefs, 24: 103–111.

Uthicke, S., Schaffelke, B. and Byrne, M. 2009. A boom-bust phylum? Ecological and evolutionary consequences of density variation in echinoderms. Ecol Monogr, 79: 3–24.

Uye, S. and Ichino, S. 1995. Seasonal variations in abundance, size composition, biomass and production rate of *Oikopleura dioica* (Fol) (Tunicata: Appendicularia) in a temperate eutrophic inlet. J Exp Mar Biol Ecol, 189: 1–11.

Vacquier, V.D. and Moy, G.W. 1977. Isolation of bindin: the protein responsible for adhesion of sperm to sea urchin eggs. Proc Natl Acad Sci USA, 74: 2456–2460.

Vadas, R.L. 1977. Preferential feeding: An optimization strategy in sea urchins. Ecol Monogr, 47: 337–371.

Vadas, R.L., Beal, B.F., Dudgen, S.R. and Wright, W.A. 2015. Spatial and temporal variability of spawning in the green sea urchin *Strongylocenytrotus droebachiensis* along the coast of Maine. J Shellfish Res, 34: 1128–1125.

Van der Plas, A.J., Koenderman, A.H.L., Deible-van Schijindel, G.J. and Voogt, P.A. 1982. Effects of estradiol-17β on the synthesis of RNA, protein and lipids in the pyloric caeca of the female starfish *Asterias rubens*. Comp Biochem Physiol, 73B: 965–970.

Vandenspiegel, D., Jangoux, M. and Flammig, P. 2000. Maintaining the life of defense: Regeneration of Cuverian tubules in the sea cucumber *Holothuria forsake* (Echinodermata: Holothuroidea). Biol Bull, 198: 34–49.

Van der Spoel, S. 1979. Strobolization in a pteropod (Gastropoda-Opisthobranchia). Malacologia, 18: 27–30.

Vasquez, J. 2001. Ecology of *Loxechinus albus*. In: *Edible Sea Urchins: Biology and Ecology*. (ed) Lawrence, J.M., Elsevier, Amsterdam, pp 161–176.

Vassetzky, S.G., Skoblina, M.N. and Sekirina, G.G. 1986. Induced fusion of echinoderm oocytes and eggs. Meth Cell Biol, 27: 359–378.

Vaughn, D. and Strathmann, R.R. 2008. Predators induce cloning in echinoderm larvae. Science, 319: 1503.

Vaughn, D. 2009. Predator-induced larval cloning in the sand dollar *Dendraster excentricus*: Might mothers matter? Biol Bull, 217: 103–114.

Vaughn, D. 2010. Why run and hide when you can divide? Evidence for larval cloning and reduced larval size as an adaptive inducible defense. Mar Biol, 157: 1301–1312.

Vickery, M.S. and Mc Clintock, J.B. 2000. Effect of food concentration and availability on the incidence of cloning in planktotrophic larvae of the sea star *Pisaster ochraceus*. Biol Bull, 199: 298–304.

Vickery, M.S., Vickery, M.C.L. and Mc Clintock, J.B. 2002. Morphogenesis and organogenesis in the regenerating planktotrophic larvae of asteroids and echinoids. Biol Bull, 203: 121–133.

Villalobos, F.B. 2005. Reproduction and larval biology of North Atlantic asteroids related to the invasion of the deep sea. Ph D Thesis, University of Southampton.

Villalobos, F.B., Avila-Poveda, O.H. and Gutierrez-Mendez, I.S. 2013. Reproductive biology of *Holothuria fuscocinerea* (Echinodermata: Holothuroidea) from Oaxaca, Mexico. Sex Early Dev Aquat Org, 1: 13–24.

Vogel, H., Gzihak, G., Chang, P. and Wolf, W. 1982. Fertilization kinetics of sea urchin eggs. Math Biosci, 58: 189–216.

Voogt, P.A. and Dieleman, S.J. 1984. Progesterone and oestrone levels in the gonads and pyloric caeca of the male sea star *Asterias rubens*: A comparison with the corresponding levels in the female sea star. Comp Biochem Physiol, 79A: 635–639.

Voogt, P.A., den Besten, P.J., Kusters, G.C. and Messing, M.W. 1987. Effects of cadmium and zinc on steroid metabolism and steroid level in the sea star *Asterias rubens* L. Comp Biochem Physiol, 86: 83–89.

Voogt, P.A., Van Ieperen, S., Van Rooyen, M.B. et al. 1991. Effect of photoperiod on steroid metabolism in the sea star *Asterias rubens* L. Comp Biochem Physiol, 100B: 37–43.

Voronina, E., Lopez, M., Juliano, C.E. et al. 2008. Vasa protein expression is restricted to the small micromeres of the sea urchin but is inducible in other lineages early in development. Dev Biol, 314: 776–786.

Voscoboynik, A., Simon-Belcher, N., Soen, Y. et al. 2007. Striving for normality: whole body regeneration through a series of abnormal generations. FASEB J, 21: 1335–1344.

Voskoboynik, A., Rinkevich, B. and Weissman, I.l. 2009. Stem cells, chimerism and tolerance: Lessons from mammals and ascidians. In: *Stem Cells in Marine Organisms*. (eds) Rinkevich, B. and Matranga, V., Springer Verlag, Dordrecht, pp 281–308.

Wahle, R.A. and Peckham, S.H. 1999. Density-related reproductive trade-offs in the green sea urchin *Strongylocentrotus droebachiensis*. Mar Biol, 134: 127–137.

Walker, C.W. and Lesser, M.P. 1998. Manipulation of food and photoperiod promotes out-of-season gametogenesis in the green sea urchin *Strongylocentrotus droebcahiensis*: implications for aquaculture. Mar Biol, 132: 663–676.

Walker, C.W., Unuma, T., McGinn, N.A. et al. 2001. Reproduction of sea urchins. In: *Edible Sea Urchins: Biology and Ecology*. (ed) Lawrence, J.M., Elsevier, Amsterdam, pp 5–26.

Walker, C.W., Harrington, L.H., Lesser, M.P. and Fagerberg, W.R. 2005. Nutritive phagocyte incubation chambers provide a structural and nutritive microenvironment for germ cells of *Strongylocentrotus droebachiensis*, the green sea urchin. Biol Bull, 209: 31–48.

Walsh, G.E., McLaughlin, L.L., Louie, M.K. et al. 1986. Inhibition of arm regeneration by *Ophioderma brevispina* (Echinodermata, Ophiuroidea) by tributyltin oxide and triphenyltin oxide. Ecotoxicol Environ Saf, 12: 95–100.

Ward, G.E., Brokaw, C.J., Garbers, D.L. and Vacquier, V.D. 1985. Chaemotaxis of *Arbacia pentactula* spermatozoa to react a peptide from the egg jelly layer. J Cell Biology, 101: 2324–2329.

Wasson, K.M., Hines, G.A. and Watts, S.A. 1998. Synthesis of testosterone and 5alpha-androstanediols during nutritionally stimulated gonadal growth in *Lytechinus variegatus* Lamarck (Echinodermata: Echinoidea). Gen Comp Endocrinol, 111: 197–206.

Wasson, K.M. and Watts, S.A. 2000. Progesterone metabolism in the ovaries and testes of the echinoid *Lytechinus variegatus* Lamarck (Echinodermata). Comp Biochem Physiol, 127C: 263–272.

Wasson, K.M., Gower, B.A. and Watts, S.A. 2000a. Responses of ovaries and testes of *Lytechinus variegatus* (Echinodermata: Echinoidea) to dietary administration of estradiol, progesterone and testosterone. Mar Biol, 137: 245–255.

Wasson, K.M., Gower, B.A., Hines, G.A. and Watts, S.A. 2000b. Levels of progesterone, testosterone and estradiol and the androstenedione metabolism in the gonads of *Lytechinus variegatus* (Echinodermata: Echinoidea). Comp Biochem Physiol, 126C: 153–165.

Wasson, K.M. and Watts, S.A. 2001. Reproductive endocrinology of sea urchins. In: *Edible Sea Urchins: Biology and Ecology*. (ed) Lawrence, J.M., Elsevier, Amsterdam, pp 43–58.

Watkins, M.J. 1958. Regeneration of buds in *Botryllus*. Biol Bull, 115: 147–152.

Watts, S.A., McClintock, J.B. and Lawrence, J.M. 2001. The ecology of *Lytechinus variegatus*. In: *Edible Sea Urchins: Biology and Ecology*. (ed) Lawrence, J.M., Elsevier, Amsterdam, pp 375–394.

Wilkie, I.C., Emson, R.H. and Mladenov, P.V. 1984. Morphological and mechanical aspects of fission in *Ophiocomella ophiactoides* (Echinodermata: Ophiuroida). Zoomorphology, 104: 310–322.

Wilson, D.P. 1978. Some observations on bipinnariae and juveniles of the starfish genus *Luidia*. J Mar Biol Ass UK, 58: 467–478.

Wyatt, T. 1973. The biology of *Oikopleura diocia* and *Fritillaria borealis* in the southern Bight. Mar Biol, 22: 137–158.

Xilin, S. 2004. The progress and prospects of studies on artificial propagation and culture of the sea cucumber, *Apostichopus japonicus*. In: *Advances in Sea Cucumber Aquaculture and Management*. (eds) Lovatelli, A., Conand, C., Purcell, S. et al., FAO Fish Tech Paper No. 463. FAO, Rome, pp 273–276.

Xiyin, L., Zhu, G., Zhau, Q. et al. 2004. Studies on hatchery techniques of the sea cucumber *Apostichopus japonicus*. In: *Advances in Sea Cucumber Aquaculture and Management*. (eds) Lovatelli, A., Conand, C., Purcell, S. et al., FAO Fish Tech Paper No. 463. FAO, Rome, pp 287–296.

Xu, R.A. and Barker, M.F. 1990a. Annual changes in steroid levels in the ovaries and pyloric caeca of *Sclerasterias molli* (Hutton, 1872) (Echinodermata: Asteroidea) during reproductive cycle. Comp Biochem Physiol, 95A: 127–133.

Xu, R.A. and Barker, M.F. 1990b. Effect of diet on steroid levels and reproduction in the starfish *Sclerasterias molli* (Hutton, 1872) (Echinodermata: Asteroidea) during reproductive cycle. Comp Biochem Physiol, 96A: 33–39.

Yagi, K., Satou, Y., Mazet, F. et al. 2003. A genomewide survey of developmentally relevant genes in *Ciona intestinalis*. III. Genes for Fox, ETS, nuclear receptors and NFκB. Dev Genes Evol, 213: 235–244.

Yamaguchi, M. and Lucas, J.S. 1984. Natural parthenogenesis, larval and juvenile development and geographical distribution of the coral reef asteroid *Ophidiaster granifer*. Mar Biol, 83: 33–42.

Yamamoto, M. and Okada, T. 1999. Origin of the gonad in the juvenile of a solitary ascidian, *Ciona intestinalis*. Dev Growth Differ, 41: 73–79.

Yaqing, C., Changqing, Y. and Songxin, F.I.R. 2004. Pond culture of sea cucumbers, *Apostichopus japonicus*, in Dalian. In: *Advances in Sea Cucumber Aquaculture and Management*. (eds) Lovatelli, A., Conand, C., Purcell, S. et al., FAO Fish Tech Paper No. 463. FAO, Rome, pp 269–272.

Yasui, K., Urata, M., Yamaguchi, N. et al. 2007. Laboratory culture of the oriental lancelet *Branchiostoma belcheri*. Zool Sci, 24: 514–520.

Yokobori, S.I., Watanabe, Y. and Oshima, T. 2003. Mitochondrial genome of *Ciona savignyi* (Urochordata, Ascidiacea, Enterogona): comparison of gene arrangement and tRNA genes with *Halocynthia roretzi* mitochondrial genome. J Mol Evol, 57: 574–587.

Yokoyama, L.Q. and Amaral, A.C.Z. 2010. Arm regeneration in two populations of *Ophionereis reticulata* (Echinodermata: Ophiuroidea). Iheringia Ser Zool, Port Alegre, 100: 123–127.

Zeng, L., Jacobs, M.W. and Swalla, B.J. 2006. Coloniality has evolved once in stolidobranch ascidians. Integ Comp Biol, 46: 255–268.

Zito, F. and Matranga, V. 2009. Secondary mesenchyme cells as potential stem cells of the sea urchin embryo. In: *Stem Cells in Marine Organisms*. (eds) Rinkevich, B. and Matranga, V., Springer Verlag, Heidelberg, pp 187–213.

Author Index

Species Index

Subject Index